T0215054

FLOOD RISK AND COMMUNITY RESILIENCE

This book details the impact of flooding on our environment, and the ways in which communities, and those that work with them, can act to manage the associated risks.

Flooding is an increasingly significant environmental hazard which inflicts major costs to the economies and livelihoods of developed countries. This book explores how local communities can identify, manage, and adapt to the ever-increasing damage flooding causes. Focusing on the future role of local communities, the benefits and challenges of their involvement, and the potential areas of transformation, this book provides insights into the efficacy of interdisciplinary and transdisciplinary working. Alongside research into similar environmental hazards, this book also draws upon the author's own knowledge of flood risk management in distinctive non-contiguous interdisciplinary settings. The chapters draw together a different and distinctive set of interdisciplinary themes in flood risk management and social resilience. In doing so, it strives to communicate the different ways of thinking that can usefully contribute to flood risk management.

This book would be ideal for those researching flood risk management, alongside scholars and non-scholars alike who are interested in finding ways of adapting to environmental hazards working with local communities.

Lindsey Jo McEwen is Professor of Environmental Management and Director of the Centre for Water, Communities and Resilience at the University of the West of England, Bristol, UK.

Earthscan Studies in Water Resource Management

Water Scarcity and Conflict in African River Basins
The Hydropolitical Landscape
Mahlakeng Khosi Mahlakeng

River Basins and International Relations
Cooperation, Conflict and Sub-Regional Approaches
Edited by Christian Ploberger

New Perspectives on Transboundary Water Governance
Interdisciplinary Approaches and Global Case Studies
Edited by Luis Paulo Batista da Silva, Wagner Costa Ribeiro, and Isabela Battistello Espíndola

The Role of Law in Transboundary River Basin Disputes
Cooperation and Peaceful Settlement
Chukwuebuka Edum

Desalination and Water Security
Chris Anastasi

Flood Risk and Community Resilience
An Interdisciplinary Approach
Lindsey Jo McEwen

For more information about this series, please visit: www.routledge.com/
Earthscan-Studies-in-Water-Resource-Management/book-series/ECWRM

FLOOD RISK AND COMMUNITY RESILIENCE

An Interdisciplinary Approach

Lindsey Jo McEwen

LONDON AND NEW YORK

Designed cover image: © Getty

First published 2024
by Routledge
4 Park Square, Milton Park, Abingdon, Oxon OX14 4RN

and by Routledge
605 Third Avenue, New York, NY 10158

Routledge is an imprint of the Taylor & Francis Group, an informa business

British Library Cataloguing-in-Publication Data
A catalogue record for this book is available from the British Library

ISBN: 978-1-138-95445-8 (hbk)
ISBN: 978-1-138-95447-2 (pbk)
ISBN: 978-1-315-66691-4 (ebk)

DOI: 10.4324/9781315666914

Typeset in Times New Roman
by KnowledgeWorks Global Ltd.

*This book is dedicated to my parents, who taught me
from an early age to value opportunities for learning.
I thank my mother who spurred me on to get
this book completed in my father's honour.
It is for both of you, with much love.*

CONTENTS

List of Figures   *ix*
List of Tables   *xiv*
List of Boxes   *xvii*
Preface   *xix*
Acknowledgements   *xxi*
List of Abbreviations   *xxii*

1 Interdisciplinary Explorations in Communities and Their
  Resilience to Changing Flood Risk: An Introduction   1

2 Different Flood Types and Diverse Communities:
  Interactions, Impacts, and Management at Local Level   36

3 Flood Hazard Perception, Awareness, and Action: From
  Individual to Community?   68

4 Different Flood Knowledges: Conflict or Integration?   99

5 Linking Flood Heritage and Local Community Resilience   131

6 Communicating Flood Science for Community Resilience   163

7 Flood Management Strategies and Their Relation to
  Community Awareness and Action   198

8 Community Participation and Agency in Local Flood
Risk Management   231

9 Community Learning for Flood Resilience: Strategies
and Pitfalls   269

10 Community Resilience to Flood Risk: Futures,
Challenges, and Opportunities   303

Index   *342*

LIST OF FIGURES

1.1   Disaster risk management conceived as a cycle with its
typical phases (adapted from Bosher, 2005; Bosher
and Chmutina, 2017; permission Lee Bosher)   6

1.2   The disaster management spiral (after Lloyd-Jones, T. (ed)
(2009) *The Built Environment Professions in Disaster Risk
Reduction and Response*. RICS, ICE, RIBA and RTPI, Max
Lock Centre, University of Westminster. Updated for this
publication by the editor/principal author).   7

1.3   Schematic representation of (a) engineering, (b) ecological
and (c) socio-ecological resilience. Resilience is illustrated
by the position (i.e., state) of a ball (i.e., system) in a basin
(i.e., conditions). Disturbances cause the ball to move (Tempels,
2016, p45, Figure 5 with permission)   15

1.4   Components of a disaster resilience framework (DFID,
2011, p7, Crown copyright)   15

1.5   The Pressure and Release (PAR) model of vulnerability to
disasters – here climate-related hazards (adapted from Blaikie
et al., 1994 with permission; simplified version edited from
Brink, 2018 by author)   21

1.6   Visualisation differentiating types of disciplinary working   23

2.1   Framework for understanding the interrelated factors
responsible for the occurrence of damaging floods (redrawn
and edited from Pielke and Downton, 2000; © American
Meteorological Society. Used with permission)   45

2.2   Physical impacts of transported sediment during the 1952 flash
floods at Lynmouth, SW UK (with permission: Peter Christie)   46

2.3   Framing sustainable flood management (concept based
      on Riley, 2007)   57
2.4   Residual risk and personal responsibility (source: Tobin pers.
      comm., based on Riley, 2007; Montz and Tobin, 2008)   59
3.1   Visualisation of the "hazard to action-chain" (Wachinger et al.,
      2013, p1054; Figure 1; Original figure was in colour; redrawn
      here in B/W)   71
3.2   Multiple influences that interact to form risk perception
      (modified from Renn and Rohrmann, 2000; source: Science
      Communication Unit, 2014, Figure 2, reprinted with permission)   74
3.3   Relationship between flood risk characteristics (Raaijmakers
      et al., 2008, Figure 1, reproduced by permission of Springer Nature)   77
3.4   Framework on the media's influence on risk perception
      (Cologna et al., 2017, Figure 1, adapted from Wachinger et al.,
      2013 , reproduced by permission of Elsevier)   83
3.5   Representation of the process of risk perception and response
      (Harries, 2007 unpublished PhD thesis, Middlesex University
      with permission; redrawn by author)   87
3.6   Westcott et al.'s (2017) expansion of Protection Motivation
      Theory (Figure 1 after Rogers, 1975 and adapted from
      Grothmann and Reusswig, 2006; permission: Rachel Westcott).
      Grey shading indicates elements of their proposed PMT expansion   89
4.1   (A) DIKW pyramid or knowledge hierarchy (based on Ackoff,
      1989); (B) the DIKW web (Spiekermann et al., 2015, Figure 2,
      reproduced by permission of Elsevier)   100
4.2   Pitfalls of, and propositions for, research to increase its
      effectiveness (source: Spiekermann et al., 2015 adapted from
      Weichselgartner and Marandino, 2012, p327; reproduced by
      permission of Elsevier). Original figure was in colour; adapted
      here to B/W with letter codes   104
4.3   Forms of flood materialisation: collective and personal
      settings. A/B, epigraphic marking of varying persistence;
      C, remembering personal flood levels; D, July 2007 'a weekend
      in history'; E, historic 1947 flood photographs; F/H, sharing
      'flood albums'; G, historic community protection measures;
      I, preserved flood water (McEwen et al., 2016, Figure 1)   109
4.4   Negotiation between active remembering and active forgetting
      within communities   112
4.5   A visual representation of the KTA process (Graham et al.,
      2006; Sudsawad, 2007, p9; permission Wolters Kluwers Health)   115
4.6   Canadian Institutes of Health Research (CIHR) research cycle
      superimposed by the six opportunities to facilitate knowledge
      translation (CIHR, 2007 with permission; Sudsawad, 2007)   116
5.1   Data on past floods collected from a variety of archive types,
      integrated into a database, statistically analysed, and climatically

modelled and interpreted (permission: PAGES Past Global
Changes Floods Working Group http://pastglobalchanges.org/
science/wg/floods/intro)   135

5.2   Examples of flood marks in different cultural settings
(composite by author): (A) 'Flommerker' showing 1789 flood
("Storofsen"), Lalm, Norway; (B) 'Wassermerke' (1594–) at
Schloss Ort Castle, Lake Traunsee, Gmunden, Austria; (C)
Watergate (1770–), Worcester, UK; (D) Upton on Severn (1947
flood marker corrected); (E) La Fleix (1728–), Dordogne,
France; (F) Exceptional high tide 1883, Chepstow, UK; (G)
Flood marks at the west entrance to King's Lynn Minster,
King's Lynn, Norfolk, England   138

5.3   During the 1927 flood: (A) A refugee camp along Mississippi
River levee, Greenville, Mississippi; (B) Depicts prisoners
brought in to place sandbags on Mississippi River levee (both
reproduced by courtesy of USACE)   143

5.4   "Houston Strong" by Houston mural artist Nicky Davis 2017
(reproduced with permission)   145

5.5   (A) "Flood in the Highlands" (1860), by Sir Edwin Landseer,
oil on canvas (reproduced courtesy of Aberdeen Art Gallery and
Museums collections); (B) "Sarpsfossen in Norwegen" (1789)
by Eric Paulsen, oil on canvas (National Gallery of Denmark)   147

5.6   Woodcut of 1607 coastal storm surge on the Severn estuary UK
(source: Title page of Lamentable newes out of Monmouthshire in
Wales, an English news-book of 1607; copyright Stephen Rippon)   148

5.7   "Flood Songs for an Unmade Album" (artwork credited to Sage
Brice, 2012; playlist credited to David Reeves, 2012; McEwen
et al., 2012; background photo credit to Richard Droker (www.
flickr.com/photos/83432882@N05/)   148

5.8   Steel bench sculpture as memorial to the Cwmcarn flood victims
(1875), Ebbw valley, Wales, UK (photograph copyright author)   150

5.9   Bringing flood heritage into conversation with other flood
knowledges (Copyright Lindsey McEwen; ESRC Sustainable
Flood Memories project)   154

6.1   Four conceptual models of risk communication (after Wardman,
2008 in Demeritt and Nobert, 2014: reproduced by permission
of Taylor and Francis Ltd)   166

6.2   Mind map of flood risk science needs in local government co-
generated with project partners, SW UK (McEwen et al., 2014;
reproduced by permission of Taylor and Francis Ltd)   170

6.3   Alternative expressions for communicating direct uncertainty
about a fact, number or scientific hypothesis (van der Bles
et al., 2019; © 2019 The Authors. Attribution 4.0 International
(CC BY 4.0))   174

6.4 Using flood maps (source: https://www.fema.gov/flood-maps)   177

6.5 Stephen Scott-Bottoms performing 'Too Much of Water' to a
community audience on the floodplain of the River Aire, UK
(Photo courtesy of Stephen Scott-Bottoms)   184

6.6 Conversations about tidal flooding while chalking during
HighWaterLine in Bristol, UK (2014) (credit: Richard
Clutterbuck)   185

6.7 Appropriated official icons from Phantom's 'Flood the Streets'
campaign (www.phantom.land, with permission)   187

6.8 The reflective cycle of good collaborative communication and
dialogue (adapted from CIRIA's Framework for communication
and engagement, Daly et al., 2015a, p22; with permission)   188

6.9 Overview of the progression of flood risk communication
strategies over the decades in the UK (source: Ping et al., 2016,
Figure 2; compiled from Parker et al., 1995, Haggett, 1998,
Handmer, 2001, McCarthy, 2007, and Parker and Priest, 2012)
(Reprinted from Journal of Water & Climate Change volume 7,
issue number 4, pages 651-664, with permission from the
copyright holders, IWA Publishing.)   188

7.1 (A) Glass flood defence wall along riverside in action
(https://www.geograph.org.uk/photo/3848842 Copyright Bob
Embleton and licensed for reuse under Creative Commons
Licence Attribution-ShareAlike 2.0 Generic (CC BY-SA 2.0));
(B) example of an earlier dismountable barrier (copyright
Environment Agency, 2009) – both Upton-upon-Severn, UK
(© Environment Agency, 2009)   201

7.2 Conceptual model of linkages between ecosystem services and
human well-being (Millennium Ecosystem Assessment, 2005)
(credit: Millennium Ecosystem Assessment; Figure 1.1)   204

7.3 Overcoming barriers to managed retreat requires diverse actors
coordinated by leadership and vision (source: Siders, 2019a,
Figure 4, p221; reproduced by permission of Elsevier and author)   208

7.4 Working with natural processes to reduce flood risk (Burgess-
Gamble et al., 2017). https://assets.publishing.service.gov.uk/
media/6036c730d3bf7f0aac939a47/Working_with_natural_
processes_one_page_summaries.pdf (permission: Lydia Burgess
Gamble)   210

7.5 Potential indicators of the effectiveness of NbS for climate
change adaptation and mitigation and associated co-benefits
(Kabisch et al., 2016, Figure 1)   213

7.6 The components of the Total Flood Warning system (TFWS)
(source: Australian Institute for Disaster Resilience, 2009;
Figure 2; CC-NC)   216

7.7 'The Flood Repairable House' (permission: Long-Dhonau)   222

8.1   A conceptual model for climate change impacts, community
      resilience, and community capitals (Kais and Islam, 2016,
      adapted from McCrea et al., 2014, with permission Md Saidul Islam)   237
8.2   Participation models and metaphors: (A) Eight rungs on a ladder
      of citizen participation (Arnstein, 1969) © American Planning
      Association, Chicago, IL. reprinted by permission of Taylor &
      Francis Ltd, http://www.tandfonline.com on behalf of American
      Planning Association, Chicago, IL; (B) the Tree of Participation
      (ToP) (Bell and Reed, 2021, Figure 2; permission: Karen Bell)   240
8.3   Model conceptualising traditional Maori approach to disaster
      risk reduction (Kenney and Phibbs, 2015, Figure 1; Permission:
      the authors; Licence: Elsevier)   249
8.4   Multiple benefits of community flood groups (source: The Flood
      Hub, Newground https://thefloodhub.co.uk/community/; with
      permission)   253
8.5   Levels of participation and engagement in citizen science
      projects (Assumpção et al., 2018, Figure 1; adapted from
      Haklay, 2013)   255
9.1   Differences between pedagogy, andragogy and heutagogy (adapted
      with permission from figure by Simon Whittemore, 2021)   273
9.2   Dimensions of ESD (UNESCO, 2014, p12, reproduced with
      permission)   275
9.3   A taxonomy of significant learning (Fink, 2005; reproduced
      with permission)   277
9.4   Scoping a framework for exploring possible Learning for
      Disaster Resilience (LfDR) content (Dufty, 2014a, Figure 2,
      permission: Neil Dufty)   283
9.5   Main content areas for disaster resilience learning (Dufty,
      2013, 2014b, Figure 2, p14, permission: Neil Dufty)   284
10.1 Some domains of possible transformation in community-focused FRM   304
10.2 The nature of CBDRM traditions (interpretations and
      worldviews) expressed through its primary features on a
      continuum (Heijmans, 2009, p27; permission: Annelies Heijmans)   306
10.3 Looming waves of destruction (copyright Graeme MacKay –
      licence Artizans.com https://mackaycartoons.net/2020/03/18/
      wednesday-march-11-2020/)   307
10.4 Dichotomies in role of civil society in extreme weather
      adaptation (source: CASCADE-NET)   309
10.5 A simplified influencing pathway of long-term health impacts
      after flooding (Zhong et al., 2018, Figure 9, p177; reproduced
      by permission of Elsevier)   315
10.6 Approaches to anticipatory governance: diverse conceptions of
      the future and actions in the present (Muiderman et al., 2020,
      Figure 1, p10; reproduced by permission of Elsevier and authors)   321
10.7 Practical Action's V2R detailed resilience framework (Pasteur,
      2011, Figure 3, p12, permission: Practical Action)   330

LIST OF TABLES

1.1 Identifying Types of Flood Losses (edited AIDR Manual 27, 2002, Table 8, with permission) 3

1.2 Differentiating "Tame" and "Wicked" Problems (Rittel and Webber, 1973; adapted from Blackman et al. 2006) 9

1.3 Definitions of Civil Society and Civic Agency 11

1.4 Definitions of Resilience, Disciplinary Context and Scale of Analysis (selected from MacKinnon and Derickson, 2013; Quinlan et al., 2016; Dubois and Krasny, 2016) 14

1.5 Selected Definitions of Community Resilience in Academia and Practice 17

1.6 Five Core Concepts in Definitions of Community Resilience That Maximise Value (CARRI, 2013, p2; Permission Meridian Institute) 18

1.7 Definitions of Further Key Terms Relevant to Local Flood Risk Management 20

1.8 Differentiating Interdisciplinary, Multidisciplinary, and Transdisciplinary Working 23

1.9 Different Cognitive Styles for Organising Knowledge Relevant to FRM 26

2.1 Examples of Different Flood Types with Their Implications for Community Resilience 38

2.2 Dimensions of Environmental Justice Applied to FRM (Drawing on Schlosberg, 2007; Svarstad et al., 2011; Bell, 2014; Alba et al., 2020) 48

2.3 Some Paradigms for Flood Management (Categorisation A–C Drawing from Tunstall et al., 2004; Johnson et al., 2005 for the UK) 53

3.1 Explanations for the Risk Perception Paradox (Figure 3.1) (Edited from Wachinger et al., 2013, p1054; permission Ortwin Renn) 71

3.2   Some Key Theories of Risk Perception and Their Disciplinary
      Setting   73
3.3   Cognitive Biases and Heuristics Reported to Affect Perceptions
      of Water Risks (Weitkamp et al., 2020, Table 1, with author
      permission)   82
4.1   Characteristics of Explicit and Tacit Knowledge (Adapted from
      Virkus, 2014 and Talisayon, 2008, cclfi.international)   102
4.2   Different Types of Local Flood Knowledge with Their
      Characteristics (See McEwen, 2023)   106
4.3   Linking Opportunities within CIHR Research Cycle to Ways
      of Undertaking Community-Focused Flood Risk Research
      (Column 1 modified from CIHR, 2005; Sudsawad, 2007)   116
5.1   International Examples of Flood Museums Promoting
      Flood Heritage   144
5.2   A Conceptualised Framework for SFM as Process-Practice
      within FRM Decision Making for Local Resilience
      (McEwen, et al., 2016, Table II)   153
5.3   Community Activities That Promote Learning about Local
      Flood Heritage   155
6.1   Different Models for Public Engagement with Science   164
6.2   Detail on Four Conceptual Models of Risk Communication in
      Figure 6.1; (*edited into a table from Demeritt and Norbert,
      2014; Wardman, 2008)   166
6.3   Purposes of Flood Risk Communications (adapted from
      Orr et al., 2015; Table 1.1; Section 2.3; Contains public sector
      information licensed under the Open Government Licence v3.0.)   169
6.4   Threshold Concepts in Flood Risk Communication   173
6.5   Six Key Issues for Practitioners to Address in Flood Risk
      Communication (adapted from Orr et al., 2015, Chapter 6,
      permission Environment Agency)   181
6.6   Principles and Advice to Inform Flood Risk Communication
      (adapted from Fisher, 2015, piv; HM Government, 2016,
      Annex 6; pp85–89)   181
6.7   Examples of Arts Practice in Communication for Flood Risk
      Awareness   184
6.8   Opportunities and Inhibitors in Flood Risk Communication
      Strategies (adapted from Ping et al., 2016, Table 2)   189
6.9   Imperatives for Effective Flood Science Communication
      (drawing from Orr et al., 2015; Ping et al., 2016 and Other
      Referenced Sources)   190
7.1   Barriers to Managed Retreat (adapted from Siders, 2019a,
      Figure 3, p221, permission from Elsevier and author)   207
7.2   Community Involvement in Early Warning Systems (EWS)
      (IFRC, 2012, Table 1)   218

7.3 Ideas for Appropriate Technology per EWS Component at
    Local/Community Level (Column A Adapted from IFRC,
    2012, Table 8; Column B by Author)   218
8.1 Defining Different Types of Social Capital (*Szreter and
    Woolcock, 2004 p654/655; $Smith, 2000–2009, np; permission:
    Michael Woolcock)   235
8.2 Flora and Flora's (2013) Community Capitals Framework
    (reproduced from Pitzer and Streeter, 2015, Table 1, p359 with
    permission)   236
8.3 Characteristics of a Resilient Community (Edited from Kais
    and Islam, 2016, p9; Author's Emphasis on Different Capitals;
    permission: Md Saidul Islam)   238
8.4 Prompt Questions about Public Participation (Australian
    Department of Sustainability and Environment, 2005 with
    permission; Edited from Krishnaswamy, 2012, p5)   243
8.5 Key Principles of Participatory Learning and Action (PLA)
    (Edited from Institute of Development Studies) (original source:
    https://www.participatorymethods.org/page/about-participatory-
    methods. Used with permission of Participation Research
    Cluster, Institute of Development Studies, University of Sussex)   244
8.6 Core Evaluation Criteria and Indicators of Successful
    Participation Tools (Adapted from Beckley et al., 2005, p21
    with permission; Explanation from Krishnaswamy, 2012)   245
8.7 'The 6Ss Framework' for Supporting Flood Group Development
    (adapted from McEwen et al., 2018)   253
8.8 Different Models of Citizen Science Data Gathering
    with FRM Examples   256
8.9 Approaches to Gathering and Sharing Soft and Thick Citizen
    Data with FRM Examples   258
9.1 Defining Different Types of Disciplinary Learning   278
9.2 Starter Questions for Community Flood Science (Modified
    from McEwen, 2011)   285
9.3 Examples of Skills Acquisition in Learning for Flood Resilience   286
9.4 Examples of Serious Games Using Low Technology: Focus and
    Target Groups   292
9.5 Examples of Interactive Physical Flood Models for Community
    Learning   293
9.6 Flood-Inspired Theatre in Community Learning for Resilience   294
10.1 A Framework for Designing Meaningful Participatory Processes
    with Potential to Transform (Drawing on Shaw, 2016; IFRC,
    2018; de Vente et al., 2016 and Author Experience)   324
10.2 Examples of Community Resilience Indices with Different
    Approaches, Scales, and Users   328

LIST OF BOXES

1.1	Useful characteristics of an interdisciplinary researcher (Lyall et al. 2011, Key advice 3.1, p35, with permission)	25
2.1	The flash flood of August 1952, Lynmouth, Devon, UK	46
3.1	Biases that act as barriers to public flood risk perception (edited from McEwen, 2011; adapted from Haley, 2008)	81
3.2	Reasons for resistance of behaviour to change (Prager, 2012, p6, after Mearns 2012, with permission)	89
4.1	Key principles and features of co-production (Hickey et al., 2021, permission from NIHR)	121
5.1	Characteristics of a place-based, flood heritage approach	151
6.1	Examples of recurrent flood science myths among the public (adapted from McEwen, 2011)	171
6.2	Pielke's (1999) flood fallacies based on scientific understanding (with permission: Roger Pielke)	172
6.3	Characteristics of maps for effective public risk communication (Minano and Peddle, 2018, p3, with permission)	178
6.4	Communicating uncertainty about 'climate impacts' (Corner et al., 2015, p9, with author permission)	180
7.1	Building Back Better (UK Environment Agency advice, undated, p20)	221
7.2	Steps in flood preparedness behaviours (Park et al., 2021, pp2–3, 12; permission: environment agency)	223
8.1	Reference to civil society and local-level DRR within Sendai Framework (UNDRR; author's emboldening)	232
8.2	CBDRM Principles (Edited from Oxfam/ADPC, 2014, p23; with permission: ADPC)	247
8.3	UK Environment Agency's Community flood plan template (open government license v3.0)	250

9.1   All-hazards household and family disaster prevention in a
       nutshell (source: International Red Cross and Red Crescent
       Societies, 2018, p23)   283
9.2   Why games? (source: Climate Centre); with permission: Red
       Cross Red Crescent Climate Centre   291
10.1 Questions in transforming community participation   323
10.2 Systemic thinking needed for transformation through
       participation in FRM   326
10.3 What should a community expect from its own resilience
       measurement tool? (Cutter, 2014, p12; permission: Susan Cutter)   331

PREFACE

Being flooded brings polarising experiences for communities – from stress and trauma to new community knowledge, connections, and cohesion that can help build stronger communities. Navigating such emotive territory with communities is challenging; principles of inclusivity, equity, relationship building, and empathy need to be at its core. Being an academic researcher in community flood risk management necessitates a strong sense of responsibility for exchanging knowledge and community voices. This sits as a backdrop to my decision back in 2015 to embark on this book – as an opportunity to share learnings from many years of research activity with the increasingly varied stakeholders involved directly or more obliquely in FRM. These include students, community groups, professionals in statutory FRM organisations, NGOs, and other sectors like culture and environment, and artists.

I owe particular thanks to the local communities and organisations, who I have worked with in my research and practice, especially those in the lower Severn valley, UK, over several research and knowledge exchange projects. These projects include ESRC Sustainable Flood Memories; AHRC Towards Hydrocitizenship; and EPSRC SESAME (business flood resilience), NERC DRY (Drought Risk and You), and ESRC CASCADE-NET (civil society and extreme weather adaptation). I gratefully acknowledge valuable collaborations with my co-researchers and their contributions to the body of work presented here.

This book has attempted the ambitious challenge of an overview – capturing a learning journey that links themes from an inevitably personal perspective. This draws from over 30 years of research and stakeholder engagement in flood risk management from different interdisciplinary perspectives. Such a task can never be exhaustive, given the burgeoning literature from community research and practice. When I started this book, 'community' was much less used as an interpretative lens.

Completing this book has sometimes felt like walking up on a downward escalator, with new and valuable material appearing rapidly in research, policy, and practice. It was an inevitable challenge to say 'stop'; however, this book has aims beyond a literature synthesis. Importantly, it aspires to offer an accessible way in for the less initiated and familiar to explore unfamiliar and distant interdisciplinary territories and possibilities, and to start making new connections between domains, interventions, and innovations and co-vision potential resilient futures. Chapter 10 culminates in an explicit focus on possible domains of transformation, that separately, but more importantly when integrated, can make a difference. Arguably, however, understandings of transformation need to start from the personal.

My research journey started formally between 1981 and 1985, when I undertook my PhD at St Andrews University, Scotland. I would like to thank particularly, Professor Alan Werritty, who initiated my passion for 'studying floods'. Since then working at different interdisciplinary interfaces has been both intellectually challenging and rewarding, particularly those collaborations with the arts and humanities. I have valued highly my Scottish educational grounding that opened up a range of natural science, social science, and humanities disciplines. This continued at St Andrews University, where I went to study English but quickly reverted to my long-standing passion for geography, along with psychology and botany. I had not appreciated until much later that my PhD was multidisciplinary – bringing together scientific and archival research to explore explanations for historical changes to Scottish rivers. Working as both a subject-based and pedagogic researcher in Higher Education during my academic career has also helped me see distinctive interdisciplinary connections in FRM and flood education, which I aspire to share here. My love of creating pottery has helped ground me – practically and metaphorically – throughout the long book writing process and in life more generally!

Producing a solo authored book has been a significant learning journey, requiring energy and determination, with inevitable highs and lows. If I knew then, what I know now! I would like to thank Routledge for sticking with me through a long writing process.

ACKNOWLEDGEMENTS

There are many people who I would like to thank wholeheartedly for their input to the production of this book.

The reviewers who gave of their time and expertise to comment on different chapters: Professor Alan Werritty (Chapters 2 and 10), Dr Sara Williams (Chapter 3), Professor Steve Poole (Chapter 5), Professor Emma Weitkamp (Chapter 6), Professor Burrell Montz (Chapter 7), Neil Dufty (Chapters 8 and 9), Dr Deepak Gopinath (Chapter 8), and Neil Dufty who contributed some Australian examples to Chapter 8.

All the numerous academic researchers, practitioners, and publishers who gave their permissions for figures, tables, and box content to be included in this book. I had some great interactions with a wide variety of interesting people in this process. These permissions are acknowledged with many thanks.

My friend and co-researcher on the DRY (Drought Risk and You) project, Antonia, with her canny Italian insight, reminding me of Voltaire's 'perfect is the enemy of the good' in my latter stages of book writing.

My long-standing friend and colleague, Caroline Mills, for offering to help proof reading and suggesting final editing and cutting some of my 'darlings' to help get this book over the line …. This kindness was much appreciated.

Finally, I would like to thank all my wonderful family and friends for their moral support and endless encouragement over a long writing journey. You know who you are! I apologise for my 'hermit-like' writing weekends – over an extended period – to get the book finished …. Celebrations will be in order when this book is finally published.

LIST OF ABBREVIATIONS

AIDR	Australian Institute of Disaster Resilience
CBDRM	Community-based Disaster Risk Management
CLDRM	Community-led Disaster Risk Management
CS	Citizen science
DRM	Disaster risk management
DRR	Disaster risk reduction
ESD	Education for sustainable development
EJ	Environmental justice
FRM	Flood risk management
FRC	Flood risk communication
IDR	Interdisciplinary research
IDL	Interdisciplinary learning
KTA	Knowledge-to-action
LfR	Learning for resilience
LfS	Learning for sustainability
NGO	Non-government organisation
TD	Transdisciplinary
TL	Transformative learning
UNDRR	United Nations Disaster Risk Reduction
WHO	World Health Organisation

1

INTERDISCIPLINARY EXPLORATIONS IN COMMUNITIES AND THEIR RESILIENCE TO CHANGING FLOOD RISK

An introduction

1.1   Introduction

In the 21st century, floods still inflict major costs to the economies and livelihoods of communities in More Economically Developed Countries (MEDCs), with flood risks and impacts set to increase further due to climate change (IPCC, 2021). It is increasingly recognised that local communities have critical roles as key actors within flood risk management and disaster risk reduction (DRR). Moreover, floods are becoming more diverse, as are the communities they affect.

Flood risk can be conceived as a combination of the magnitude (size) and frequency of flooding, with the degree of exposure of human activities and their vulnerability. Complex interactions occur between physical and societal factors, over different spatial and temporal scales, which determine the flood impacts in any location. In some settings (e.g., Europe; European Environment Agency, EAA[1]), extreme river floods are among the most damaging climate-induced events. In other areas, floods overlay or cascade in complex ways with environmental hazards (e.g., droughts, heatwaves, and wildfires in the USA and Australia). Hence, effective flood risk management and wider DRR are inter-connected local, national, and international priorities.

The United Nations Office for Disaster Risk Reduction's *Sendai Framework for Disaster Risk Reduction* (2015–2030) (hereafter 'Sendai Framework'[2,3]) outlines targets and priorities for action to prevent new, and reduce existing, disaster risks. It strongly advocates community involvement in DRR. Over 15 years, it has aimed to achieve

> The substantial reduction of disaster risk and losses in lives, livelihoods and health and in the economic, physical, social, cultural and environmental assets of persons, businesses, communities, and countries.
>
> *UNDRR (2015, p6)*

DOI: 10.4324/9781315666914-1

The Sendai Framework highlights concern for community roles in terms of their knowledge, representation, health, ways of working, and empowerment.

Efforts to deal with flooding at local scale can involve an increasingly wide range of actors, intersecting policy areas, and management activities that overlay within river catchments, coastal settings, and cities. There is potential for multi-stakeholder learning across risk settings that have different environmental and governance histories, and across communities with varied social and cultural characteristics which involve distinct mixes of stakeholders (e.g., communities, statutory and non-statutory organisations).

This book poses four overarching questions:

- What is the past, present, and potential future role of local communities in the strategy, policy, and practice of local flood risk management?
- How can the benefits and challenges of community involvement be addressed to increase local resilience to changing flood risk?
- What insights can be brought from increased interdisciplinary and transdisciplinary working and systemic thinking in research and practice within community-focused flood risk management?
- What are the possible areas of leverage and transformation within local community-focused flood risk management?

This introductory chapter highlights background contexts to flood risk management and explores key concepts like 'community' and 'resilience' when dealing with environmental risk. One challenge for multi-stakeholder working is use of terminology and language, with many contested terms with different disciplinary histories and professional uses that lead to communication issues. The chapter then considers opportunities and challenges of interdisciplinarity and transdisciplinarity for research and practice, and different ways of thinking that can contribute to flood risk management.

1.2 Aims

This book aims to provide an accessible synthesis of a burgeoning research, policy, and practice literature on flood risk management (FRM), considered explicitly through a 'community lens'. Its primary focus is on the MEDCs. However, it also explores understandings from wider community engagement with flood risk, for example, from community-led approaches in Less Economically Developed Countries (LEDCs). It aspires to promote cultural exchanges about policies and practices that develop community resilience. The book draws creatively on research into challenges unrelated to flooding (e.g., other environmental hazards, public health, community development) that provide transferable insights for FRM contexts in MEDCs. Spanning the boundary between the natural and social sciences, the book also considers co-working with the arts and humanities.

In writing this book, the author draws on over 20 years of engaged interdisciplinary research into local FRM and experience of working with diverse at-risk communities in the UK. The intended readership is diverse, encompassing environmental and social scientists, community groups, professionals in statutory and non-statutory FRM and the cultural sector and artists, to promote collaborative learning and support development of a community of practice with potential to increase local flood resilience.

1.3 Background contexts

The challenge of reducing flood risk, managing floods better and decreasing societal impacts remains high on political agendas within many countries globally. A strong sense of urgency exists among flood risk communities and statutory FRM agencies to try and think differently about how to manage floods. In scene-setting for this book, nine contextual questions to FRM are briefly considered.

1.3.1 How are flooding and its impacts experienced locally?

Floods are frequently experienced as highly sensory and emotive events – but that is not always the case, and not all floods are disasters. The occurrence and sequence of severe floods versus more routine (chronic) nuisance events experienced, and the alternation between floods and other hazards like droughts, can be critical in determining local perception and impacts. However, extreme floods can have major socio-economic consequences – tangible and intangible – for individuals and communities (Table 1.1).

TABLE 1.1 Identifying Types of Flood Losses (Edited AIDR Manual 27, 2002, Table 8, with permission)

Can the lost item be bought and sold for [money]?	Direct loss: Loss from contact with flood water	Indirect loss: No contact; loss as consequence of flood water
Yes – monetary (tangible)	For example, • Buildings and contents • Vehicles • Livestock • Crops • Infrastructure	• Disruption to transport, etc. • Loss of value added in commerce and business interruption (where not made up elsewhere) • Legal costs associated with lawsuits
No – non-monetary (intangible)	For example, • Lives and injuries • Loss of memorabilia • Damage to cultural or heritage sites • Ecological damage	• Stress and anxiety • Disruption to living • Loss of community • Loss of cultural and environmental sites and collections • Ecosystem resource loss

Trends in numbers of people affected by flooding are increasing (e.g., Paprotny et al., 2018 for Europe – 1870–2016). It is widely recognised that the traditional engineering approach to managing floods solely by increasing investment in large-scale physical infrastructure has failed to reduce losses (Zevenbergen et al., 2020). Dykes can fail; 'protective' structures tend to encourage development and can put more at risk (Tierney, 2014). As MunichRE[4] reflects, "there is no such thing as complete protection".

1.3.2   What are past, present, and future flood patterns?

Mapping recent global hazard distribution evidences some flood incidence in over a third of the world's land area – where around 82% of the world's population lives (Dilley et al. 2005; *Prevention Web*). Research on past extreme floods has highlighted the catastrophic and differential impacts on communities and their economic, social, and cultural assets (e.g., 1927 Mississippi floods, the USA; 1966 floods, Italy; 1953 storm surge, Netherlands/UK). Within the MEDCs, high-profile, severe floods occurring since the 1990s include Hurricane Katrina, New Orleans, US (in 2005); Queensland, Australia (in 2011, 2022); and Europe (e.g., Central Europe, in 2009; Elbe/Danube, in 2013; Germany, in 2021). Several extreme floods have occurred in the UK (e.g., in 1998, 2000, summer 2007, winters 2013/2014, 2015/16, 2019/20) – a 'flood rich' period compared to the 1970s and 1980s when research and policy attention was more focused on managing drought.

Future flood risks are also set to increase in many settings in Europe and globally (Alfieri et al., 2017, 2018). Climate change projections of global rainfall variability suggest more extreme and variable precipitation and flooding is our future (e.g., Penderglass et al., 2017; Tabari, 2020). This has significant implications for human security (IPCC, 2022), and how floods are planned for, and mitigated, locally. For example, despite uncertainties in modelling, annual flood losses in Europe can be expected to increase five-fold by 2050 and up to 17-fold by 2080 (Alfieri et al., 2015; EEA, 2016).

1.3.3   What is the role of changing society in increasing flood impacts?

Tierney (2014) emphasised "the social roots of risk", in determining what and who is exposed and vulnerable. This is not just about growing differential impacts of climate change but also increases in the economic value of assets on floodplains. As EEA (2016, p53) reflected:

> The major share of this increase [in calculated future flood losses] (70–90%) is estimated to be attributable to socio-economic development, and the remainder (10–30%) to climate change.

Alfieri el. (2015) predicted these annual flood damages in Europe – under a 'high end' climate scenario – could rise from EUR 5.3 billion to 40 billion by

2050, with numbers of people affected increasing from 200 thousand to over 0.5 million. A critical concern is how such projections might play out for communities in specific physical, social-economic, and cultural settings. At the hyperlocal scale, community flood risk varies with floodplain topography and the spatial distribution of vulnerable land uses, such as residential developments on exposed lower-cost land.

1.3.4 How have approaches to flood management changed?

Several paradigm shifts have occurred in approaches to flood management (e.g., Van Ruiten and Hartmann, 2016 for EU; Werritty, 2019 for Scotland; Chapter 2). Importantly, each has implications for the responsibilities and required agency of local communities. A key shift has been from engineering approaches to more integrated and sustainable FRM. The latter recognises that investment in defensive infrastructure alone, with its costs and design limits, can only be part of the solution. Risk management involves *all* stakeholders, including the public, taking some responsibility for residual flood risk and their own protection. Residual risk is risk remaining after any FRM measures have been implemented. It involves "attaching responsibility for risky choices at the level where the decision is made" (Montz and Tobin, 2008, p3).

In the UK, such local community action is a key expectation in the delivery of the *Flood and Water Management Act* (2010). However, many issues in local FRM relate to these paradigm shifts, and variable public awareness or acceptance, as individuals and communities, of these changes and the implications for their role in local flood risk planning. If individuals are unaware that management of residual risk is their responsibility, then they are unlikely to prepare to minimise personal or collective losses. Even when people do perceive risk, they still may not act rationally to protect themselves against future flooding (Baker, 2007). They also may not know how to prepare or have the resources necessary to act.

A further paradigm shift has been towards involving communities in risk management. It is this 'community-focused FRM' model that is explored in this book (Chapters 2, 8, and 10). This recognises the changing roles of citizens – their individual and collective agency – in developing local resilience, within calls for people-centred, participatory approaches to disaster risk management (DRM; e.g., Wolff, 2021). In exploring how to develop greater resilience for *all* community members (e.g., Riley, 2007), increased emphasis is now being placed on community learning for resilience to ensure better understanding of residual risk, effective inputs into local FRM, and informed adoption of preparedness measures (Chapter 9). Parallel shifts in focus have occurred in other aspects of sustainable development (see UN Sustainable Development Goals[5]), with attention placed on resilience building, water security, reducing inequalities, climate change adaptation, and linking local and global (Kelman et al., 2015).

1.3.5 How has citizen agency within the timeline for flood management changed?

Post-flood reviews (e.g., Cabinet Office, 2008 after UK 2007 floods) have led to increased awareness of the valuable roles that communities can play during acute events. However, another shift in FRM has seen a reduced focus on solely improving emergency response during floods, and increased attention given to citizen agency within the full DRM cycle and wider resilience planning. The disaster 'cycle' or 'wheel' is a well-established, simple metaphor for thinking about FRM. It highlights relationships between five phases: pre-disaster preparedness activities, the event itself and disaster relief, shorter-term recovery and rehabilitation, longer-term reconstruction, and mitigating actions (Figure 1.1).

Preparedness is "in large part, a function of the capacity of society to reorganize itself to reduce risks" (Gober and Wheater, 2015, p4784). Numerous potential points of intervention or leverage (e.g., in assumptions and decisions) exist in the physical and socio-cultural 'idea-spaces' for innovation within local FRM systems. Approaching these differently could challenge the status quo and potentially lead to changed individual and collective behaviours and actions that promote better recovery, while also strengthening community resilience.

An elaboration of the model of the DRM cycle conceives it as a complex spiral building upwards (Figure 1.2), while "residual risks keep the spiral on the loop, despite continuous improvements in management practices" (Aubrecht et al., 2013, p2; Lloyd-Jones et al., 2016; Bosher et al., 2021). Here, emphasis is on better preparedness where communities, through learning and adapting, become more

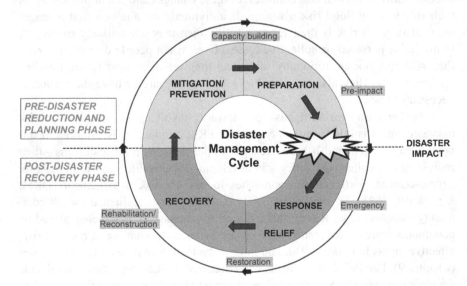

FIGURE 1.1   Disaster risk management conceived as a cycle with its typical phases (adapted from Bosher, 2005; Bosher and Chmutina, 2017; permission Lee Bosher)

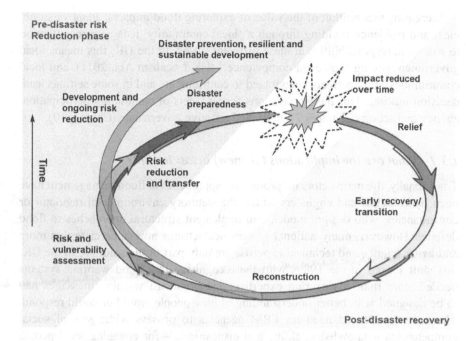

FIGURE 1.2   The disaster management spiral (after Lloyd-Jones, T. (ed) (2009) *The Built Environment Professions in Disaster Risk Reduction and Response*. RICS, ICE, RIBA and RTPI, Max Lock Centre, University of Westminster. Updated for this publication by the editor/principal author).

resilient rather than returning to pre-event conditions. For these aspirational reasons, the notion of a DRM spiral is adopted in this book.

1.3.6   What is happening with flood governance?

Another key consideration is the governance culture in risk management in different national settings, with distinct roles of state and citizen linked to political histories. Approaches to flood and wider governance in many countries are also evolving. This poses questions about the best scale at which governance for FRM should operate, on a continuum that runs from reliance on a centralised state, to increased devolved responsibility to citizens and 'the local'. In settings like the UK, a retreat of centralised state direction intertwines with localism agendas and the impacts of austerity.[6] Statutory agencies are increasingly concerned with ensuring and supporting increased roles of civil society (defined below) in extreme weather adaptation and wider disaster resilience. This agenda involves risk planning and action at different scales – from individual households and businesses, local and regional government, right up to the state and its relationships with its citizens, and international directives like the Sendai Framework.

Increasing recognition of the value of exploring flood impacts, adaptive strategies, and resilience building through a 'local community' lens requires attention to roles and responsibilities at these different scales. In the UK, this means local government having 'power of competence'[7] (UK Localism Act, 2011), and local communities being actively encouraged to contribute to, and in some settings lead, decision-making. This has implications for all facets of community participation, agency and action (Chapters 7 and 8), and adaptive governance (Chapter 10).

1.3.7   What are the implications for (new) actors in FRM?

Traditionally, the main actors in 'protective' approaches to flood management have been hydrologists and engineers within the statutory environmental regulator or consultancies, who design, model, and implement structural approaches to flood defence. However, many national governance settings now appreciate that more and better science and technical expertise are only part of the solution. In the UK, this shift started in the 1980s with the recognition that flood warning systems needed more than engineering expertise, and broadcast warning messages had to be designed with better understanding of how people would or could respond. There was evidenced need for FRM agencies to possess wider sets of social competencies – knowledge, skills, and dispositions – for engaging local people and understanding their behaviours. Locally focused, place-based FRM can now involve a wide range of stakeholders, statutory and non-statutory, including non-government organisations (NGOs), community flood groups, small businesses, and local media actors. This brings challenges.

> In this sense, efforts to confront flooding touch on a wide range of activities, policy fields, and stakeholders within river basins. No wonder, then, that flood risk management is typically characterised by high stakes, competing interests, and conflict!
>
> *Challies (2016, np)*

Relationships between academic institutions and local FRM practice are also being rethought, with hybridity of flood knowledge and differing perspectives increasingly valued. It is timely to revisit the roles of key actors in local FRM and actively explore potential for newer players to contribute community resilience thinking. This includes other guardians of local knowledge and promoters of safe spaces for dialogue (e.g., the cultural sector; Chapters 5, 6, and 10).

1.3.8   What is happening to hazard research?

A large literature exists on floods as environmental hazards – exploring interactions between human and physical domains and the role of social and cultural construction in how these are understood (Tierney, 2014). This work dates to the Chicago

School in the 1960s (Chapter 2), with its strong focus on individual risk perception (Chapter 3). More recently, increased research attention has been given to the roles of communities, their capitals or assets, capacities, and social networks in resilience building, alongside increasing the agency of individual citizens. Methods of research and practice are increasingly concerned about co-production of knowledge, and effective and meaningful participation (Chapters 4, 8, and 10). Again, strong potential exists for learning across different social, cultural, environmental, and governance contexts, as evidenced in EU-funded, cross-national research[8] on community flood resilience (e.g., WESENSIT[9]; STAR-FLOOD[10]). In addition, there is growing recognition of the value and complementarity of different types of data as evidence – science *and* story – reflecting different expertise in FRM (Chapters 4, 8, and 10).

1.3.9 In what ways is flooding a wicked problem?

Dealing with floods as hazards within climate change can be conceived as embodying many qualities of a 'wicked problem', characterised by subjectivity, complexity, interconnection, uncertainty, and divergence in terms of both floods *and* people and their behaviours. A 'wicked problem', first coined by Rittel and Webber (1973), is one that is ill-defined, involving many stakeholders with shared power relations. Christensen et al. (2009, piv) warned of the temptation to "succumb to [problem] complexity … by offering simplistic solutions to incorrectly defined problems". Differentiating a problem as tame (conventional) or wicked is an initial step (Table 1.2).

TABLE 1.2 Differentiating "Tame" and "Wicked" Problems (Rittel and Webber, 1973; Adapted from Blackman et al. 2006)

Dimension	A "Tame" or more routine problem	A "Wicked" problem
The problem	Clearly defined and stable	Lack of agreement about what the problem is
Thought processes	Linear	Systemic and complex
Solutions	Known when solution is reached	Attempts at solution change problem
Outcomes	Solutions are true or false	Solutions are good or bad
Approach to problem	Scientific; uses knowledge protocols	Draws on stakeholder views. Solutions are subjective and judgemental
Solving processes	Belongs to a group of similar problems solvable in similar ways	No principles of solution fit all members of a group of problems
Causes and boundaries	Causes are clear. Solutions can be tried and abandoned	Many causal levels exist. Problems are symptoms of other problems

Collaboration is critical in solving wicked problems; they benefit significantly from creative interdisciplinary approaches, relevant practical expertise, and transdisciplinary imagination (Brown et al., 2010), which draws on multiple knowledge cultures, and different skills and dispositions.

Participatory design processes that involve co-working with communities to deal with tame problems need rethinking for wicked problem-solving. Critical elements include agility of process; ability to undertake and support relationship building (e.g., trust, commitment); key roles (e.g., facilitating leadership[11]); engagement skills (e.g., active listening) and dispositions (e.g., empathy, propensity for authentic conversations, suspended judgement); promotion of specific ways of thinking (e.g., creative mindmapping, lateral and systems thinking); and alternative communication (e.g., visual representation). Scaffolding for participation may be needed, recognising participant diversity and different competencies. Solving wicked problems also needs to build "ongoing, adaptive learning into the process", with awareness of responsible power relations (Manville, 2016, np) and what constitutes effective communication (Chapter 6).

1.4 Framing key concepts in local FRM: Navigating language differences

So, what are the enablers and inhibitors of effective dialogue in community-focused FRM? An essential starting point is to recognise the challenges of terminology, where interpretations of key concepts might differ across academic disciplines and between the multiple stakeholders involved in FRM.

1.4.1 Defining 'civil society' and citizenship

Civil society as a concept is much discussed in academia (sociology, politics) and political arenas (Fowler, 2010); equally critical is what it means in practice, and its relationship to voluntary action locally. Civil society comprises individual civilians and groups of citizens but also organisations that work with civilians in resilience building (Table 1.3). Other related thematic concepts are civic agency and active citizenship, both with implications for learning and capacity-building needs (Chapters 8 and 9).

A Position Statement of US National Council for the Social Studies[12] (2001, p1) defined an effective citizen, among its characteristics, as someone who:

- "Accepts responsibility for the well-being of oneself, one's family, and the community"
- "Seeks information from varied sources and perspectives to develop informed opinions and creative solutions"
- "Uses effective decision-making and problem-solving skills in public and private life"
- "Actively participates in civic and community life"

Such citizenship can be critical to local FRM.

TABLE 1.3 Definitions of Civil Society and Civic Agency

Term	Definition	Source
Civil society	"'Third sector' of society, along with government and business. It comprises civil society organizations and non-governmental organizations"	UNESCO[a]
	"A community of citizens linked by common interests and collective activity"	Oxford 'Living' Dictionaries
Civic agency	"Capacity of human communities and groups to act cooperatively and collectively on common problems across their differences of view"	Boyte (2007, np)
	"Civic agency can also be understood in cultural terms, as practices, habits, norms, symbols and ways of life that enhance or diminish capacities for collective action"	
Active citizenship	"Participation that requires respect for others and that does not contravene human rights and democracy"	Hoskins (2014, np)

[a] https://whc.unesco.org/en/glossary/203.

1.4.2   Defining 'community' and 'communities'

An awareness of the potential heterogeneity of at-risk groups is essential to effective collective FRM. The term 'community' (from Latin *communis* meaning 'common, public, shared by all or many') is complex, contested, and used in diverse ways by different stakeholder groups (Shaw, 2012; Titz et al., 2018). Media coverage frequently attends to its demise or return (e.g., Niven, 2013[13]). The National Institute for Health and Clinical Excellence (NICE, 2008) defined 'community' as

> an umbrella term, to cover groups of people sharing a common characteristic or affinity, such as living in a neighbourhood, or being in a specific population group, or sharing a common faith or set of experiences.

The term is also used as a surrogate for other constructs like solidarity (Yerbury, 2011). Drawing on Yerbury (2011), South (2015, p7) acknowledged the word 'community' is used "as shorthand for the relationships, bonds, identities, and interests that join people together or give them a shared stake in a place, service, culture or activity". Its scale is a social unit for planning and action larger than an individual household. Complicating matters, spatial or geographical, place-based communities can be overlaid with communities of identity or interest (by vocation/workplace, hobby, faith, culture, values, and world view), communities of supporters (organisational and voluntary), and temporary communities (like students, tourists, or travellers) (Burns et al., 2004; UK Cabinet Office, 2011).

Researchers have attempted to classify community in other ways. For example, Colclough and Sitaraman (2005, p478/9) distinguished between simple, complex, and large communities. Simple communities are smaller and place-based where participants choose to be active "in a single dimension of their everyday lives" (e.g., hobby or sport). In contrast, complex communities comprise "many more groups or divisions, and typically include numerous activities in the lifeworlds of their members" like work, family, and friendships. This involves multiple communities in the same locale so individuals can be members of several communities simultaneously. However, "this does not necessarily ensure consensus or integration". Their third category involved larger, more permanent communities like immigrant communities. Such classifications highlight the dynamics of community development, and the potential complexities in relationships between place-based communities.

Within this book, the idea and implications of 'community' will be explored from internal and external perspectives. Coates' (2010) research on FRM in Yorkshire, UK, found that the term 'community' was variously understood as simply a group of individuals; residents of a defined area; inhabitants of a rural area; a group of at-risk people for whom flood managers have responsibilities; people within a government-defined political or administrative category (e.g., parish communities); or all people who lack 'expert' knowledge. She found that people felt part of a local community most intensely when:

> three aspects [of community], the social, spatial, and mental, were understood to be present [simultaneously]. That is, there was an attachment or sense of belonging to the locality; residents felt a shared, place-based identity and dense localised networks existed within the community boundaries.
>
> *Coates (2010, p198)*

Additionally, where a group of individuals find themselves affected by the same event, communities of circumstance can develop out of their shared experience of flood risk (UK Cabinet Office, 2011). Communities are not "neatly bounded" or "static" (Barrios, 2014, p330), and this sense of community may be short-lived as flood memory fades. However, the growth of online (virtual) communities for sharing memories and support (Garde-Hansen, 2015), and the development via social media of national and global communities of interest, can build and link communities of risk. For Twigger-Ross et al. (2011, p6), alongside spatial elements, community involves

> social relations and structures such as networks; and cognitive or psychological elements such as local or group identities and the creation of belonging/ exclusion.

Coates (2015, p55) further proposed the idea of the 'conscious community'; this builds on conceptualisation of community as "a structure of meaning, to show

how the cultural, spatial and social elements of local community creation are inextricably linked".

In sum, there are many dimensions to 'community', with different connotations and trajectories in different cultural settings. Households, as constituent units of these communities, can also have varied composition and functioning, for example, in the extent of intergenerational connection within families and in community functioning. Yet, as Dinnie and Fischer (2019, p243) have noted:

> policy narratives seldom acknowledge the multiple meanings associated with the term community; they, therefore, fail to make a distinction between community as it is experienced in everyday settings, and more formal community organising.

Both meanings are important in local FRM. This book uses the term 'community' inclusively to embrace connected groups of citizens while also recognising involvement of other local stakeholders (e.g., government, NGOs, business, academic institutions, and cultural sector), tasked to, or with the potential to, work with and for local communities in FRM.

1.4.3 Defining 'resilience'

Resilience (derived from Latin *resalire*, 'springing back' or 'rebounding') is also a contested and multi-faceted concept, with different disciplinary origins and uses, and an extensive and growing interdisciplinary literature (e.g., Adger, 2000; Evans and Reid, 2014; Kelman et al., 2015; Ungar, 2018; Rogers, 2020). The idea has been applied at wide-ranging scales and in different contexts – landscapes, ecosystems, species, cities, disasters, communities, organisations, businesses, and individuals. Understanding, defining, and promoting resilience (Norris et al., 2008; Keating et al., 2014; Davidson et al., 2016) are all key concerns, with resilience framed as both process and outcome (Table 1.4). Visualisation via ball-and-cup heuristics helps to clarify the different concepts of system stability and change applicable within different disciplinary domains (Figure 1.3; Tempels, 2016; Rölfer et al., 2022).

Recent definitions have highlighted potential for more flexible evolutionary senses of resilience, involving "transformative capacity" in response to changing and uncertain stresses (Béné et al., 2012, p21). Practically, resilience can be envisaged holistically as a spectrum of aspiration that requires flexibility in definition depending on context (McClymont et al., 2020). Whittle et al. (2010) made distinctions between observing resilience as

- *"Resistance – 'holding the line'*
- *Bounce-back – 'getting back to normal'*
- *Adaptation – 'adjusting to a new normal'*
- *Transformation – 'owning a need to change'".*

TABLE 1.4 Definitions of Resilience, Disciplinary Context and Scale of Analysis (Selected from MacKinnon and Derickson, 2013[a]; Quinlan et al., 2016[b]; Dubois and Krasny, 2016[c])

Resilience (disciplinary context)	Definition	Scale	Emphasis	Key references
Engineering resilience (engineering)	"System's speed of recovery to equilibrium following a shock"[b]	Physical system	Return time to recovery, efficiency, equilibrium[c]; constancy	Pimm (1984)
Ecosystem resilience (ecology)	"Ability of a system to withstand shock and maintain critical relationships and functions"[b]	Ecological system	Buffer capacity, withstand shock, persistence, robustness[c]	Holling (1996)
Social resilience (geography)	"Ability of groups or communities to cope with external stresses and disturbances as a result of social, political, and environmental change"[b]	Community	Social dimensions, heuristic device	Adger (2000, p347)
Socio-ecological resilience (varied disciplines)	Capacity of a socio-economic system to continually change, adapt, or transform so as to maintain ongoing processes in response to gradual and small-scale change or transform in the face of devastating change.[c]	Ecosystems comprised of ecological and social sub-systems	Adaptive capacity, learning, innovation	Berkes et al. (2003)
Psychological resilience (psychology)	"The capacity for successful adaptation and functioning despite high risk, stress and trauma"[a]	Individual	Adaptation	Egeland et al. (1993)

in dealing with emergencies (Twigger-Ross et al., 2011, p5). This thinking is evident in DFID's (2011) Disaster Resilience Framework (Figure 1.4), with its common elements of context, disturbance, capacity to deal with disturbance, and different reactions to disturbance. This framework is used to examine different kinds of resilience and establish existing levels.

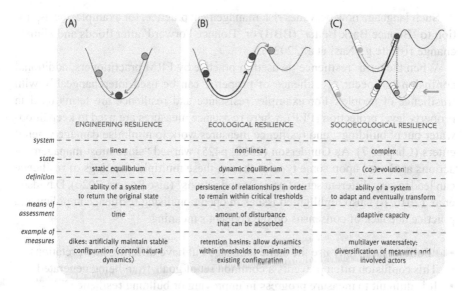

FIGURE 1.3   Schematic representation of (A) engineering, (B) ecological and (C) socio-ecological resilience. Resilience is illustrated by the position (i.e., state) of a ball (i.e., system) in a basin (i.e., conditions). Disturbances cause the ball to move (Tempels, 2016, p45, Figure 5 with permission)

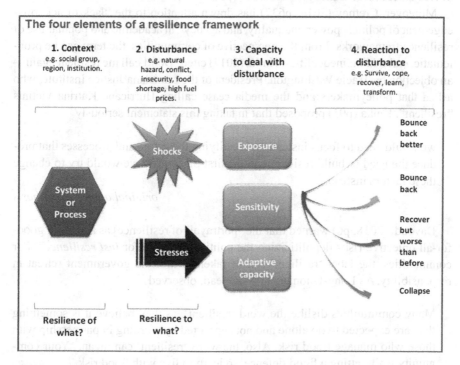

FIGURE 1.4   Components of a disaster resilience framework (DFID, 2011, p7, Crown copyright)

Such language now pervades risk management practice, for example, the aspiration to 'Bounce Back Better' (BBB) or 'Bounce Forward' after floods and climate change risks (e.g., Nasi et al., 2023).

When the term 'resilience' is used in practice by FRM practitioners, additional confusion can occur. 'Resilience of property' can be used interchangeably with 'resilience of people'. For example, resistance and resilience are terms used in property level protection (PLP), where resistance measures are used to keep floodwater out of buildings; and resilience measures work to minimise damage when it enters (Chapter 7). As Gunderson (2000, p425) warned "since most management actions are based upon some type of theory, these multiple meanings of resilience can lead to very different sets of policies and actions" (also Rogers, 2020). Davidson et al. (2016, p1) also highlighted issues in operationalising resilience in policy and practice due lack of consensus on the concept's meaning:

- "Policymakers may use the same language but have differing interpretations;
- This confusion often prevents a common set of goals from being generated;
- It is difficult to measure progress in improving or building resilience."

Arguing for "conceptual clarity" and "effective practice", they warned of "likelihood that the resilience concept will become nothing more than a rhetorical device unless these issues are resolved" (p9).

Moreover, Cretney (2014, p637) has drawn attention to the "lack of acknowledgement of politics, power, inequality, and agency" in academic and popular use of resilience frameworks. From the perspective of communities, the term can be problematic, embedding inequalities. Kaika (2017) cited 'Don't call me resilient again' – an objection by Tracie Washington, President of the Louisiana Justice Institute, who asked that policymakers and the media cease calling Hurricane Katrina victims "resilient". Kaika (p95) proposed that in taking this statement seriously

we would need to focus instead on identifying the actors and processes that produce the <u>need</u> to build resilience in the first place. And we would try to change these factors instead.

original author's emphasis

Davoudi (2018, p6) warned that the "portrayal [of resilience] as a neutral, good-for-all objective risks depoliticizing the political struggle for *just resilience*". For communities, the label 'resilient' can be taken to indicate government retreat in responsibility. As Long-Dhonau, an NGO lead, observed

Many communities dislike the word 'resilient', as they believe it is something they are expected to do alone and not supported by working in partnership with those who manage flood risk. Also, the word 'resilient' can mean, 'Your community isn't getting a flood defence' so learn to live with flood risk.

pers. comm.

1.4.4   Bringing 'community' and 'resilience' together

As Cutter et al. (2014, p65) stated:

> Resilience, especially the concept of community resilience is becoming the de facto framework for enhancing community-level disaster preparedness, response, and recovery in the short term, and climate change adaptation in the longer term.

The way community resilience is defined affects attempts to measure and enhance it (Patel et al., 2017). When the two contested terms 'community' and 'resilience' are put together, a multitude of potential meanings exist (Table 1.5), depending on whether resilience is considered "a metaphor, theory, set of capacities, and [or] strategy for disaster readiness" (Norris et al., 2008, p127).

TABLE 1.5 Selected Definitions of Community Resilience in Academia and Practice

Definition	Source
"A process linking a network of adaptive capacities (resources with dynamic attributes) to adaptation after a disturbance or adversity"	Norris et al. (2008, p127)
"Communities and individuals harnessing local resources and expertise to help themselves in an emergency, in a way that complements the response of the emergency services"	UK Cabinet Office (2011, p10)
"The capability to anticipate risk, limit impact, and bounce back rapidly through survival, adaptability, evolution, and growth in the face of turbulent change"	Community and Regional Resilience Institute (CARRI, 2013, p10)

Useful reviews of community resilience include Community and Regional Resilience Institute – CARRI (2013), Twigger-Ross et al. (2014), Patel et al. (2017), and Ntontis et al. (2019). CARRI (2013), in classifying definitions of community resilience, distinguished ontological definitions of "being" as an attribute versus phenomenological definitions of "becoming" as a process. Reviewing over 25 definitions, they proposed five core concepts within "definitions that are most valuable in terms of improving the ability of communities to recover after disasters explicitly or implicitly" (p2; Table 1.6).

In contrast, Patel et al. (2017) proposed a focus on nine core elements of community resilience: "local knowledge, community networks and relationships, communication, health, governance and leadership, resources, economic investment, preparedness, and mental outlook". They argued that breaking down the concept like this may be more productive than working to characterise and research it as a distinct entity. In another approach, Cutter et al. (2014) used different frames or lenses for appraising inherent aspects of place-based resilience, specifically flood resilience. These included social, economic, environmental, housing and infrastructure, institutional and community capitals (Section 8.3 of Chapter 8).

TABLE 1.6 Five Core Concepts in Definitions of Community Resilience That Maximise Value (CARRI, 2013, p2; Permission Meridian Institute)

Core concept	Explanation
Attribute	Resilience is an attribute of the community
Continuing	A community's resilience is an inherent and dynamic part of the community
Adaptation	The community can adapt to adversity
Trajectory	Adaptation leads to a positive outcome for the community relative to its state after the crisis, especially in terms of its functionality
Comparability	The attribute allows communities to be compared in terms of their ability to positively adapt to adversity

Evidence suggests socio-ecological resilience, dealing with complexity and change, is emerging as the dominant resilience interpretation within community resilience. Such systemic thinking has important implications for planning and practice, with the need to develop and implement adaptive, flexible management systems. Some strategic documents conceive community resilience in narrow ways, linking solely to the acute event (emergency) response phase, and relationships with specific groups like emergency services. Perspectives on community resilience within the DRM spiral influence what is understood by 'flood resilience practice' (e.g., whether short-term responsive or long-term adaptive; Wenger, 2017).

A concern can be difference in rhetoric and reality within practice of FRM organisations for developing community resilience. To quote the independent review of "the extreme weather event South Australia 2016" (Burn et al., 2017, pxiii)

> Resilience is often talked about; there is a national strategy agreed by the Council of Australian Governments (COAG), and whilst there is mention of it in plans and documents, it would appear that at this time, it is insufficient and requires a concerted effort to develop strategies, operationalise them and achieve government, community, and public 'buy-in'.

Ntontis et al. (2019, p11) also cautioned about the implications of different conceptualisations of community resilience for its operationalisation by policymakers and practitioners. They argued "a focus on enhancing some specific core aspects of community resilience might be more fruitful than attempts to enhance resilience in abstract".

In practice, there are also increasing concerns about fairness and equity in recovery particularly in the framings of 'urban resilience' – building "back better so that no one is left behind" (Collodi et al., 2021). Meerow and Newell (2019) in their articulation of the '5Ws', went beyond "just resilience for whom and what, but also where, when, and why". Indeed, DeVerteuil and Golubchikov (2016, p143) argued for a heterogeneous understanding of "critical resilience" linked to issues including social justice and power relations. Recognition of the need for social justice is made explicit in definitions by some government organisations. For example, the United States Agency for International Development (USAID) defines 'resilience' as

> the ability of people, households, communities, countries, and systems to miti-gate, adapt to, and recover from, shocks and stresses in a <u>manner that reduces chronic vulnerability and facilitates inclusive growth</u>.

> *author's emphasis*

Changing community resilience at a place over time also needs consideration – the notion of historic resilience, with resilience not necessarily increasing over time due to climate change, development, and increasing inequalities. Also more frequently in practice, consideration of community resilience is not specific to flooding but also to an increasing range of stresses that might occur at-a-place. The interdependency of spatial scales of focus is also important (McClymont et al., 2020). Community and individual resilience are affected by different factors, with Béné et al. (2016) arguing for social construction of resilience at household level.

There are alternative approaches proposed for community groups to foster. MacKinnon and Derickson (2013) offered the concept of resourcefulness in place of resilience. They identified four key elements as an initial framework: resources, skill sets and technical expertise, indigenous and folk knowledge and recognition (sense of confidence, self-worth, and self- and community-affirmation, group status).

1.4.5 'Exposure', 'vulnerability', 'environmental justice', and 'sustainability'

Physical exposure involves the assets located in flood-prone areas. Several additional terms are important in risk management, with meanings that have developed from different disciplines (Table 1.7).

Work on relationships between resilience and social vulnerability needs early reference (Miller et al., 2010), with their different disciplinary and theoretical origins. Like resilience, vulnerability has various definitions; care is needed in the temptation to define one in terms of the other. According to UNDRR Terminology (2017),[16] vulnerability can be defined as

> The characteristics determined by physical, social, economic and environmental factors or processes which increase the susceptibility of an individual, a community, assets or systems to the impacts of hazards.

> It directly affects disaster preparation, response, and recovery.

TABLE 1.7 Definitions of Further Key Terms Relevant to Local Flood Risk Management

Term	Definition	Source
A. Social vulnerability	"Characteristics of a person or group in terms of their capacity to anticipate, cope with, resist and recovery from the impact of a natural hazard"	Wisner et al. (2004, p11), http://svi.cdc.gov
B. Temporal vulnerability	Vulnerabilities which are inherently connected to the issue of time, timing, or temporality	De Vries (2011)
C. Environmental justice	Embraces notions of justice, based in recognition, capabilities, and participation (Schlosberg, 2007) Equitable exposure to environmental good and harm (Wolch et al., 2014)	Scholsberg (2007); Wolch et al. (2014)
D. Flood disadvantage	"Shows how flood-related social vulnerability combines with the potential for exposure to flooding. It accounts for both the likelihood of coming into contact with a flood and also the severity of negative impacts on the health and wellbeing of local communities that could occur as a result of that contact"	Climate Just[14]
E. Sustainability	Meeting our own needs without compromising the ability of future generations to meet their own needs	UNESCO[15]

A. Social vulnerability: Not all people or communities are affected by flooding to similar extent. Social vulnerability combines three sets of factors (England and Knox, 2015, p2): personal factors like age and health status (sensitivity); societal factors (adaptive capacity) including "income, tenure, mobility, social isolation, access to information and insurance"; and environmental factors that increase exposure (e.g., housing and neighbourhood characteristics). Blaikie et al. (1994), in their Pressure and Release (PAR) model, highlighted the progression of vulnerability determined by factors with increasing distance in space and time from the acute disaster event (simplified and focused on flooding in Figure 1.5). Unsafe conditions are caused by dynamic pressures and underlying social and political causes – with uneven distribution.

The progression of vulnerability

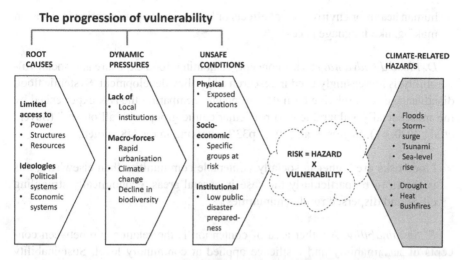

FIGURE 1.3 The Pressure and Release (PAR) model of vulnerability to disasters – here climate-related hazards (adapted from Blaikie et al., 1994 with permission; simplified version edited from Brink, 2018 by author)

As De Vries (2011, p163) pointed out, *(B) Temporal vulnerability* is also a critical concept in understanding response to flood warnings and other preparedness actions. This

> not only relates to the historical dimension of social vulnerability, not only to the dynamic multiscalarity of temporality in models of vulnerability, but also to how time itself provides bias in our understandings of social-history of vulnerability.

C. Environmental justice: Inequality is the "unequal and/or unjust distribution of resources and opportunities among members of a given society" (Koh, 2020). Environmental justice (EJ), a concern for patterns of environmental inequality and injustice, has its origins in US social movements, linked to civil rights. This recognised that environmental pollution and technological risk were disproportionately in areas occupied by Black, Asian and Minority Ethnic (BAME) communities.

Over recent years, there has been a strong emphasis on bringing ideas of EJ and inequalities of burden to flooding (e.g., Cutter, 2006; Schlosberg, 2007; Walker, 2012; Chapter 2). As Morse (2008, p1) reflected on the goal of EJ in context of Hurricane Katrina (in 2005):

> in the legal realm, … to secure for all communities and persons the same degree of protection from environmental and health hazards, and the same opportunity to influence the decision-making process. This objective is not met when low-income or minority communities are burdened disproportionately by adverse

human health or environmental effects or by barriers to participation in decision making, like language access.

D. Flood disadvantage: This concept, combining flood exposure and social vulnerability, is increasingly used in research and policy development. Systemic flood disadvantage is "a relative term determined by comparing the risk experienced by the most socially vulnerable or a particular ethnic group with all others" (Sayers et al., 2020, p1). Sayers et al. (2018, p339), working in the UK, noted:

> Flood risks are higher in socially vulnerable communities than elsewhere; this is shown to be particularly the case in coastal areas, economically struggling cities, and dispersed rural communities.

E. Sustainability: Another area of contention is the relationship between concepts of sustainability and resilience applied at community level. Sustainability links concern for environmental, economic, and social resources. As Lew et al. (2016, p23) articulated, communities need to keep asking two goal-focused questions that may be challenging to negotiate collectively:

- *What do we want to protect and conserve, and to keep from changing? (sustainability)*
- *What do we want to adapt and change into something new, and maybe better? (resilience)*

This distinction applies in brokering between Learning for Sustainability (LfS) and Learning for Resilience (LfR) within FRM and in recognising their potential areas of intersection and difference (Chapter 9). In delivering on both, interdisciplinary focus is critical.

1.5 Interdisciplinary working in risk management: Research and practice

Navigating this complex conceptual territory requires awareness of the opportunities and challenges of interdisciplinary and transdisciplinary working in research and practice (Table 1.8). These are differentiated from disciplinary and multidisciplinary approaches to problem-solving (Figure 1.6). The chapter now appraises what these approaches offer to increase understanding within different domains of community FRM.

1.5.1 Interdisciplinary approaches

"Interdisciplinary research and collaborations are essential to disentangle complex and wicked global socio-ecological challenges" (Kelly et al., 2019, p149). Whereas traditionally under a flood defence paradigm, hydrology and engineering were

TABLE 1.8 Differentiating Interdisciplinary, Multidisciplinary, and Transdisciplinary Working

Term	Definition
Interdisciplinary research	"A mode of research by teams or individuals that integrates information, data, techniques, tools, perspectives, concepts, and/or theories from two or more disciplines ... to solve problems" (US Institute of Medicine, 2005, p2)
Multidisciplinary research	"Involves researchers from two or more disciplines working collaboratively on a common problem, without modifying disciplinary approaches or developing synthetic conceptual frameworks" (Graybill et al., 2006, p757)
Transdisciplinary research	"Different academic disciplines working together with non-academic collaborators to integrate knowledge and methods, to develop and meet shared research goals ..." (Kelly et al., 2019, p150)

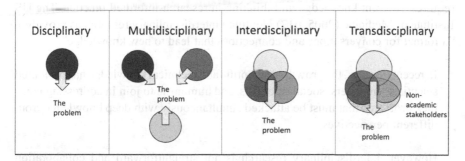

FIGURE 1.6   Visualisation differentiating types of disciplinary working

dominant disciplines in flood management, increased recognition exists that other disciplines have important collaborative roles to play. Hence people and organisations, working in FRM, need to become aware of what interdisciplinary working means (or could mean), with the imperative to "embrace complexity" and "push your boundaries" (Kelly et al., 2019, p149). To quote Popper (1963, p88),

> We are not students of some subject matter, but students of problems. And problems may cut right across the borders of any subject matter or discipline.

Selby (2006, p59) also reflected – in context of LfS – but applicable here:

> We have been great at thinking the world apart; what we need to do now is to think it back together.

However, interdisciplinary working requires awareness that different academic disciplines are characterised by their own theories of knowledge, disciplinary concepts, methodologies, languages and means of communication, identity and culture, learning styles, and worldviews (Becher and Trowler, 2001).

Definitions for research: Interdisciplinary working involves negotiating and integrating approaches and methods between academic disciplines, contrasting with multidisciplinary working (Figure 1.6; Table 1.6). These different terms can, however, be used without precision. Selby (2006) usefully differentiated between 'contiguous' (or proximate) and 'non-contiguous' (or non-proximate) interdisciplinarity in terms of closeness of academic disciplines (e.g., sciences working with sciences, cf. sciences working with arts). The potential value and challenge of non-contiguous interdisciplinarity should be noted – when disciplines are more distant and without histories of co-working to tackle a 'wicked' problem.

Application: Traditionally research approaches to flood management have tended to sit in disciplinary or professional silos. So, engineers have focused on structural measures, hydrologists on modelling while social scientists focused on individual risk perceptions and behaviours. Investigating 'wicked' problems requires different knowledge and thinking in research-informed practice. The US Institute of Medicine (2005, p17) identified interdisciplinary research as providing "a format for conversations and connections that lead to new knowledge".

In recent decades, the growth of scientific and technical knowledge has prompted scientists, engineers, social scientists, and humanists to join in addressing complex problems that must be attacked simultaneously with deep knowledge from different perspectives.

However, interdisciplinary research is not straightforward and collaboration with researchers from diverse backgrounds involves risk and challenge. Karlqvist (1999, p379) cautioned about significant steps between "appealing idea" and "operational method", requiring "understanding of the disciplines themselves" and "how to connect disciplinary knowledge". Interdisciplinarity, as aspiration, needs to be made explicit as a process and goal; time and space are required for dialogue, imagination, and making connections alongside early surfacing of issues. US Institute of Medicine (2005 p2/3) also noted "It may take extra time for building consensus and for learning new methods, languages, and cultures". However, as Tait and Lyall (2007, p1) observed, "the rewards can be substantial, in terms of advancing the knowledge base and helping to solve complex societal problems".

Some disciplines are well placed to act as hubs for interdisciplinary working. For example, geographers and psychologists already communicate across the natural-social science interface within their disciplines. Lyall et al. (2011) proposed valuable characteristics in interdisciplinary researchers (Box 1.1).

Challenges and opportunities of research that involves non-contiguous or 'unexpected' interdisciplinarity include different understandings of what research might constitute (e.g., between sciences and arts); how a creative and emergent methodology might be developed; what is evaluated (e.g., the artefact or people's reactions to it); what are considered as research outcomes; and what comprises

---

BOX 1.1   USEFUL CHARACTERISTICS OF AN INTERDISCIPLINARY RESEARCHER (LYALL ET AL. 2011, KEY ADVICE 3.1, P35, WITH PERMISSION)

In interdisciplinary research, personality and attitudes of researchers are at least as important for success as discipline base and specialisation. Useful characteristics are:

- flexibility, adaptability, creativity;
- curiosity about, and willingness to learn from other disciplines;
- an open mind to ideas coming from other disciplines and experiences;
- good communication and listening skills;
- an ability to bridge the gap between theory and practice;
- a good team worker;
- a willingness to tolerate ambiguity.

'impact' (e.g., its scale – individual transformation or policy change). This also requires a rethinking of relationships between research, knowledge exchange, and community engagement. Finally, pursuing interdisciplinary territory can be challenged in some national settings where historically interdisciplinary research is perceived as less safe in terms of reward in national, peer-reviewed, research excellence frameworks that determine individual academic progression. Transformation of academic research cultures can be slow.

1.5.2   Transdisciplinary working

This discussion leads to post-disciplinary thinking through transdisciplinary research (TR), defined internationally in different ways. These range from a "diffuse conceptual term located above individual disciplines to any research that involves stakeholders" (Hadorn et al., 2008, pvii). Wickson et al. (2006) distinguished TR through three characteristics: its problem focus, evolving methodology, and collaboration. Transdisciplinarity brings valuable opportunities for co-working with communities, using participatory methods with impact (e.g., knowledge exchange) through their processes, as well as in their outcomes. Two developmental areas in TR explored in this book are co-production and participatory methods (Chapters 4, 8, and 10; Polk, 2015; McEwen, 2023).

1.6   Different types of thinking: Application to researching floods and communities

Combining interdisciplinary and transdisciplinary research approaches to study interactions between risk and communities for increased resilience has potential to draw on, and promote, different ways of thinking. Western and Chinese

TABLE 1.9 Different Cognitive Styles for Organising Knowledge Relevant to FRM

Category of thinking	Character	Links to practice	Examples	References
(A) Holistic thinking (opposite of analytical thinking)	Focuses on context to a situation. Recognises interconnectedness and interdependencies of disparate elements that form larger patterns	Key skill in LfS	"Interactive planning for flood resilience" (Vojinović, 2015)	Kasser (2015)
(B) Systems thinking (branch of holistic thinking)	Examines links, interactions, feedbacks, and causality between components	Seeing world in a particular way affects how situations are approached or specific tasks undertaken	Systems thinking using river catchment as its spatial unit A system of systems approach to managing complex interdependent risks (Cavallo and Ireland, 2014)	Ackoff (2018)
(C) Conception as hydrosocial cycle	Involves flows beyond physical water of narratives, knowledge, responsibility, power, and resources	Focus on wider interrelationships with water (human value systems)	Potential to understand relatedness of downstream communities through and with floodwater flows (Krause, 2016)	Swyngedouw (2009); Linton and Budds (2014)

philosophies around knowledge contrast: the former underpinned by assumptions that it is possible to abstract an element or event from its environment and study it in isolation; the latter emphasising interconnectedness and interdependency of scales (Wong, 2020). Several different frames for organising knowledge are relevant to FRM (Table 1.9; Chapter 4), including holistic thinking (A).

Systems thinking (Table 1.9B): Hydrological cycling and systems thinking emphasise connections, connectivity, and flows. Thinking and talking in systems is the traditional language of environmental scientists (e.g., hydrologists, ecologists, and geomorphologists) but this concept now possesses wider reach (Senge, 2006). Key concepts of systems thinking (Twhink.org[17]) are inter-connected parts, links between system structure and dynamic and emergent behaviours, and controlling feedback loops. Complex social systems can exhibit counterintuitive behaviour. Systems thinking allows identification of leverage points, where interventions can be made influencing system functioning. Notions of trigger and tipping points within human and physical systems are also useful in transforming FRM.

Conception as hydrosocial cycle (Table 1.9C): The hydrological cycle is concerned with physical water flows. In contrast, Linton and Budds (2014, p170) defined the hydrosocial cycle as attending

> to the social nature of these flows as well as the agential role played by water, while highlighting the dialectical and relational processes through which water and society interrelate.

They proposed this cycle as "both an analytical tool for investigating hydrosocial relations, and as a broader framework for undertaking critical political ecologies of water" (p170). More recently, the value of such thinking has been critiqued in its contribution to understanding relationships between flooding and society, and human experience of (flood)water and risky rivers (e.g., Camargo and Cortesi, 2019). This includes questioning of a hydrosocial emphasis on the human (cf. ecology), and the notion that everything associated with water cycles. They argued that floods provide valuable opportunities to rethink analytical tools for studying water and for increased interdisciplinary working.

1.7 Scope and structure of this book

Against this backdrop, this book now explores several domains within FRM through a 'community lens'. This is an extensive and rapidly developing territory; the book aims to provoke thought but cannot be exhaustive. It asks questions and provides an accessible way into interdisciplinary territories, aspiring to take readers on a personal journey that connects diverse research and practice for new or increased understandings. Importantly, this process aims to underpin improved community resilience to changing flood risk, recognising the major, often emotional, challenge this represents for at-risk communities, and the diverse stakeholders that work with them. It aspires to promote holistic, systemic, and creative thinking in local FRM, with inter-linking of themes between its chapters.

Notes

1 https://www.eea.europa.eu/data-and-maps/indicators/river-floods-3/assessment.
2 https://www.undrr.org/publication/sendai-framework-disaster-risk-reduction-2015–2030.
3 The Framework was adopted at the Third UN World Conference on DRR in Sendai, Japan, in 2015.
4 https://www.munichre.com/topics-online/en/climate-change-and-natural-disasters/natural-disasters/floods/no-complete-protection-flooding.html.

5 https://www.undp.org/sustainable-development-goals.
6 "Difficult economic conditions created by government measures to reduce public expenditure" (Oxford Languages).
7 Gives local government power to do anything an individual can do provided it is not prohibited by other legislation.
8 https://ec.europa.eu/environment/water/flood_risk/links.htm.
9 https://www.wesenseit.com/.
10 www.starflood.eu.
11 https://www.forbes.com/sites/brookmanville/2016/05/15/six-leadership-practices-for-wicked-problem-solving/.
12 https://www.socialstudies.org/sites/default/files/publications/se/6505/650511.html.
13 https://www.theguardian.com/voluntary-sector-network/2013/may/03/community-spurs-fans.
14 https://www.climatejust.org.uk/glossary/.
15 https://www.un.org/en/academic-impact/sustainability.
16 https://www.undrr.org/terminology/vulnerability.
17 https://www.thwink.org/sustain/glossary/SystemsThinking.htm.

References

Ackoff, R. (2018) A lifetime of systems thinking. https://thesystemsthinker.com/a-lifetime-of-systems-thinking/

Adger, W.N. (2000) Social and ecological resilience: Are they related? *Progress in Human Geography*, 24, 347–364.

Alfieri, L., Bisselink, B., Dottori, F., Naumann, G., de Roo, A., Salamon, P., Wyser, K. and Feyen, L. (2017) Global projections of river flood risk in a warmer world. *Earth's Future*, 5, 171–182. https://doi.org/10.1002/2016EF000485

Alfieri, L., Burek, P., Feyen, L. and Forzieri, G. (2015) Global warming increases the frequency of river floods in Europe. *Hydrology and Earth System Sciences*, 19(5), 2247–2260. https://doi.org/10.5194/hess-19-2247-2015

Alfieri, L., Dottori, F., Betts, R., Salamon, P. and Feyen, L. (2018) Multi-model projections of river flood risk in Europe under global warming. *Climate*, 6(1), 6.

Aubrecht, C., Ozceylan Aubrecht, D., Klerx, J. and Freire, S. (2013) Future-oriented activities as a concept for improved disaster risk management. *Disaster Advances*, 6(12), 1–10.

Australian Institute of Disaster Resilience (2002) *Disaster loss assessment guidelines*, Manual 27, Australian Emergency Manual Series, Attorney-General's Department, Australian Government, Canberra. https://knowledge.aidr.org.au/media/1967/manual-27-disaster-loss-assessment-guidelines.pdf

Baker, V.R. (2007) Flood hazard science, policy, and values: A pragmatist stance. *Technology in Society*, 29, 161–168.

Barrios, R.E. (2014) 'Here, I'm not at ease': Anthropological perspectives on community resilience. *Disasters*, 38, 329–350. https://doi.org/10.1111/disa.12044

Becher, T. and Trowler, P. (2001) *Academic Tribes and Territories: Intellectual Enquiry and the Culture of Disciplines* (2nd edn). Open University Press/SRHE, Buckingham.

Béné, C., Al-Hassan, R.M., Amarasinghe, O., Fong, P., Ocran, J., Onumah, E., Ratuniata, R., Van Tuyen, T., Allister McGregor, J. and Mills, D.J. (2016) Is resilience socially constructed? Empirical evidence from Fiji, Ghana, Sri Lanka, and Vietnam. *Global Environmental Change*, 38, 153–170. https://doi.org/10.1016/j.gloenvcha.2016.03.005

Béné, C., Godfrey-Wood, R., Newsham, A. and Davies, M. (2012) *Resilience: New Utopia or New Tyranny? Reflection about the Potentials and Limits of the Concept of Resilience in Relation to Vulnerability Reduction Programmes*. IDS working Paper 405, Institute of Development Studies, Brighton, pp. 61.

Berkes, F., Colding, J. and Folke, C. (2003) *Navigating Social-Ecological Systems: Building Resilience for Complexity and Change*. Cambridge University Press, Cambridge, UK.

Blackman, T., Greene, A., Hunter, D.J., McKee, L., Elliott, E., Harrington, B., Marks, L. and Williams, G. (2006) Performance assessment and wicked problems: The case of health inequalities. *Public Policy and Administration, 21*(2), 66–80. https://doi.org/10.1177/095207670602100206

Blaikie, P., Cannon, R., Davis, I. and Wisner, B. (1994) *At Risk: Natural Hazards, People's Vulnerability and Disasters*. Routledge, New York, 284pp.

Bosher, L. (2005) *The Divine Hierarchy: the Social and Institutional Elements of Vulnerability in South India*. Unpublished PhD, University of Loughborough, University of Loughborough.

Bosher, L.S. and Chmutina, K. (2017) *Disaster Risk Reduction for the Built Environment*, Wiley, London.

Bosher, L., Chmutina, K. and van Niekerk, D. (2021) Stop going around in circles: Towards a reconceptualisation of disaster risk management phases *Disaster Prevention and Management*, 30(4/5), 525–537. https://doi.org/10.1108/DPM-03-2021-0071

Boyte, H.C. (2007) Building civic agency: The public-work approach. https://www.opendemocracy.net/en/building_civic_agency_the_public_work_approach/

Brink, E. (2018) *Adapting Cities: Ecosystem-Based Approaches and Citizen Engagement in Municipal Climate Adaptation in Scania, Sweden. Unpublished PhD, Lundt University.*

Brown, V.A., Harris, J.A. and J.Y, R. (eds.) (2010) *Tackling Wicked Problems through the Transdisciplinary Imagination*. Earthscan, London.

Burn, G., Adams, L. and Buckley, G. (2017) *Independent review of the extreme weather event South Australia, 28 September–5 October 2016*. Report presented to Premier of South Australia. Available at: https://www.dpc.sa.gov.au/__data/assets/pdf_file/0003/15195/Independent-Review-of-Extreme-Weather-complete.pdf

Burns, D., Heywood, F., Taylor, M., Wilde, P. and Wilson, M. (2004) *Making Community Participation Meaningful: A Handbook for Development and Assessment*. Published for Joseph Rowntree Foundation, The Policy Press, Bristol.

Camargo, A. and Cortesi, L. (2019) Flooding water and society. *Wiley Interdisciplinary Reviews: Water*, 6. https://doi.org/10.1002/wat2.1374

Cavallo, A. and Ireland, V. (2014) Preparing for complex interdependent risks: A system of systems approach to building disaster resilience. *International Journal of Disaster Risk Reduction*, 9, 181–193.

Challies, E. (2016) Participation and collaboration for sustainable flood risk management? https://sustainability-governance.net/2016/03/07/participation-and-collaboration-for-sustainable-flood-risk-management/, 7th March 2016.

Christensen, C.M., Grossman, J.H. and Hwang, J. (2009) *The innovator's Prescription: a Disruptive Solution for Health Care*. McGraw Hill Medical, New York.

Coates, T. (2010) Conscious community: belonging, identities and networks in local communities' response to flooding. PhD thesis, Middlesex University. https://eprints.mdx.ac.uk/6592/

Coates, T. (2015) Understanding local community construction through flooding: The 'conscious community' and the possibilities for locally based communal action: Understanding local community construction through flooding. *Geo: Geography and Environment*, 2, 55–68. https://doi.org/10.1002/geo2.6

Colclough, G. and Sitaraman, B. (2005) Community and social capital: What is the difference? *Sociological Inquiry*, 75(4), 474–496.

Collodi, J., Pelling, M., Fraser, A., Borie, M. and Di Vicenz, S. (2021) How do you build back better so no one is left behind? Lessons from Sint Maarten, Dutch Caribbean, following Hurricane Irma. *Disasters*, 45(1), 202–223.

Community and Regional Resilience Institute (2013) *Definitions of Community Resilience: An Analysis*. CARRI Report. Available at https://www.merid.org/wp-content/uploads/2019/08/Definitions-of-community-resilience.pdf

Cretney, R. (2014) Resilience for whom? Emerging critical geographies of socio-ecological resilience. *Geography Compass*, 8, 627–640. https://doi.org/10.1111/gec3.12154

Cutter, S. L. (2006) *Hazards Vulnerability and Environmental Justice*. Routledge, London. https://doi.org/10.4324/9781849771542

Cutter, S.L., Ash, K.D. and Emrich, C.T. (2014) The geographies of community disaster resilience. *Global Environmental Change*, 29, 65–77. https://doi.org/10.1016/j.gloenvcha.2014.08.005

Davidson, J.L., Jacobson, C., Lyth, A., Dedekorkut-Howes, A., Baldwin, C.L., Ellison, J.C., Holbrook, N.J., Howes, M.J., Serrao-Neumann, S., Singh-Peterson, L. and Smith, T.F. (2016) Interrogating resilience: Toward a typology to improve its operationalization. *Ecology and Society*, 21(2), 27. http://dx.doi.org/10.5751/ES-08450-210227

Davoudi, S. (2018) Just resilience. *City & Community*, 17(1). https://doi.org/10.1111/cico.12281

De Vries, D.H. (2011) Temporal vulnerability in hazardscapes: Flood memory-networks and referentiality along the North Carolina Neuse River (USA). *Global Environmental Change*, 21, 154–164. http://dx.doi.org/10.1016/j.gloenvcha.2010.09.006

DeVerteuil, G. and Golubchikov, O. (2016) Can resilience be redeemed? *City*, 20(1), 143–151. https://doi.org/10.1080/13604813.2015.1125714

DFID (2011) *Defining disaster resilience: A DFID approach paper*. DFID. https://assets.publishing.service.gov.uk/government/uploads/system/uploads/attachment_data/file/186874/defining-disaster-resilience-approach-paper.pdf

Dilley, M., Chen, R.S., Deichmann, U., Lerner-Lam, A.L. and Arnold, M. (2005) *Natural Disaster Hotspots: A Global Risk Analysis*. World Bank, Washington, DC.

Dinnie, E. and Fischer, A. (2019) The trouble with community: How 'sense of community' influences participation in formal, community-led organisations and rural governance. *Sociologia Ruralis*, 60(1), 243–259.

DuBois, B. and Krasny, M. (2016) Educating with resilience in mind: Addressing climate change in post-Sandy New York City. *The Journal of Environmental Education*, 47, 255–270.

Egeland, B., Carlson, E. and Sroufe, L. (1993) Resilience as process. *Development and Psychopathology*, 5, 517–528.

England, K. and Knox, K. (2015) *Targeting Flood Investment and Policy to Minimise Flood Disadvantage*. Joseph Rowntree Foundation, UK.

European Environment Agency (2016) Flood risks and environmental vulnerability. Exploring the synergies between floodplain restoration, water policies and thematic policies. Report No 1/2016. https://www.eea.europa.eu/publications/flood-risks-and-environmental-vulnerability

Evans, B. and Reid, J. (2014) *Resilient Life: the Art of Living Dangerously*. Wiley, Chichester.

Fowler, A. (2010) Civic agency. In H.K. Anheier and S. Toepler (eds.) *International Encyclopedia of Civil Society*. Springer, New York. https://doi.org/10.1007/978-0-387-93996-4_69

Garde-Hansen, J. (2015) *Digital Memories and media of the Future*. The Routledge Companion to British Media History, The Routledge Companion to British Media History, pp600–611.

Gober, P. and Wheater, H.S. (2015) Debates – perspectives on sociohydrology: Modeling flood risk as a public policy problem. *Water Resources Research*, 51, 4782–4788. https://doi.org/10.1002/2015WR016945.

Graybill, J.K., Dooling, S., Shanda, V., Withey, J., Greve, S. and Simon, G.L. (2006) A rough guide to interdisciplinarity: Graduate student perspectives. *BioScience*, 56(9), 757–763. https://doi.org/10.1641/0006-3568(2006)56[757:ARGTIG]2.0.CO;2

Gunderson, L.H. (2000) Ecological resilience – in theory and application. *Annual Review of Ecology and Systematics*, *31*, 425–439. http://www.jstor.org/stable/221739

Hadorn, G.H. et al. (2008) The emergence of transdisciplinarity as a form of research. In G.H. Hadorn et al. (eds.) *Handbook of Transdisciplinary Research*. Springer, Dordrecht. https://doi.org/10.1007/978-1-4020-6699-3_2

Holling, C.S. (1996) Engineering resilience versus ecological resilience. In P. Schulze (ed.) *Engineering within Ecological Constraints*. National Academies Press, Washington, DC, pp31–44.

Hoskins, B. (2014) Active citizenship. In A.C. Michalos (ed.) *Encyclopedia of Quality of Life and Well-Being Research*. Springer, Dordrecht. https://doi.org/10.1007/978-94-007-0753-5_16

Institute of Medicine (2005) *Facilitating Interdisciplinary Research*. The National Academies Press, Washington, DC. https://doi.org/10.17226/11153.

IPCC (2021) Climate Change 2021: The Physical Science Basis. Contribution of Working Group I to the Sixth Assessment Report of the Intergovernmental Panel on Climate Change [Masson-Delmotte, V., P. Zhai, A. Pirani, S.L. Connors, C. Péan, S. Berger, N. Caud, Y. Chen, L. Goldfarb, M.I. Gomis, M. Huang, K. Leitzell, E. Lonnoy, J.B.R. Matthews, T.K. Maycock, T. Waterfield, O. Yelekçi, R. Yu, and B. Zhou (eds.)]. Cambridge University Press, Cambridge, United Kingdom and New York, NY, USA, 2391 pp. doi:10.1017/9781009157896.

IPCC (2022) Summary for policymakers. In H.-O. Pörtner, D.C. Roberts, E.S. Poloczanska, K. Mintenbeck, M. Tignor, A. Alegría, M. Craig, S. Langsdorf, S. Löschke, V. Möller, and A. Okem (eds) *Climate Change 2022: Impacts, Adaptation, and Vulnerability. Contribution of Working Group II to the Sixth Assessment Report of the Intergovernmental Panel on Climate Change*. Cambridge University Press, Cambridge and New York, USA, pp. 3–33.

Kaika, M. (2017) *'Don't call me resilient again!'*: The New Urban Agenda as immunology … or … what happens when communities refuse to be vaccinated with 'smart cities' and indicators. *Environment & Urbanization*, 29(1), 89–102. https://doi.org/10.1177/0956247816684763

Karlqvist, A. (1999) Going beyond disciplines: The meanings of interdisciplinarity. *Policy Sciences*, 32(4), 379–383.

Kasser, J.E. (2015) *Holistic Thinking: Creating Innovative Solutions to Complex Problems* (2nd edn). The Right Requirement, Cranfield, UK.

Keating, A., Campbell, K., Mechler, R. et al. (2014) *Operationalizing Resilience against Natural Disaster Risk: Opportunities, Barriers, and a Way Forward*. Zurich Flood Resilience Alliance. https://pure.iiasa.ac.at/id/eprint/11191/1/zurichfloodresiliencealliance_ResilienceWhitePaper_2014.pdf

Kelly, R., Mackay, M., Nash, K.L. et al. (2019) Ten tips for developing interdisciplinary socio-ecological researchers. *Socio-Ecological Practice Research*, 1, 149–161 https://doi.org/10.1007/s42532-019-00018-2

Kelman, I., Gaillard, J.C. and Mercer, J. (2015) Climate change's role in disaster risk reduction's future: Beyond vulnerability and resilience. *International Journal of Disaster Risk Science*, 6, 21–27. https://doi.org/10.1007/s13753-015-0038-5

Koh, S. Y. (2020) Inequality. *International Encyclopedia of Human Geography* (2nd edn). https://www.sciencedirect.com/topics/social-sciences/inequality

Krause, F. (2016) "One man's flood defense is another man's flood": Relating through water flows in Gloucestershire, England. *Society & Natural Resources*, 29(6), 681–695. https://doi.org/10.1080/08941920.2015.1107787

Lew, A.A., Ng, P.T., Ni, C. and Wu, T. (2016) Community sustainability and resilience: Similarities, differences and indicators. *Tourism Geographies*, 18(1), 18–27. https://doi.org/10.1080/14616688.2015.1122664

Linton, J. and Budds, J. (2014) The hydrosocial cycle: Defining and mobilizing a relational-dialectical approach to water. *Geoforum*, 57, 170–180. https://doi.org/10.1016/j.geoforum.2013.10.008

Lloyd-Jones, T. (ed.) (2009) *The Built Environment Professions in Disaster Risk Reduction and Response*. RICS, ICE, RIBA and RTPI, Max Lock Centre, MLC Press, University of Westminster.

Lloyd-Jones, T., Davis, I. and Steele, A. (2016) *Topic Guide: Effective Post-disaster Reconstruction Programmes*. *Evidence on Demand*, UK. xiv, 93p. Department for International Development. Available from: https://www.gov.uk/dfid-research-outputs/topic-guide-effective-post-disaster-reconstruction-programmes

Lyall, C., Bruce, A., Tait, J. and Meagher, L. (2011) *Interdisciplinary Research Journeys: Practical Strategies for Capturing Creativity*. Bloomsbury Academic, London.

Mackinnon, D. and Derickson, K. (2013) From resilience to resourcefulness: A critique of resilience policy and activism. *Progress in Human Geography*, 37. https://doi.org/10.1177/0309132512454775

Manville, B. (2016) Six Leadership Practices for 'Wicked' Problem Solving. *Forbes* https://www.forbes.com/sites/brookmanville/2016/05/15/six-leadership-practices-for-wicked-problem-solving/

McClymont, K., Morrison, D., Beevers, L. and Carmen, E. (2020) Flood resilience: A systematic review. *Journal of Environmental Planning and Management*, 63(7), 1151–1176. https://doi.org/10.1080/09640568.2019.1641474

McEwen, L.J. (2023) Co-production and the role of lay knowledge in community resilience: Learnings for local flood risk management. Chapter 23. In J. Lamond, D. Proverbs and N. Bhattacharya-Mis (eds.) *Research Handbook on Flood Risk Management*. Edward Elgar Publishing, Cheltenham, UK.

Meerow, S. and Newell, J.P. (2019) Urban resilience for whom, what, when, where, and why? *Urban Geography*, 40(3), 309–329. https://doi.org/10.1080/02723638.2016.1206395

Miller, F., Osbahr, H. and Boyd, E. (2010) Resilience vulnerability: Complementary or conflicting concepts? *Ecology and Society*, 15. https://doi.org/10.5751/ES-03378-150311

Montz, B.E. and Tobin, G.A. (2008) From false sense of security to residual risk: Communicating the need for new floodplain development models. *Geograficky Casopis (Geographical Journal – Slovak Academy of Sciences)*, 60(1), 3–14.

Morse, R. (2008) *Environmental Justice through the Eye of Hurricane Katrina*. Washington, DC, Joint Center for Political and Economic Studies, Health Policy Institute.

Nasi, V.L., Jans, L. and Steg, L. (2023) Can we do more than "bounce back"? Transilience in the face of climate change risks. *Journal of Environmental Psychology*, 86, 101947.

National Institute for Health and Clinical Excellence (NICE) (2008) *Community Engagement to Improve Health*. NICE, London.

Niven, R. (2013) The complexity of defining community. *The Guardian*. https://www.theguardian.com/voluntary-sector-network/2013/may/03/community-spurs-fans

Norris, F.H., Stevens, S.P., Pfefferbaum, B., Wyche, K.F. and Pfefferbaum, R.L. (2008) Community resilience as a metaphor, theory, set of capacities, and strategy for disaster readiness. *Journal of Community Psychology*, 41, 127–50.

Ntontis, E., Drury, J., Amlôt, R., Rubin, G.J. and Williams, R. (2019) Community resilience and flooding in UK guidance: A critical review of concepts, definitions, and their implications. *Journal of Contingencies and Crisis Management*, 27(1), 2–13.

Paprotny, D., Sebastian, A. and Morales-Nápoles, O. et al. (2018) Trends in flood losses in Europe over the past 150 years. *Nature Communications*, 9, 1985. https://doi.org/10.1038/s41467-018-04253-1

Patel, S.S., Rogers, M.B., Amlôt, R. and Rubin, G.J. (2017). What do we mean by 'community resilience'? A systematic literature review of how it is defined in the literature. *PLoS Currents*, 9. https://doi.org/10.1371/currents.dis.db775aff25efc5ac4f0660ad9c9f7db2

Penderglass, A.G., Knutti, R., Lehner, F., Deser, C. and Sanderson, B.M. (2017) Precipitation variability increases in a warmer climate. *Scientific Reports*, 7, 17966. https://doi.org/10.1038/s41598-017-17966-y

Pimm, S.L. (1984) The complexity and stability of ecosystems. *Nature*, 307, 321–326.

Polk, M. (2015) Transdisciplinary co-production: Designing and testing a transdisciplinary research framework for societal problem solving. *Futures*, 65, 110–122. https://doi.org/10.1016/j.futures.2014.11.001

Popper, K.R. (1963) *Conjectures and Refutations. The Growth of Scientific Knowledge*. Routledge and Kegan Paul, New York.

Quinlan, A., Berbés-Blázquez, M., Haider, L. and Peterson, G. (2016) Measuring and assessing resilience broadening understanding through multiple disciplinary perspectives. *Journal of Applied Ecology*, 53, 677–687.

Riley, D. (2007) Improving public safety – from federal protection to shared risk reduction. *Gilbert F. White National Flood Policy Forum, Floodplain Management 2050* Washington, DC, November 6–7, 2007, 90–92.

Rittel, H.W.J. and Webber, M.M. (1973) Dilemmas in a general theory of planning. *Policy Sciences*, 4(2), 155–169.

Rogers, P. (2020) The evolution of resilience. *Connections – The Quarterly Journal*, 19(3), 13–32. https://doi.org/10.11610/Connections.19.3.01

Rölfer, L., Celliers, L. and Abson, D.J. (2022) Resilience and coastal governance: Knowledge and navigation between stability and transformation. *Ecology and Society*, 27(2), 40. https://doi.org/10.5751/ES-13244-270240

Sayers, P.B., Carr, S., Moss, C. and Didcock, A. (2020) *Sayers - Flood Disadvantage – Social Vulnerable and Ethnic Minorities*. Research undertaken by Sayers and Partners for Flood Re. Sayers and Partners, London.

Sayers, P., Penning-Rowsell, E.C. and Horritt, M. (2018) Flood vulnerability, risk, and social disadvantage: Current and future patterns in the UK. *Regional Environmental Change*, 18, 339–352. https://doi-org.ezproxy.uwe.ac.uk/10.1007/s10113-017-1252-z

Schlosberg, D. (2007) *Defining Environmental Justice: Theories, Movements, and Nature*. Oxford University Press, Oxford.

Selby, D. (2006) The catalyst that is sustainability: Bringing permeability to disciplinary boundaries. *Planet*, 17, 57–59. https://doi.org/10.11120/plan.2006.00170057

Senge, P. (2006) *The Fifth Discipline: the Art and Practice of the Learning Organization* (2nd edn). Random House Business, London.

Shaw, R. (2012) Chapter 1: Overview of community-based disaster risk reduction. In R. Shaw (ed.) *Community-Based Disaster Risk Reduction (Community, Environment and Disaster Risk Management, Vol. 10)*, Emerald Group Publishing Limited, Bingley, pp3–17. https://doi.org/10.1108/S2040-7262(2012)0000010007

South, J. (2015) *A guide to community-centred approaches for health and wellbeing*. Project Report. Public Health England/NHS England.

Swyngedouw, E. (2009) The political economy and political ecology of the hydro-social cycle. *Journal of Contemporary Water Research & Education*, 142, 56–60. https://doi.org/10.1111/j.1936-704X.2009.00054.x

Tabari, H. (2020) Climate change impact on flood and extreme precipitation increases with water availability. *Science Reports*, 10, 13768. https://doi.org/10.1038/s41598-020-70816-2

Tait, J. and Lyall, C. (2007) *Short Guide to Developing Interdisciplinary Research Proposals* ISSTI Briefing Note 1.

Tempels, B. (2016) *Flood Resilience: A Co-Evolutionary Approach: Residents, Spatial Developments and Flood Risk Management in the Dender Basin*. PhD, University of Gent. InPlanning.

Tierney, K. (2014) *The Social Roots of Risk: Producing Disasters, Promoting Resilience*. Stanford University Press, Standford.

Titz, A., Cannon, T. and Krüger, F. (2018) Uncovering 'community': Challenging an elusive concept in development and disaster related work. *Societies, Basel*, 8(3). https://doi.org/10.3390/soc8030071

Twigger-Ross, C., Coates, T., Deeming, H., Orr, P., Ramsden, M. and Stafford, J. (2011) *Community Resilience Research: Final Report on Theoretical Research and Analysis of Case Studies*. Report to Cabinet Office and Defence Science and Technology Laboratory. Collingwood Environmental Planning Ltd, London.

Twigger-Ross, C., Kashefi, E., Weldon, S., Brooks, K., Deeming, H., Forrest, S., Fielding, J., Gomersall, A., Harries, T., McCarthy, S., Orr, P., Parker, D. and Tapsell, S. (2014) *Flood Resilience Community Pathfinder Evaluation: Rapid Evidence Assessment*. Defra, London.

UK Cabinet Office (2008) *The Pitt Review: Lessons Learned from the 2007 Floods*. https://webarchive.nationalarchives.gov.uk/ukgwa/20100807034701/http:/archive.cabinetoffice.gov.uk/pittreview/_/media/assets/www.cabinetoffice.gov.uk/flooding_review/pitt_review_full%20pdf.pdf.

UK Cabinet Office (2011) *Strategic National Framework on Community Resilience*. HMG, London. https://www2.oxfordshire.gov.uk/cms/sites/default/files/folders/documents/fireandpublicsafety/emergency/StrategicNationalFramework.pdf

UK Flood and Water Management Act (2010) https://www.legislation.gov.uk/ukpga/2010/29/contents. Accessed 2 August 2014

UK Localism Act (2011) https://www.legislation.gov.uk/ukpga/2011/20/contents/enacted

UNDRR (2015) *Sendai Framework for Disaster Risk Reduction (2015–2030)* https://www.undrr.org/publication/sendai-framework-disaster-risk-reduction-2015-2030

Ungar, M. (2018) Systemic resilience: Principles and processes for a science of change in contexts of adversity. *Ecology and Society*, 23(4), 34. https://doi.org/10.5751/ES-10385-230434

US National Council for Social Studies (2001) Creating effective citizens: A position statement of national council for the social studies. *Social Education*, 65(5), 319.

van Ruiten, L.J. and Hartmann, T. (2016) The spatial turn and the scenario approach in flood risk management – Implementing the European floods directive in the Netherlands. *AIMS Environmental Science*, 3(4), 697–713. https://doi.org/10.3934/environsci.2016.4.697

Vojinović, Z. (2015) Flood Risk: The Holistic Perspective: *From Integrated to Interactive Planning for Flood Resilience IWA Publishing*. https://doi.org/10.2166/9781780405339

Walker, G.P. (2012) *Environmental Justice: Concepts, Evidence and Politics*. Routledge. London, p. 256.

Wenger, C. (2017) The oak or the reed: How resilience theories are translated into disaster management policies. *Ecology and Society*, 22(3), 18. https://doi.org/10.5751/ES-09491-220318

Werritty, A. (2019) 'How can we learn to live with floods? Challenges for science and management': Guest editorial introduction. *Scottish Geographical Journal*, 135(1–2), 1–4. https://doi.org/10.1080/14702541.2019.1667625

Whittle, R., Medd, W., Deeming, H., Kashefi, E., Mort, M., Twigger-Ross, C., Walker, G. and Watson, N. (2010) After the rain – learning the lessons from flood recovery in Hull. Final project report for *'Flood, vulnerability and urban resilience: a real-time study of local recovery following the floods of June 2007 in Hull.'* Lancaster University, Lancaster.

Wickson, F., Carew, A.L. and Russell, A.W. (2006) Transdisciplinary research: Characteristics, quandaries and quality. *Futures*, 8(9), 1046–1059. https://doi.org/10.1016/j.futures.2006.02.011

Wisner, B., Blaikie, P., Cannon, T. and Davis, I. (2004) *At Risk: Natural Hazards, People's Vulnerability and Disasters* (2nd edn). Routledge, New York.

Wolch, J. R., Byrne, J., and Newell, J. P. (2014) Urban green space, public health, and environmental justice: The challenge of making cities 'just green enough.' *Landscape and Urban Planning*, 125, 234–244. https://doi.org/10.1016/j.landurbplan.2014.01.017

Wolff, E. (2021) The promise of a "people-centred" approach to floods: Types of participation in the global literature of citizen science and community-based flood risk reduction in the context of the Sendai Framework. *Progress in Disaster Science*, 10, 100171. https://doi.org/10.1016/j.pdisas.2021.100171

Wong, D. (2020) Comparative Philosophy: Chinese and Western. In E.N. Zalta (ed.) *The Stanford Encyclopedia of Philosophy* (Fall 2020 Edition), https://plato.stanford.edu/archives/fall2020/entries/comparphil-chiwes/.

Yerbury, H. (2011) Vocabularies of community. *Community Development Journal*, 47(2), 184–98.

Zevenbergen, C., Gersonius, B. and Radhakrishan, M. (2020) Flood resilience. *Phil. Trans. R. Soc. A*, 378, 20190212. http://dx.doi.org/10.1098/rsta.2019.0212

2

DIFFERENT FLOOD TYPES AND DIVERSE COMMUNITIES

Interactions, impacts, and management at local level

2.1 Introduction

Understanding a flood risk situation and its local impacts requires an appraisal of the intersections between 'different types of flooding' and 'different kinds of community'. Many studies – from the 1970s onwards – already outline causes and intensifying factors for river (fluvial) flooding with its magnitude, frequency, and physical characteristics. Flood modellers are also now advancing our understanding of other flood types beyond fluvial (e.g., surface water, groundwater, coastal etc.) and their interactions over time and space. This chapter brings diversity of both floods and communities together. Building on Chapter 1, it sets the scene for other domains explored within this book. It asks:

- What are the different types of floods – their character and distinctiveness – as experienced by impacted communities, and what are the implications for building community flood resilience?
- How do diversity and change in both floods and communities manifest in these interactions, and what are the consequences for environmental justice (EJ)?
- What happens when different paradigms (ways of thinking) about flood management are viewed through a 'community' lens? What are the implications for community empowerment, agency, and ownership of these changing strategies?

Exploring this territory requires contributions from academic disciplines including geography, hydrology, and sociology. The chapter integrates research evidence with examples from specialist guidance targeted at at-risk citizens and communities by statutory flood risk management (FRM) agencies.

DOI: 10.4324/9781315666914-2

2.2   Flood science and lived experience

2.2.1   Looking back: Communities establishing local river flood histories

In considering flood risk, vulnerability of communities, and flood management strategies *at-a-place*, it is important to take a longer-term view (e.g., Paprotny et al., 2018). Reconstruction of local flood histories can be undertaken by hydrologists and insurance companies drawing on archival sources beyond shorter-term gauged river flow records, *and* individual citizens motivated to research their local flood chronologies through oral histories and archival evidence (government; community; Chapters 4 and 5).

The flood time-series in a river drainage basin or catchment can intersperse 'flood rich' periods with those termed by hydrologists as 'flood poor' – a term problematic for at-risk communities. 'Routine' and 'catastrophic' flood events form part of the same hydrological series, and episodic occurrence of the most extreme river floods as outliers means that (not all) people at flood risk may have had experience of such events. The inter-arrival times of floods of different magnitudes (sizes) and frequencies, and associated flood experiences, have important implications for what is considered 'extreme' by communities, the extent of social flood memory, and experiential flood knowledge (McEwen et al., 2016). What is perceived as routine or severe is also culturally determined (Chapter 3). Historic extreme floods may only be remembered by older people, or recorded in archives, such as in physical flood marking and documentary evidence (Chapter 5). Marking of instantaneous maximum flood levels can act as a valuable source of local environmental knowledge that helps in reconstructing the magnitude and frequency of historic flood patterns over the past 100–400 years. Comparing past flood patterns to present occurrence and future projections benefits both hydrologists and communities (e.g., Blöschl et al., 2019 for Europe).

2.2.2   Different causes of flooding and challenges for communities: A brief resume

Flooding can be caused by wide-ranging, climate-related agents, as outlined in Table 2.1. Different flood types vary in their physical characteristics (e.g., specific climatic and physiographic settings) and their implications for communities (e.g., sensory experience; ability of scientists and local people to forecast events; effectiveness of adaptive strategies). The lived experience of people can be greatly affected by a flood's hydrological characteristics (depth, frequency, speed of rise, velocity, duration; spatial extent); its visibility, sound, and smell; its seasonality and timing; occurrence during day or night, and under grey or sunny skies. Different types can therefore be associated with variations in public hazard perception, attribution, and psychological effects, and therefore affecting the likelihood of individuals and communities undertaking adaptive measures (Chapters 3 and 8). Each flood type can occur individually or potentially

TABLE 2.1 Examples of Different Flood Types with Their Implications for Community Resilience

Flood cause	Physical character	Character as experienced	Factors affecting community impact	Sources of information	Potential for warning communities	Implications for community preparedness	Indicative reference
A. Fluvial or river floods	• Occurs when rivers and streams flow overbank • Flood hydrograph can vary in terms speed of rise; maximum flow, duration, etc. • River confluences particularly vulnerable	• Sensory – on larger rivers, flooding is normally highly visible and frequently audible • Emotive – fear and trauma but sometimes also awe	• Proximity to water courses • Height of property	• Official flood forecasting on main rivers • Local knowledge of signs of rising water and critical levels for impacts	• Formal warnings possible on larger water courses	• Action needed during events and in wider resilience planning • Needs catchment thinking	See Chapter 1
B. Rain-related (pluvial): flash flooding Problematic in small steep catchments or 'rapid response' catchments	• Intense local rainfall falls on a ground surface at rates which do not allow infiltration • Increases effective drainage density of catchments	• Flood front commonly reported as a 'wall of water' • Surges as bridges and other obstructions are breached	• Force (power) of water • Size of sediment available for transportation in upstream water courses	• Urban pluvial flood modelling, forecasting, and management tools • Weather radar	• Flooding can occur in settings not on fluvial flood risk maps • Flood peaks with very short lag time and steep rising limb mitigating against formal warning	• Immediate heavy structural property damage making home uninhabitable • Can carry away large pieces of property like vehicles	Archer and Fowler (2021)
C. Rain-related (pluvial): surface water flooding	• Generated by high-intensity rainfall falling on impermeable urban surfaces	• Rapid accumulation of surface water	• Extent of impermeable surfaces	• Weather radar	• Limited potential for official warning	• Depends on scale – direct impacts to property; indirect 'nuisance' impacts	Fortunato et al. (2014)

(Continued)

TABLE 2.1 (Continued)

Flood cause	Physical character	Character as experienced	Factors affecting community impact	Sources of information	Potential for warning communities	Implications for community preparedness	Indicative reference
C. Rain-related (pluvial): surface water flooding (Continued)	• When storm sewer drains surcharged in their capacity	• High velocities on slopes	• Anomalies of flow pathways and ponding over urban surfaces • Extent of sustainable drainage measures	• UK Environment Agency		• Temptation to wade/drive through	
D. Groundwater flooding	• Water levels rise within soil or rock or water flows from abnormal springs • Water flowing through aquifers (chalks, sands, and gravels) • Tends to occur after longer periods of sustained high rainfall	• Communities may not be aware; may not be initially visible, slow flowing, and silent • Not obvious in buildings if below ground floor	• Proximity of groundwater system to pollution sources. These sources can be located	• Monitored groundwater levels; alerts and warnings For example, USGS Groundwater Watch; UK Groundwater Forum • May not be flood risk maps for groundwater	• Variable	• Practical advice available	Groundwater UK[a]
E. High tide flooding Also 'supertides'	• Daily rhythm means increased risk at certain times of tidal cycle (e.g., high spring tides at full and new moons)	• Inundation of exposed coastal property and transport links	• Described as 'sunny day flooding' or 'blue sky flooding'	• Statutory authorities	• Flood risk predictable in terms of time of day; repeat flooding possible	• Needs event preparedness and wider resilience planning	US National Oceanic and Atmospheric Administration[b]

(Continued)

TABLE 2.1 (Continued)

Flood cause	Physical character	Character as experienced	Factors affecting community impact	Sources of information	Potential for warning communities	Implications for community preparedness	Indicative reference
E. High tide flooding Also 'supertides' (Continued)	• Can come up through urban sewer systems • Increased risk occurs when coincident with exacerbating factors (F)		• Occurs when least expected; without rainfall	• Tidal predictions			
F. Coastal storm surge	• Strong onshore winds (e.g., hurricane or typhoons) lead to rise in sealevel • Water is pushed onto land, leading to flooding • Combined with low atmospheric pressure • Can be caused by European weather systems (extra-tropical cyclones) and tropical cyclones (hurricanes)[c]	• Awaiting possible impending inundation in unprotected settings or whether water overtops or breaches coastal defences	• Pressure, wind strength, and direction; width and slope of continental shelf • Can last from hours to days and affect hundreds of km[b]	• Requires coastal flood warning system monitoring wind, wave, tides, and weather conditions • Examples: US National Hurricane Center (Historical SLOSH Simulations) UK Coastal Monitoring and Forecasting (UKCMF)	• Intersection of conditions can be forecast	• Population density on coastal strip • Height above mean sea level • Economic productivity of coastal zone • Positioning of key infrastructure	Muis et al. (2016)

(Continued)

TABLE 2.1 (Continued)

Flood cause	Physical character	Character as experienced	Factors affecting community impact	Sources of information	Potential for warning communities	Implications for community preparedness	Indicative reference
G. Tsunami	• "Wall of water" • Triggered by tectonic or seismic activity, and submarine or terrestrial landslides	• Rapid onset • Difficult to escape if without high ground • Warning signs known by indigenous cultures	• Height of wave	• Landslide hazard Norway – monitoring of cracks with seismic sensors; webcams; meteorological stations • Expectation to give 72 hours warning	• If failure site monitored	• Length of any warning may be limited	Harbitz et al. (2014)
H. Dam burst floods Influenced by dam scale and construction	• Strongly place-specific • Catastrophic failure • Rapid large-scale flood flows • Wall of water • Very high stream powers	• Rapidity and noise • Limited time for escape without warning	• Historically lack of warning • Now efficiency of alarm systems • With warning, evacuation	• Risk assessments	• Efficiency of alarm systems • "Probability of huge rains is known days in advance and that of possible overtopping hours in advance, and because modern telecommunication is efficient worldwide" p148)	• Evacuation escape routes required	Lempérière (2017)

(Continued)

TABLE 2.1 (Continued)

Flood cause	Physical character	Character as experienced	Factors affecting community impact	Sources of information	Potential for warning communities	Implications for community preparedness	Indicative reference
I. Ice-jam floods River dam formed by blocks of ice	• Seasonal (spring) caused by melting snow and ice	• Deep, unusually fast, rushing water (cf. river flooding)	• Complex flows of ice blocks and water • Extent of warning • Proximity to river	• Satellite imagery important due to large survey areas (e.g., Joint Polar Satellite System) •	• Work on improving warnings for ice-bound rivers – without extensive false alarms[d]	• Also community observation	Das et al. (2018)
J. Jökulhlaup (Icelandic literally "glacial run") – type of glacial outburst flood	• Catastrophic failure • Wall of water • Very high stream powers	• Visual and sound • Large debris (boulders; ice) transported	• Proximity	• "Coupling early warning systems with organized response including active involvement of local populations"	• "EWS effective in reducing losses and increasing the resilience of communities against [...] jökulhlaups" (p971)	• Rural Icelandic communities historically affected • Needs co-ordinated response with active community involvement	Gudmundsson (2015)

a http://www.groundwateruk.org/FAQ_groundwater_flooding.aspx.

b https://coast.noaa.gov/states/fast-facts/recurrent-tidal-flooding.html.

c https://ntslf.org/storm-surges/about-storm-surges.

d https://eomag.eu/improved-flood-warnings-on-ice-bound-rivers/.

in combination, with complex interactions having different socio-economic impacts in specific settings.

Over time, the floods in a particular locality can change in character and complexity, with implications for the extent of exposed activities, local community flood experiences, and the geography of directly and indirectly affected communities within a locale. This has implications for risk communication (Chapter 6) and adaptive learning strategies involving communities (Chapter 9).

Each flood type provides specific challenges for communities. Historically, inland flooding is typically equated by citizens and communities with river flooding of exposed floodplains (Table 2.1A). Fluvial flood risk on large water courses in more developed countries now tends to be routinely modelled and mapped by environmental regulators and insurance companies, so risk information is normally publicly available, if not accessed (Chapter 6). However, potential risks from smaller unmonitored water courses can be more hidden and forgotten by communities (Table 2.1B). The full extent of the drainage network of a river catchment may only become visible locally when it is episodically switched on during intense localised rainfall and associated flash floods. Surface water flooding (Table 2.1C) is increasing for urban communities, with both nuisance and episodic extreme impacts. For example, only after major impacts of surface water flooding to property and road infrastructure during the UK Summer 2007 floods did its management move higher up political agendas, with subsequent mapping and modelling undertaken by statutory agencies (UK Cabinet Office, 2008). In contrast, groundwater flooding (Table 2.1D) is much less visible and the impacts can be delayed: affected households and businesses may think they have escaped flooding but have water below floorboards or in cellars within their properties. For example, in the same 2007 floods in Hull, UK, secondary flood damage was only identified later when households began to smell damp in their houses (Medd et al., 2015).

Communities occupying exposed coastal areas can be subject to both tidal flooding and storm surges (Table 2.1E/F). 'Sunny day flooding' is counter-intuitive in terms of risk, but a coincidence of certain weather and astronomical effects can lead to conditions sometimes referred to as 'super tides' where communities experience repeated inundation, twice-daily, during high tide cycles. The catastrophic 1953 North Sea storm surge – combining gale-force winds, low atmospheric pressure, and high tides – badly affected low-lying areas of the Netherlands and the UK, and the high loss of life (over 1,800 and 307 deaths, respectively) demonstrated the impacts of such conditions coinciding, without severe flood warnings in place.

Other types of coastal flooding are strongly tied to the physiography of place, for example, tsunami risk (Table 2.1G). This is seen, for example, when major landslides occur in Norwegian fiords impacting their exposed shoreline communities. In 1934, a 64-m high tsunami killed 40 people in three villages in Tafjorden, Norway. Established fiord-side communities may have to be prepared to live with such risk. When the monitored Åkerneset landslide fails, it will trigger the largest peacetime evacuation in Norwegian history (Harbitz et al., 2014).

While artificial dams can store floodwater, the potential for catastrophic dam failure (Table 2.1H) also poses a major location-specific flood risk for downstream communities. This can occur because of elderly infrastructure, heavy rainfall, or earthquake incidence (Lempérière, 2017), and historically, there has been limited capacity to issue warnings. Dam burst floods have occurred at contrasting scales with devastating impacts on downstream communities, including high loss of life, large-scale destruction of property and displacement. Examples include the Great Sheffield flood 1864, UK (at night – ca. 240 deaths[1]) and the Teton Dam failure 1976, Idaho, USA (Arthur, 1977; 10 deaths, 30,000 displaced). Critical is any awareness of impending failure, warning technologies at the time, and whether the event happened by day or night (Lempérière, 2017).

Interaction of water and ice blocks within cold environments can bring risk of extensive community impacts (Table 2.1I), as evidenced by the Badger flood, Newfoundland (2003) with its rapid river-level rise (2.3 m in 1 hour) and large ice chunks pushed into the town.[2] As Das et al. (2018, p1) noted, "due to high water levels, unusual velocities, and complex formation mechanisms of ice-induced floods, damage, injuries, and deaths to humans occur more frequently in these types of floods than in open-water floods".

Jökulhlaups or 'megafloods' (Table 2.1J), caused by breaching of glacial ice dams, can lead to very large and sudden increases in discharge of glacier meltwater streams. Such episodic flooding had major impacts on exposed 19th-century farming communities in Southern Iceland (Sigurmundsson et al., 2014). In the 20th century, however, the effects of jökulhlaups on Icelandic communities were more typically indirect, with frequent breaches to the main highway that traverses the coastal outwash plain.

2.2.3 Understanding flood intensifying factors: Differentiating evidence and myth

The non-specialist may struggle to understand the nature and magnitude of the flood risks they are exposed to. Factors that intensify fluvial flooding operate at different scales (e.g., basin, network, and channel conditions) – both natural and those linked to human activity (Pielke and Downton, 2000; Figure 2.1).

Examples of human impacts on hydrology include construction of bridges that obstruct flood flows, and river or coastal management strategies such as 'channel improvement'. Myths can pervade in the public's attribution of, and blame for, extreme floods to human agency, for example, the impact of lack of river dredging on extreme floods (Chapters 3 and 6). Scientific evidence underpinning potential management solutions requires careful unpicking with flood-affected communities[3] (see CIWEM's 2014 '*Floods and dredging: a reality check*').

> That is not to say dredging has no role to play. It can reduce water levels on a local scale and may be critical to flood risk management in key locations…. However, dredging cannot hope to prevent flooding caused when heavy rainfall results in flows that vastly exceed the capacity of the river channel.
>
> *p1*

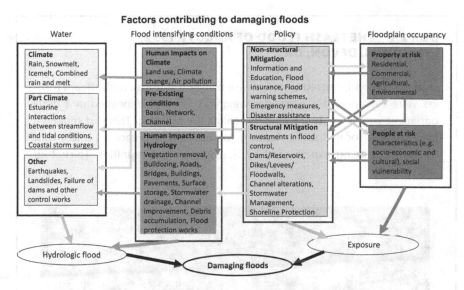

Factors contributing to damaging floods

FIGURE 2.1   Framework for understanding the interrelated factors responsible for the occurrence of damaging floods (redrawn and edited from Pielke and Downton, 2000; © American Meteorological Society. Used with permission)

Changes in both climate and land use can impact flood risk (Chapter 1), but myths can persist within communities about their relative impacts on the range and scale of river floods. One myth is that land use change causes extreme floods. However, science indicates catchment land use is likely a more significant influencing factor in moderate floods. This has implications for the efficacy of FRM strategies that involve land use manipulation. For example, Natural Flood Management (NFM) measures (Chapter 7) are far from a 'silver bullet' and may take decades to be effective. Also, the most extreme floods on record will tend to be extreme independent of land use (Newson, 1992), with tarmac and saturated ground behaving similarly hydrologically.

2.2.4   Communities and risk from polluted floodwaters

Flood impacts on communities do not relate solely to water quantity. Rivers transport load (dissolved; suspended; bedload; organic and inorganic) from erosion, and from present and past human activities within their catchments. This has implications for local flood impacts, adding to demands for clear-up in flood recovery. Human impacts of transported sediment load can be large; smaller steep, high-energy, river catchments, combined with large calibre sediment, can operate in a binary fashion, switching on during flash floods. Such high sediment loadings lead to major direct damage to property, infrastructure, and casualties (Alexander and Cooker, 2016), one example of which was the catastrophic Lynmouth 1952 flood, Devon, UK (Dobbie and Wolf, 1953) (Box 2.1; Figure 2.2). This event emphasised

BOX 2.1 THE FLASH FLOOD OF AUGUST 1952, LYNMOUTH, DEVON, UK

During this flood, caused by an intense convective storm, high stream powers were evidenced by a boulder weighing 7.5 tons deposited in a hotel basement with ca. 6 m of boulders against the building. This vast sediment load blocked local bridges, leading to subsequent breaching and major flood pulses being experienced by downstream communities. Thirty-four people died.

FIGURE 2.2   Physical impacts of transported sediment during the 1952 flash floods at Lynmouth, SW UK (with permission: Peter Christie)

the importance of both scientists and communities being aware of high upstream sediment availability in local flood risk assessments.

Not all river load is visible. Lower-energy, river environments tend to be associated with finer sediment transport, with different implications for flood impacts on property. Organic and inorganic load in floods, as point source and diffuse pollution, can vary with local factors like agricultural type in upstream fields (McEwen et al., 2002), or location of sewage treatment plants, sometimes overtopped in floods.

Organic deposits can include human and animal waste (with *E. coli* bacteria), leading to property-level, public health impacts. Hence, it is important to limit human and animal contact with floodwater, which may contain "chemical spills, biotoxins, invasive species, waste, sewage and debris" (Erickson and Brooks, 2017, np). This risk is often overlooked by residents when evacuating from, or returning to, flooded areas.

Polluted flood water is not just a risk at the time of the flood itself. Contaminated flood water can also bring pathogens including *E. coli*, salmonella, hepatitis A and norovirus into private, school, and community locations growing vegetables for food. For example, US food science researchers[4] provided safety guidance to growers.

How concerned gardeners have to be about using garden produce after a flood depends, to a large degree, on how 'clean' the flood water was or whether it was likely to have been contaminated with sewage, river or creek water, farm runoff, or industrial pollutants. The most conservative answer—one that eliminates any and all risks—is that gardeners should discard all produce that was touched by flood water.

Ingham and Ingham (2007, p1)

Flooding in areas with histories of heavy metal mining can bring additional risks to communities through mobilisation and re-deposition of contaminated sediment, with longer-term implications for soil health. Toxicity of sediment represents a major issue in the clear-up of flooded buildings and community open spaces. A scientific challenge is in knowing pre-flood levels of heavy metals relative to those found in flood deposits. For example, after the 2005 New Orleans floods, USA, there was still concern over persistence of elevated arsenic levels in flooded schoolyards and playgrounds after an 18-month recovery period (Rotkin-Ellman et al., 2010).

Understanding the pathways and interactions of surface and subsurface flood flows through a catchment can be critical in working out community exposure to water quality issues. Groundwater flooding can also carry significant dissolved chemical load and bacteria, depending on underlying geology and whether the area has an industrial heritage. This can determine toxicity of water quality in seasonally and episodically flooded cellars and basements, with health risk implications for both human contact with floodwater and private water supplies from groundwater (Waller, 1994). Such impacts on communities are more frequently cited in Less Economically Developed Countries (LEDCs), for example Iran and Pakistan, but also apply in More Economically Developed Countries (MEDCs) where water-bearing permeable rocks underground (aquifers) are used for water supply. High tide flooding can also bring pollutants (pollutants, nutrients) off the land into the sea,[5] with health and well-being implications for people and nature.

2.3 Intersections between diverse communities and flood types

Alongside variety in flood type, communities living in exposed flood risk areas can also be diverse. Degrees of flood experience can intersect with varied socio-economic, cultural, and scientific capital (or assets; Chapter 8). Some settings (e.g., coastal and riverside properties with high aesthetic value) can be considered highly desirable in communities with higher economic capital, and worth a degree of risk-taking alongside ability to bear losses. However, historically urban floodplains have tended to be lower cost but exposed land; lower socio-economic, transient, or intermittent residents, as more vulnerable societal groups are likely to occupy these zones. This leads to unequal flood impacts and issues of environmental and climate justice (Section 1.4.5 of Chapter 1). As Cutter (2006, pxxvii) observed, "disasters are income- and gender-neutral and color-blind. Their impacts, however, are not", and "of course, not all communities have the same capacity to adapt" (Global Commission on Adaptation, 2019, p7).

EJ as a concept has several dimensions: distributive, procedural, as recognition and sense of justice, differentiated in Table 2.2.

TABLE 2.2 Dimensions of Environmental Justice Applied to FRM (drawing on Schlosberg, 2007; Svarstad et al., 2011; Bell, 2014; Alba et al., 2020)

Dimension of EJ	Issue and aims
Distributive	Distribution of flooding or impact of decisions in FRM differentially affecting particular groups. EJ needs to consider politics and values that determine local intervention strategies (Alba et al., 2020). Aims for "an equitable distribution of environmental 'goods' and protection from environmental harms for all socioeconomic groups" (Bell, 2014, p1)
Procedural	Inequalities of participation occurring within processes of decision-making in FRM. EJ aims for "fair, participatory and inclusive structures and processes of environmental decision making" (Bell, 2014, p1)
As recognition	"The range of social and cultural values and practices that impede the full recognition of a group as an accepted member of the moral and political community" (Schlosberg, 2007, p16)
Sense of justice/ legitimacy	"When studying social impacts of changes such as those imposed by sets of policy instruments"; "How people affected by changes themselves perceive and evaluate the changes" (Svarstad et al., 2011, p11)

In making the case for framing flooding in the UK as an EJ problem, Walker and Burningham (2011, p217) identified two key questions:

> to ask what patterns of social inequality exist in relation to the environmental good or bad that is at issue, a distributional question that seeks to reveal patterns of unevenness and difference.
>
> to evaluate these patterns in the light of what is expected to be a just and fair situation, in respect of, for example, how the inequalities are being produced, who is responsible for them, how relevant decisions have been made and how government policy and practice are enacted.

Bell (2014) also asked "Do we just all want to be polluted the same?", arguing that EJ should aspire to a healthy environment for all.

EJ issues interweave with the temporality of flooding in complex ways. In socially disadvantaged settings, sequences of flood impacts can lead to a ratcheting down of resilience through inability of households to recover before the next flood hits. They can also manifest in cascading (secondary) risks and vulnerabilities from floods (e.g., increased likelihood of disease and other health and well-being issues; Chapter 10). As Chakraborty et al. (2019, p244) noted, such "unequal social consequences" have a strong potential to increase as climate changes, with significant public health implications.

Hurricanes hitting the east coast, USA, are frequently cited as examples of differential impacts among particular racial, ethnic, and socio-economic groups. During Hurricane Katrina (in 2005) in New Orleans, with storm surge flooding (25–28 feet above normal tide levels), African Americans living at low elevations had high exposure to back swamp flooding and poor access to transportation. Ways of recovering from natural disasters may also solidify structural racism (Morse, 2008). In another example, Hurricane Harvey (in 2017) caused catastrophic flooding in Greater Houston, Texas, USA, with differential impacts on areas with higher proportions of non-Hispanic Black and socio-economically deprived residents (Chakraborty et al., 2019). These groups had constrained access to resources necessary for response, recovery, and medical care.

Hurricane Katrina (2005) represented "a classic example of inadequate and inequitable preparation and response to flooding, based on racism and classism" (Bell, 2020, p71; see "Understanding Katrina"[6]). As Eugene Robinson ("Never Again" National Forum, Washington, DC, 2006) noted: "environmental injustice began long before Hurricane Katrina ever hit, in the basic pattern of settlement in the city" (Morse, 2008, p1).

Similar issues are evident in other geographical settings. Working in the context of Canadian First Nations in Manitoba, Thomson (2015, p222) acknowledged "environmental risk is distributed inequitably along race, class, and indigenity [sic] differences". Here, floodwater, during a 'superflood', was diverted intentionally towards First Nation communities to protect downstream settler cities. Alba et al. (2020, p364) also noted in Venice, Italy:

> sea walls, artificial islands and mobile dams not only create uneven geographies of vulnerability but also reflect specific approaches to CCA [climate change adaptation] and incorporate specific views and claims to the coast.

Additionally, "access to information and capacity for self-protection are typically distributed unevenly within populations" (Eiser et al., 2012, p5). Ethnicity and language can act as a barrier, leading to differential ability to participate in local decision-making. Some groups are 'less heard' not because they are 'hard to reach', but due to assumptions of vulnerability or incapability, and hence are not brought into the conversation. This involves lack of understanding of their cultures and constraints (Chapter 10; Bell, 2014, 2020).

However, this complex territory has risks of stereotyping. Grineski et al. (2015), comparing flooding and air pollution, argued for nuance and against sweeping characterisation of different demographics of affected people and their capital (Chapter 8). It is important to understand and include different local environmental and cultural knowledge in adaptation planning for future coastal flooding through inclusive participation, and to openly discuss values (Douglas et al., 2012 in Boston, Massachusetts; Alba et al., 2020 in Italy; Chapters 4 and 10).

2.4 Compound, overlaid, sequential, and cascading risks and communities

While communities are increasingly diverse, risks are becoming ever more complex. Floods can occur from the same cause but in quick succession so affecting community recovery. For example, a sequence of major storms will wet up a catchment, with insufficient time for sustained lower flows before the next storm-generated flood peak. By contrast, compound flooding is "when different sources occur concurrently or in close succession" (Hendry et al., 2019, p3117); its potential impacts on communities can be greater than if each extreme were to occur separately. Sequential independent risks occur in succession, cascading flood risks have some sort of secondary causality, while overlaid risks occur simultaneously. All have implications for flood impacts and speed of community recovery.

2.4.1 Compound flooding: Interactions between different causes of flooding

The most extreme historic floods on record – sometimes before instrumented gauging of river flows – in river catchments can relate to interactions between different causal factors over space and time (e.g., tidal influence, snowmelt, and rainfall distribution). For communities, understanding these different causes can give a sense of the maximum possible flood if several potential factors were to coincide. For example, the extreme November 1770 flood on the lower River Severn, England,

UK, the highest flood within the historic record, combined severe snowmelt, rainfall, and tidal causes – a situation that has not recurred in later centuries.

Risk of compound flood events is increasing due to climate change (e.g., Wahl et al., 2015; Bevacqua et al., 2019; Ganguli and Merz, 2019; Hendry et al., 2019; Gori et al., 2020; Rahimi et al., 2020; Del-Rosal-Salido et al., 2021; Hsiao et al., 2021). Hence, specialist, single-cause, flood risk maps used to promote flood preparedness in communities can underestimate actual risk. A current challenge for hydrologists is modelling and mapping compound flood risk, with flows difficult to predict in combination with other meteorological factors. This poses questions about how compound risks are perceived and experienced by communities, with floodwater coming in potentially different timings and from multiple directions. For example, on Canada's coasts with exposed communities, "spatial and temporal trends and variability of compound flooding", of total water level, storm surge, precipitation, and streamflow, are a particular cause for concern (Jalili Pirani and Najafi, 2020, p1). Establishing local flood risk at a particular geographical location can usefully combine modelling appraisal by specialist hydrologists, and observations at 1:1 by local citizens of interactions between multiple causes during extreme events (Chapter 4).

2.4.2   Complex hazards and communities

Flood risks – compound or otherwise – to communities do not exist in isolation but can be combined with many other stressors through sequential, cascading, and overlaid events. Increased interest exists in how physical processes across different environmental risks (extreme weather – flooding, drought, heatwaves, wildfires, cyclones – and disease, e.g., pandemics) interact across spatial and temporal scales in multi-risk environments (Gill and Malamud, 2014). This leads to multiplier effects on risks, and complexity in risk sources and their consequences for local communities.

Complex causal chains can lead to extreme impacts, while unusual (low probability) combinations of processes make these difficult to foresee. Many national governments establish registers of individual risk types against likelihood, impact, and confidence (e.g., HM Government UK National Risk Register, 2020), but not normally multiple risks together. However, Australian National Emergency Risk Assessment Guidelines (AIDR, 2020, p43) state that "identified risks need to consider the broadest range of potential consequences. This includes cascade, cumulative and 'knock-on' (secondary) effects". Such overlays present challenges for scientific modelling and risk assessment, infrastructural design, and community risk perception and awareness, as illustrated below in three scenarios: floods and droughts ('rapid visual' versus 'slow onset and hidden'); floods and heat (both 'intense' and 'sensory'); and floods and pandemics ('visual' versus 'hidden').

2.4.2.1   Floods and droughts

There is increased recognition among statutory agencies of the need for communities to possess better understanding of relationships between floods (more localised) and droughts (regional) within the same catchment. Guerreiro et al. (2018) assessed future changes in flood, heatwave, and drought impacts for all 571 European cities and found 100 particularly vulnerable to two or more climate impacts, with previous underestimation of risk. High-intensity rainfall on soils which have already been compacted due to drought can increase the likelihood of local flash flooding at times when flooding "may be far from people's minds" (UK Environment Agency[7]). Engaging communities with the need for flood preparedness during periods of underlying drought or alerting them to drought risk when floods have a high media profile can be challenging (Weitkamp et al., 2020; Chapter 6).

2.4.2.2 Sequential flood-heatwave events

Sequential extremes include trends in consecutive flood-heatwave events leading to significant local impacts that impede community recovery. For example, Zhang and Villarini (2020) found a high percentage of floods in the central USA were preceded by a heat stress event. Significant potential impacts included overwhelming critical infrastructure (e.g., damaged roads, hospitals, and power grid), with high numbers of fatalities and economic damage.

2.4.2.3 Floods and pandemics

Overlaid risk from flooding and disease (infection risk and its mitigation) is increasingly being researched, given impacts of the COVID-19 pandemic (Chapter 10). Pei et al. (2020) explored impacts of hurricane evacuation during this pandemic in southeast Florida, USA, finding that hurricane evacuation increased the total number of COVID-19 cases in both origin and destination locations. Overlaid or sequential flood and COVID-19 risks have socio-economic implications for community resilience. Negatives include further challenges to health and well-being in flood recovery; positives include ability to draw on existing social and volunteer networks within communities (Section 8.6 of Chapter 8).

The likelihood, scenarios, and impacts of such multiple risks will alter with climate change (e.g., Sadegh et al., 2018), with urban communities particularly vulnerable due to amassed people and structures. Better understanding of such risks needs to feed into improved projections of "potential high-impact events" to support close working between "climate scientists, engineers, social scientists, impact modellers and decision-makers" (Zscheischler et al. 2018, p469).

2.4.3 Compound risks: Implications for community resilience

It is well established that community resilience can be increased through prior personal and collective knowledge of how to cope and adapt, gained from previous floods. However, the changing nature of flood risk, with increasingly compound events, has implications for the role of social flood memory and intergenerational local knowledge, and their value as a resource which can be drawn upon for community resilience (Chapter 4). This also poses questions about public risk perception of compound risks, given that these are becoming more frequent and perception is strongly influenced by recent experience (alongside other social processes; Chapter 3). Such complicated interweaving of risks and impacts brings additional challenges for FRM professionals tasked with communicating risk to, and with, communities. This includes how best to visualise key elements of multi-hazard interaction and their impacts on different stakeholders (Gill and Malamud, 2014; Chapter 6).

Systemic thinking (Chapter 1) is also important in considering compound risk, externalities, and implications for community resilience. For example, International Alert (2015, p5) emphasised this approach "to analysing and measuring compound risks" for working in complex settings "with slow- and rapid-onset emergencies, violent conflict, climate change, and economic shocks". The implications of considering such whole-system interconnection – "'multiple exposure' to multiple threats" due to "deep rooted vulnerability processes" (cf. Kelman et al., 2015, p21) – are explored in Chapter 10.

2.5 The changing role of community in flood management strategies

Given these challenges at the interface of hydrology and sociology, this section introduces flood management from historic and community perspectives. Flood management has evolved through several paradigm shifts in moves from rural to urban living, for example, from land drainage to flood defence to FRM (Table 2.3A–C; e.g., Butler and Pidgeon, 2011).

Table 2.3 summarises these historical paradigms.

Paradigm A: Early rural communities in MEDCs had little alternative but to bear losses, adapt to living with local flooding through individual or collective action (e.g., through land drainage) or relocate. For example, in the UK, the internal drainage

TABLE 2.3 Some Paradigms for Flood Management (Categorisation A–C drawing from Tunstall et al., 2004; Johnson et al., 2005 for the UK)

Paradigm	Responsibility	Approach/ measures	Issues	Changing role of citizens and communities
A. Agricultural land drainage – rural focus	Farmers/ rural land managers	Drainage ditches to speed the flow	• Increases effective drainage density of floodplain or catchment so water runs off quicker	Local ditch implementation and maintenance in rural settings

(Continued)

TABLE 2.3 (Continued)

Paradigm	Responsibility	Approach/ measures	Issues	Changing role of citizens and communities
B. Flood defence – urban focus. Emphasises resilience as 'resistance' or 'protection' and a 'keep floodwater out' approach to reducing flood probabilities	State or riparian landowner responsibility	Suite of structural measures including embankments or levees, channel enlargement, interception channels, and dams	• Engineering measures only effective up to design limits • Not normally viable in cost-benefit terms to protect up to maximum probable flood • New schemes costly • Leaves a residual (unprotected) risk to those communities living in flood risk zones • Can have effect of placing more property at risk • Negative hydrological and environmental consequence	Need to deal with residual risk when infrastructure is exceeded or fails but frequently unaware
C. Flood risk management	Distributed responsibility for dealing with residual risk	Integrated strategic multi-methods – diverse range of interventions	• Variable effectiveness for different types and scales of flooding	Different roles of state and citizen with each intervention. Citizens need to be aware of residual risk and act to prepare. Can involve rural-urban relationships in same catchment
D. Community-focused local FRM (see Chapters 2 and 10)	Communities have key role in resilience planning. Different models: community-based and -led)	Range of local measures	• Requires community agency • May require statutory authorities to undertake different roles (including resourcing but also stepping back)	Requires communities to participate and potentially undertake lead roles

boards[8] (IDBs; 1930–) and predecessor organisations had important roles in managing local rural water levels for flood protection so supporting agricultural production.

Paradigm B: Here flood management is subdivided into structural or engineering measures ('flood control') and those involving 'adjusting people'. The structural paradigm, favoured historically by statutory actors, privileges engineering solutions for flood defence over non-structural (or behavioural) measures. Smith and Tobin (1979) provided an accessible resumé of perceived advantages and disadvantages of different measures at that time. A critical concern in engineering approaches is determining the design flood limits for failure – "reflecting the protection standard or safety level of FRM" (Nakamura and Oki, 2018, p5504). They reflected on historic increases to this parameter in Japan "to reflect the sociohydrological situation" of an era. However, performance of infrastructure relative to its design flood size (i.e., whether it achieves that level of protection during floods) is also critical for community experience.

Paradigm C: FRM focuses on managing risk systemically, as opposed to the management of actual flood events. This involves a suite of distributed or devolved approaches for dealing with residual flood risk (Chapter 1). However, risk and uncertainty management implies negotiation of socially acceptable levels of risk and exposure between the state and society (Challies et al., 2016). As Challies (2016, np) mused, this "high stakes" territory is "inevitably sensitive and often controversial". Success in FRM can be evaluated on various criteria (e.g., individual and community security, loss of life, property damage, speed of recovery, business interruption, and social disruption; Sayers et al., 2013). Transparency and objectivity are critical in its appraisal.

Paradigm D: It can be argued that community-focused FRM is a further paradigm, with its central attention to community engagement and participation in local risk management (Chapter 1; Table 2.3D). This is the central focus of this book and is explored further below.

So, approaches to managing floods can be envisaged as a dynamic and evolving process, controlled by wide-ranging local, regional, and international factors and triggers (Plate, 2002). Each paradigm has different implications for the roles, responsibilities, capitals (or assets), and capacities needed for engagement and participation of individual citizens, communities, and other local actors in managing risk. For citizens and communities, these 'colour shifts' from 'green' to 'grey' to 'blue-green' can be challenging to navigate. Similarly, the shift from more targeted pin-pointed measures, and the optics of 'doing something', to more systemic and distributed approaches with co-benefits, needs careful brokering with different socio-economic and cultural groups. This involves concern not just for public perception of actual risk but also the efficacy of different FRM approaches and measures and their agency within these (Chapters 3 and 7). Critical is where responsibilities for mitigation of residual risk sit within the public psyche.

Drivers for change have varied. While structural engineering approaches to river flood management form an important part of an integrated suite of measures (e.g., Montz and Tobin, 2008 in the USA; Kelman and Rauken, 2012 in Norway; Serra-Llobet et al., 2018; Tariq et al., 2020), rising costs of flood impacts alongside expense and potential failure of structural measures, increased risks due to climate change and requirements of higher environmental standards, prompted a major re-think towards FRM. The precise nature, emphasis, and timing of these paradigm shifts vary in different MEDCs globally but with dominant patterns. For example, Johnson et al. (2005, p562) articulated three paradigms and their timings in the UK. Each reflects "a changing set of beliefs, values, and attitudes towards the flood prob-lem" that in turn influenced multi-stakeholder attitudes to different mitigation strate-gies. During the World War II period to the 1970s, rural land drainage was driven by agricultural production and food security as priorities. The focus changed in the 1980s/1990s to flood defence and hard engineering or structural approaches to con-trolling urban river floods, driven by "economic growth, national security, and welfare standards" (p564). This in turn shifted to a more integrated multi-method, risk-based approach (FRM) that encompassed concern for ecological and environmental values in the 2000s (Tunstall et al., 2004; Johnson and Priest, 2008; Krieger et al., 2012). The European Commission's Floods Directive (2007/60/EC; implemented 2007) was critical here, forcing shifts in thought, policy, and practice within the UK and other member states about how best to manage floods. Scott et al. (2013, p3) reflected on the changing landscape of risk-based flood governance in many countries

> towards a more strategic, holistic and long-term approach characterised by miti-gating both flood risk and adaptation, or increasing resilience to flooding events. This is typified by the Dutch "room for the river" approach ...

However, the extent and speed of uptake of, and preference for, different non-structural measures to mitigate risk varies by country. For example, the USA was relatively slow in uptake due to economic drivers in decision-making (Buss, 2005) while Australia has invested heavily and early in flood warning system develop-ment (Chapter 7).

The nature and sequencing of paradigm shifts can also contrast in different cultural and socio-hydrological contexts. For example, Nakamura and Oki (2018) identified three eras of flood management in Japan, with diverse triggers. These are "Era 1: 1910–1935, changing society", "Era 2: 1935–1970, response to mega floods", and "Era 3: 1970–2010, response to economic growth" in shifts from a "green" to a "technological" society.

One key framing of FRM is Sustainable Flood Management (SFM; Werritty, 2006), integrating FRM explicitly with sustainability principles (social, economic, and environmental; Chapter 1). Scottish Executive (2005) stated that SFM

> provides the maximum possible social and economic resilience against flooding, by protecting and working with the environment, in a way which is fair and affordable both now and in the future.

Its objective is to integrate "strategies that benefit communities, reduce economic damage, protect environments, and meet the needs of future generations" (Das et al., 2018, p3 working in cold regions). This wider policy driver of promoting sustainability emphasises delivery of FRM through participatory and collaborative governance involving communities (Challies et al., 2016).

2.5.1   From FRM to community-focused FRM

FRM, with its systemic view, puts a strong emphasis on multi-stakeholder networks, collaboration and communication, including increasingly participation of local communities as key stakeholders. Scott et al. (2013, p13) also noted this growing diversity of actors, along with the "development of new roles and different forms of both horizontal and vertical collaboration". Importantly, shifts to distributed FRM in dealing with residual flood risk integrate changed responsibilities, multi-methods, and flood education (e.g., Riley, 2007 in the USA; Figure 2.3), bringing community learning into recovery and preparedness phases (Chapters 1 and 9). Community agency – locally attuned – is required for many elements of distributed FRM to be effective.

This poses questions about societal positioning along a continuum from: "Where are we now – *The government will protect us*" and "Where do we need to be – *We are all responsible for our safety*"? (Riley, 2007, p90). A community's

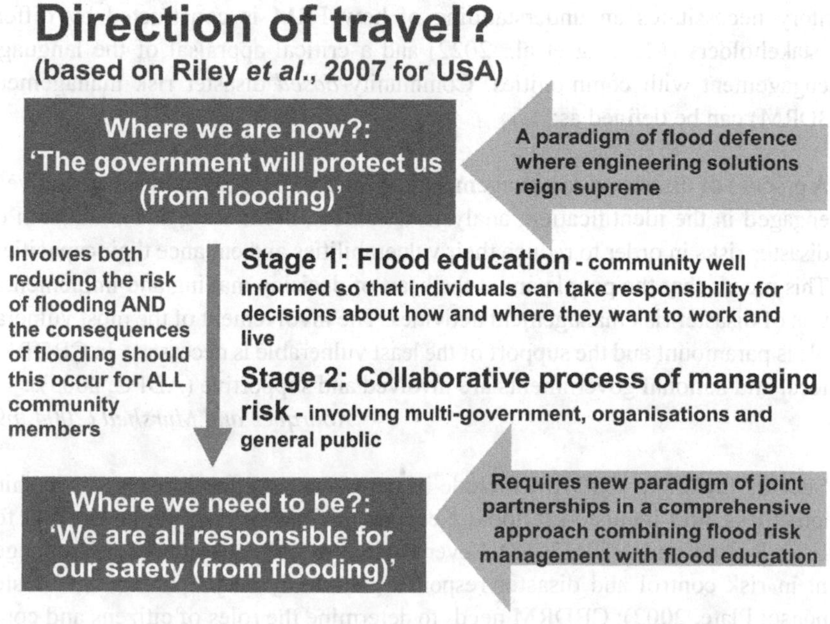

Direction of travel?
(based on Riley *et al.*, 2007 for USA)

Where we are now?:
'The government will protect us
(from flooding)'

A paradigm of flood defence where engineering solutions reign supreme

Involves both reducing the risk of flooding AND the consequences of flooding should this occur for ALL community members

Stage 1: Flood education - community well informed so that individuals can take responsibility for decisions about how and where they want to work and live

Stage 2: Collaborative process of managing risk - involving multi-government, organisations and general public

Where we need to be?:
'We are all responsible for
our safety (from flooding)'

Requires new paradigm of joint partnerships in a comprehensive approach combining flood risk management with flood education

FIGURE 2.3   Framing sustainable flood management (concept based on Riley, 2007)

ability to act to minimise losses does not necessarily increase over time at-a-place, as flood risks and communities change.

Sayers et al. (2013) provided a valuable summary of economic, social, and ecological sustainability objectives and outcome measures of success for good FRM. This requires public participation in establishing needs and goals (Chapter 8). A key question is the possible roles of communities in such systemic approaches, and how to build community resilience throughout the FRM spiral (Chapter 1). There may be differences between strategic and community measures in determining understanding of 'success' and what is valued in terms of processes and outcomes. As Sayers et al. (2013, p5) summarise:

> Flood risk management [...] embeds a continuous process of adaptation that is distinct from the 'implement and maintain' philosophy of a traditional flood defence approach. Taking a longer-term, whole-system view places a much higher demand upon those affected by flooding and those responsible for its mitigation. It involves collaborative action across governments, the public sector, businesses, voluntary organizations, and individuals. This places an increasing emphasis on effective communication of the residual risks and actions to be taken.

A significant impetus in research and policy is to develop principles and practices for different models of community-focused FRM (Table 2.3D). This inclusive term is used in this book to describe the broad approach to strong community involvement, whether community-based or community-led (Chapter 8). Such territory necessitates an understanding of how FRM is constructed by different stakeholders (Mehring et al., 2022) and a critical appraisal of the language of engagement with communities. Community-*based* disaster risk management (CBDRM) can be defined as:

> A process of disaster risk management in which at risk communities are actively engaged in the identification, analysis, treatment, monitoring and evaluation of disaster risks in order to reduce their vulnerabilities and enhance their capacities. This means that the people are at the heart of decision making and implementation of disaster risk management activities. The involvement of the most vulnerable is paramount and the support of the least vulnerable is necessary. In CBDRM, local and national governments are involved and supportive (ADPC, 2003).
>
> *Abarquez and Murshed (2004, p9)*

Success factors in CBDRM include its permanence, effectiveness, ownership, adaptiveness, and inclusion (Global Network of Civil Society Organisations for Disaster Reduction GNDR, 2018). Several stages exist in operational risk management in risk control and disaster response (risk analysis, preparedness, disaster response; Plate, 2002); CBDRM needs to determine the roles of citizens and communities in each. Such thinking brings both challenges and opportunities.

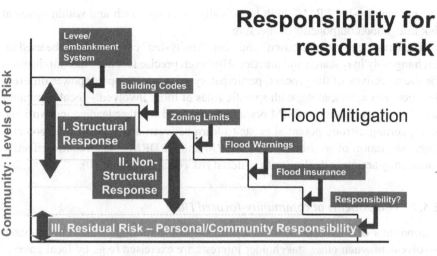

Hypothetical depiction of residual risk. Note: (i) Vertical intervals will not necessarily be identical; (ii) Model will be place specific; and (iii) The system is dynamic and will change over time.

FIGURE 2.4   Residual risk and personal responsibility (Source: edited from Tobin pers. comm., concept based on Riley, 2007; Montz and Tobin, 2008)

Figure 2.4 (Riley, 2007; Montz and Tobin, 2008) highlights how different adaptation measures – each involving different types and degrees of agency from local communities – can be integrated to reduce flood risk incrementally.

In some settings, structural measures, often implemented by government, offset design risk but residual risk remains. Residual risk may also be increasing with climate change (Eisenberg, 2021). Other adaptive options to deal with this risk are non-structural (e.g., land use zoning), led by government or other actors; many require community awareness and agency to be effective (e.g., early flood warning systems). With these measures, residual flood risk remains to be dealt with by households at risk (e.g., through property-level protection), as personal and/or community responsibility. However, building social capacities for effective community-based FRM involves attention to strategies for capacity building not just for at-risk communities, but also the wider organisational and institutional environment (Kuhlicke et al., 2011; Scott et al., 2013) (Chapters 4, 8, and 10).

Community-*led* DRM (CLDRM) has a different knowledge and power dynamic to CBDRM, with strongest inputs to adaptation options from local citizens working collectively. However, not all communities have the resources to lead but in some remoter settings, communities can have little alternative. This poses questions about balance and changing responsibility in any transitioning between community-based and community-led FRM – both holistically as an approach and within different risk mitigation components or measures.

The terms "community-based" and "community-led" can sometimes be used interchangeably in research and practice. However, precise labelling has implications for the objectives of the process, participatory principles adopted, power differentials between actors, along with specific roles of those involved – local communities, NGOs, and statutory DRM organisations – and whether leading, co-working, or supporting. Strong potential exists to learn between different national contexts from integration of measures – in CBDRM and CLDRM – that require effective community action in dealing with residual risk (Chapters 7 and 10).

2.5.2 Components of community-focused FRM

Components of FRM can exist independently of community- and citizen-based involvement when other stakeholder interests are exercised (e.g., by local government and spatial planners). However, components also need to be developed to reflect community interests and concerns.

2.5.2.1 Natural flood management

One configuration of SFM is NFM (Cook et al., 2016), a catchment-based approach, involving manipulations of catchment land use – rural or urban – and engagement of local communities. Long-standing controversies exist about the relationships between floods, forests, and land management (Alila et al., 2009). Scotland provides a good example of policy preceding scientific evidence. While NFM may reduce impacts of less than 1- in 10-year floods, this is not yet validated for major floods in Scotland. However, NFM is located positively within a larger context of multiple benefits – with habitat creation, biodiversity protection, carbon sequestration, and sediment capture. Communities need to be aware of the strengths and limitations of NFM measures. A common question at public meetings promoting NFM can be 'How many trees and when will they be effective?' This complexity highlights the importance of building positive community attitudes to NFM, promoted by information, collaboration, and trust-building (Marshall et al., 2019; Chapter 7).

2.5.2.2 Spatial planning and place-making

Another component of integrated FRM is the spatial turn in planning (van Ruiten and Hartmann, 2016, p698; Busscher et al., 2019), referring to "the need for land in flood risk management". As Scott et al. (2013, p3) noted in the enactment of EU legislation "spatial planning has increasingly moved centre stage as part of a 'whole catchment' framework to risk management" – an approach widely adopted (e.g., in the UK and Netherlands). This component focuses on good place-making as a key element of local urban flood resilience (Foley, 2021[9]), systemically linking several agendas – FRM, effective spatial planning, community development, and climate adaptation. Good place-making for local communities provides significant quality of life and health benefits, particularly for vulnerable and marginalised groups.

This integrative thinking offers "an essential tool for making land use choices that help to achieve greater flood and coastal resilience, as well as wider environmental benefits" (Ran and Nedovic-Budic, 2016; Foley, 2021 np). Embedding place-making into spatial planning also emphasises the importance of inclusive collaborative planning processes that attend to community participation and EJ issues (Hudson and Slavíková, 2022).

2.5.2.3 Recovery-based approach to urban infrastructure

Changes in assessments of urban infrastructural performance are also occurring in FRM. In surface water flood management, there is impetus for changes in approach from using arbitrary deterministic design thresholds for failure of infrastructure to a risk-based system that mirrors those of river flood engineering (Haghighatafshar et al., 2020). Once design limits are exceeded, recovery work shifts to "the affected" in communities and "first responders, insurance policies, and long-term government recovery processes" (p6). This thinking integrates a recovery-based approach to impacted urban infrastructure, and its relationships to community functioning after floods. For example, this involves determining an acceptable time for local schools to be out of service due to future flooding, given their role in different stages of recovery and return to normal community functioning. In Boulder County, the USA, this time was appraised as four weeks. With community recovery time placed at the centre, replacement infrastructure needs to meet a minimum standard.

2.6 Conclusions

Increasing diversity of both floods and communities intersect to present major challenges for local FRM and equitable climate resilience. Many current issues in integrated FRM in MEDCs relate to limited awareness, or acceptance by communities, of their changing roles within the different paradigm shifts articulated above. While effective community-focused FRM requires action from communities and a wider group of stakeholders, communities have variable knowledge, empowerment or resources to participate in local flood risk planning. Community-focused FRM needs to give strong attention to systemic EJ issues for marginalised groups in exposed flood risk settings. This poses questions about what such FRM looks like – or could look like, how it is constructed by different stakeholders, and its research and resourcing needs.

This urgent imperative for increased community engagement and participation in community-focused FRM sets an important backdrop to this book. Various interrelated domains impact this socio-hydrological territory: flood perception, flood knowledge, flood heritage, community participation within local FRM, flood learning, and in identifying pivotal areas for innovation and transformation. Critical throughout is knowledge exchange and co-generation *with* communities *and* social learning *within* communities to build capital and capacities (Chapter 8). This affects the roles of both communities and other local stakeholders in FRM; the following chapters explore this territory further.

Notes

1 https://sheffieldfloodclaimsarchive.shu.ac.uk/aboutFlood.cfm.
2 https://www.gov.nl.ca/ecc/files/waterres-flooding-situation-report.pdf.
3 https://www.therrc.co.uk/news/floods-and-dredging-reality-check.
4 https://eu.redding.com/story/life/home-garden/2019/04/05/avoid-contaminants-after-your-garden-floods/3343402002;https://www.wiscontext.org/10-tips-safely-using-produce-flooded-garden.
5 https://e360.yale.edu/features/as-high-tide-flooding-worsens-more-pollution-is-washing-to-the-sea.
6 https://items.ssrc.org/category/understanding-katrina/.
7 https://www.bbc.co.uk/news/uk-17679723.
8 https://www.ada.org.uk/wp-content/uploads/2017/12/IDBs_An_Introduction_A5_2017_web.pdf.
9 https://environmentagency.blog.gov.uk/2021/09/22/planning-for-flood-resilient-places/.

References

Abarquez, I. and Murshed, Z. (2004) *Community-Based Disaster Risk Management: Field Practitioners Handbook*. Asian Disaster Preparedness Center (ADPC), Thailand.

AIDR (2020) Australian National Emergency Risk Assessment Guidelines.

Alba, R., Klepp, S. and Bruns, A. (2020) Environmental justice and the politics of climate change adaptation – the case of Venice. *Geographica Helvetica*, 75, 363–368. https://doi.org/10.5194/gh-75-363-2020

Alexander, J. and Cooker, M.J. (2016) Moving boulders in flash floods and estimating flow conditions using boulders in ancient deposits. *Sedimentology*, 63, 1582–1595. https://doi.org/10.1111/sed.12274

Alila, Y., Kuras, P.K., Schnorbus, M. and Hudson, R. (2009) Forests and floods: A new paradigm sheds light on age-old controversies. *Water Resources Research*, 45, W08416. https://doi.org/10.1029/2008WR007207

Archer, D. and Fowler, H. (2021) A historical flash flood chronology for Britain. *Journal of Flood Risk Management*, 14, e12721. https://doi.org/10.1111/jfr3.12721

Arthur, H.G. (1977) Teton Dam Failure. In *The Evaluation of Dam Safety*, Engineering Foundation Conference Proceedings, Asilomar, 1976, ASCE, pp61–71.

Bell, K. (2014) *Achieving Environmental Justice: A Cross-National Analysis*. Policy Press, University of Bristol, Bristol, UK.

Bell, K. (2020) *Working-Class Environmentalism: An Agenda for a Just and Fair Transition to Sustainability*. Palgrave Macmillan, Cham.

Bevacqua, E., Maraun, D., Vousdoukas, M.I., Voukouvalas, E., Vrac, M., Mentaschi, L. and Widmann, M. (2019) Higher probability of compound flooding from precipitation and storm surge in Europe under anthropogenic climate change *Science Advances*, 5(9) EAAW5531. https//doi.org/10.1126/sciadv.aaw5531

Blöschl, G., Glendell, M. et al. (2019) Twenty-three unsolved problems in hydrology (UPH) – a community perspective. *Hydrological Sciences Journal*, 64(10), 1141–1158. https://doi.org/10.1080/0262666.2019.1620507

Buss, L.S. (2005) Nonstructural flood damage reduction within the U.S. Army Corps of Engineers. *Journal of Contemporary Water Research & Education*, 130, 26–30.

Busscher, T., van den Brink, M. and Verweij, S. (2019) Strategies for integrating water management and spatial planning: Organising for spatial quality in the Dutch "Room for the River" program. *Journal of Flood Risk Management*, 12, e12448. https://doi.org/10.1111/jfr3.12448

Butler, C. and Pidgeon, N. (2011) From "Flood Defence" to "Flood Risk Management": Exploring governance, responsibility, and blame. *Environment and Planning C, Government & Policy*, 29(3), 533–547.

Chakraborty, J., Collins, T.W. and Grineski, S.E. (2019) Exploring the environmental justice implications of Hurricane Harvey flooding in Greater Houston, Texas. *AJPH Climate Change*, 109(2), 244–250.

Challies, E. (2016) Participation and collaboration for sustainable flood risk management? https://sustainability-governance.net/2016/03/07/participation-and-collaboration-for-sustainable-flood-risk-management/

Challies, E., Newig, J., Thaler, T., Kochskämper, E. and Levin-Keitel, M. (2016) Participatory and collaborative governance for sustainable flood risk management: An emerging research agenda. *Environmental Science and Policy*, 55(2), 275–280.

CIWEM (2014) *Floods and dredging: A reality check*. CIWEM. https://www.ciwem.org/assets/pdf/Policy/Reports/Floods-and-Dredging-a-reality-check.pdf

Cook, B., Forrester, J., Bracken, L.J., Spray, C.J. and Oughton, E. (2016) Competing paradigms of flood management in the Scottish/English Borderland. *Disaster Prevention and Management*, 25(3), 1–15. https://doi.org/10.1108/DPM-01-2016-0010

Cutter, S.L. (2006) *Hazards, Vulnerability and Environmental Justice*. Earthspan, Abingdon, UK.

Das, A., Reed, M. and Lindenschmidt, K.-E. (2018) Sustainable ice-jam flood management for socio-economic and socio-ecological systems. *Water*, 10(2), 135. http://dx.doi.org/10.3390/w10020135

Del-Rosal-Salido, J., Folgueras, P., Bermúdez, M., Ortega-Sánchez, M. and Losada, M.A. (2021) Flood management challenges in transitional environments: Assessing the effects of sea-level rise on compound flooding in the 21st century. *Coastal Engineering*, 167, 103872. https://doi.org/10.1016/j.coastaleng.2021.103872

Dobbie, C.H. and Wolf, P.O. (1953) The Lynmouth flood of August 1952. *Proceedings of the Institution of Civil Engineers*, 2, 522–546. https://doi.org/10.1680/ipeds.1953.12369

Douglas, E.M., Kirshen, P.H., Paolisso, M. et al. (2012) Coastal flooding, climate change and environmental justice: Identifying obstacles and incentives for adaptation in two metropolitan Boston Massachusetts communities. *Mitigation and Adaptation Strategies for Global Change*, 17, 537–562. https://doi.org/10.1007/s11027-011-9340-8

Eisenberg, D. (2021) The need to consider residual risk. *Nature Climate Change*, 11, 803–804. https://doi.org/10.1038/s41558-021-01129-z

Eiser, J.R., Bostrom, A., Burton, I., Johnston, D.M., McClure, J., Paton, D., van der Pligt, J. and White, M.P. (2012) Risk interpretation and action: A conceptual framework for responses to natural hazards. *International Journal of Disaster Risk Reduction*, 1, 5–16. https://doi.org/10.1016/j.ijdrr.2012.05.002

Erickson, T.B. and Brooks, J. (2017) After a disaster, contaminated floodwater can pose a threat for months to come. *The Conversation.* https://theconversation.com/after-a-disaster-contaminated-floodwater-can-pose-a-threat-for-months-to-come-84247

European Commission (2007) Directive 2007/60/EC of the European Parliament and of the Council of 23 October 2007 on the assessment and management of flood risks. (Text with EEA relevance) Official Journal of the European Union L 288, 6/11/2007, 27–34.

Foley, J. (2021) Planning for flood-resilient places. Environment Agency. https://environmentagency.blog.gov.uk/2021/09/22/planning-for-flood-resilient-places/

Fortunato, A., Oliveri, E. and Mazzola, M.R. (2014) Selection of the optimal design rainfall return period of urban drainage systems. *Procedia Engineering*, 89, 742–749.

Ganguli, P. and Merz, B. (2019) Extreme coastal water levels exacerbate fluvial flood hazards in Northwestern Europe. *Scientific Reports*, 9(1), 13165. https://doi.org/10.1038/s41598-019-49822-6

Gill, J.C. and Malamud, B.D. (2014) Reviewing and visualizing the interactions of natural hazards. *Reviews of Geophysics*, 52, 680–722. https://doi.org/10.1002/2013RG000445

Global Commission on Adaptation (2019) Adapt now: A global call for leadership on climate resilience. Global Center on Adaptation/World Resources Institute. https://gca.org/wp-content/uploads/2019/09/GlobalCommission_Report_FINAL.pdf

Global Network of Civil Society Organisations for Disaster Reduction GNDR (2018) *Institutionalising sustainable CBDRM: Key success factors.* https://www.gndr.org/resource/cbdrm/success-factors-for-sustainable-community-based-disaster-risk-management/

Gori, A., Lin, N. and Smith, J. (2020) Assessing compound flooding from landfalling tropical cyclones on the North Carolina coast. *Water Resources Research*, 56, e2019WR026788. https://doi.org/10.1029/2019WR026788

Grineski, S., Collins, T.W., Chakraborty, J. et al. (2015) Hazardous air pollutants and flooding: A comparative interurban study of environmental injustice. *GeoJournal*, 80, 145–158. https://doi.org/10.1007/s10708-014-9542-1

Gudmundsson, M.T. (2015) Chapter 56 – Hazards from Lahars and Jökulhlaups. In H. Sigurdsson (ed.) *The Encyclopedia of Volcanoes* (2nd edn). Academic Press, Cambridge, pp971–984. https://doi.org/10.1016/B978-0-12-385938-9.00056-0

Guerreiro, S.B., Dawson, R.J., Kilsby, C., Lewis, E. and Ford, A. (2018) Future heat-waves, droughts and floods in 571 European cities. *Environmental Research Letters*, 13, 034009.

Haghighatafshar, S., Becker, P., Moddemeyer, S., Persson, A., Sörensen, J., Aspegren, H. and Jönsson, K. (2020) Paradigm shift in engineering of pluvial floods: From historical recurrence intervals to risk-based design for an uncertain future. *Sustainable Cities & Society*, 61, 102317. https://doi.org/10.1016/j.scs.2020.102317

Harbitz, C.B., Glimsdal, S., Løvholt, F., Kveldsvik, V., Pedersen, G.K. and Jensen, A. (2014) Rockslide tsunamis in complex fjords: From an unstable rock slope at Åkerneset to tsunami risk in western Norway. *Coastal Engineering*, 88, 101–122. https://doi.org/10.1016/j.coastaleng.2014.02.003

Hendry, A., Haigh, I.D., Nicholls, R.J., Winter, H., Neal, R., Wahl, T., Joly-Laugel, A. and Darby, S.E. (2019) Assessing the characteristics and drivers of compound flooding events around the UK coast. *Hydrology and Earth System Sciences*, 23, 3117–3139. https://doi.org/10.5194/hess-23-3117-2019

HM Government (2020) National Risk Register, 2020 Edition. https://www.gov.uk/government/publications/national-risk-register-2020

Hsiao, S.-C., Chiang, W.-S., Jang, J.-H., Wu, H.-L., Lu, W.-S., Chen, W.-B. and Wu, y-T. (2021) Flood risk influenced by the compound effect of storm surge and rainfall under climate change for low-lying coastal areas. *Science of The Total Environment*, 764, 144439. https://doi.org/10.1016/j.scitotenv.2020.144439

Hudson, P. and Slavíková, L. (2022) The role of risk transfer and spatial planning for enhancing the flood resilience of cities. In T. Hartmann, L. Slavíková and M.E. Wilkinson (eds.) *Spatial Flood Risk Management: Implementing Catchment-Based Retention and Resilience on Private Land*. Elgar Publishing, Cheltenham, UK, pp148–163.

Ingham, B. and Ingham, S. (2007) Safely Using Produce from Flooded Garden. University of Wisconsin Extension. https://pddc.wisc.edu/wp-content/blogs.dir/39/files/Fact_Sheets/LC_PDF/Safely_Using_Produce_from_Flooded_Gardens.pdf

International Alert (2015) *Compounding risk: Disasters, fragility and conflict*. Policy brief. May 2015. https://www.international-alert.org/publications/compounding-risk/

Jalili Pirani, F. and Najafi, M.R. (2020) Recent trends in individual and multivariate compound flood drivers in Canada's coasts. *Water Resources Research*, 56, e2020WR027785. https://doi.org/10.1029/2020WR027785

Johnson, C.L. and Priest, S.J. (2008) Flood risk management in England: A changing landscape of risk responsibility? *International Journal of Water Resources Development*, 24(4), 513–525. https://doi.org/10.1080/07900620801923146

Johnson, C.L., Tunstall, S.M. and Penning-Rowsell, E.C. (2005) Floods as catalysts for policy change: Historical lessons from England and Wales. *International Journal of Water Resources Development*, 21(4), 561–575. https://doi.org/10.1080/07900620500258133

Kelman, I., Gaillard, J.C. and Mercer, J. (2015) Climate change's role in disaster risk reduction's future: Beyond vulnerability and resilience. *International Journal of Disaster Risk Science*, 6, 21–27. https://doi.org/10.1007/s13753-015-0038-5

Kelman, I. and Rauken, T. (2012) The paradigm of structural engineering approaches for river flood risk reduction in Norway. *Area*, 44(2), 144–151. https://doi.org/10.1111/j.1475-4762.2011.01074.x

Krieger, K., Broekhans, B. and Correlje, A.F. (2012). Flood protection policies in Germany UK and The Netherlands: Towards a risk-based approach? In F. Klijn and T. Schweckendiek (eds.) *Comprehensive Flood Risk Management: Proceedings of the 2nd European Conference on Flood Risk Management FLOODrisk2012*. Taylor & Francis, Abingdon, pp1245–1250.

Kuhlicke, C., Steinführer, A., Begg, C. et al. (2011) Perspectives on social capacity building for natural hazards: Outlining an emerging field of research and practice in Europe. *Environmental Science & Policy*, 14(7), 804–814. https://doi.org/10.1016/j.envsci.2011.05.001

Lempérière, F. (2017) Dams and floods. *Engineering*, 3(1), 144–149. https://doi.org/10.1016/J.ENG.2017.01.018

Marshall, K., Waylen, K. and Wilkinson, M. (2019) *Communities at Risk of Flooding and Their Attitudes towards Natural Flood Management*. CRW2018_03. Scotland's Centre of Expertise for Waters (CREW), Scotland's Centre of Expertise for Waters (CREW).

McEwen, L.J., Garde-Hansen, J., Holmes, A., Jones, O. and Krause, F. (2016) Sustainable flood memories, lay knowledges and the development of community resilience to future flood risk. *Transactions of the Institute of British Geographers*, 42(1), 14–28.

McEwen, L.J., Hall, T., Hunt, J., Harrison, M. and Dempsey, M. (2002) Flood warning, warning response and planning control issues associated with caravan parks: The April 1998 floods on the lower Avon flood plain, Midlands Region, UK. *Applied Geography*, 22, 271–305.

Medd, W., Deeming, H., Walker, G., Whittle, R., Mort, M., Twigger-Ross, C., Walker, M., Watson, N. and Kashefi, E. (2015) The flood recovery gap: A real-time study of local recovery following the floods of June 2007 in Hull, North East England. *Journal of Flood Risk Management*, 8, 315–328.

Mehring, P., Geoghegan, H., Cloke, H.L. and Clark, J.M. (2022) Going home for tea and medals: How members of the flood risk management authorities in England construct flooding and flood risk management. *Journal of Flood Risk Management*, 15(1), e12768. https://doi-org.ezproxy.uwe.ac.uk/10.1111/jfr3.12768

Montz, B.E. and Tobin, G.A. (2008) From false sense of security to residual risk: Communicating the need for new floodplain development models. *Geograficky Casopis (Geographical Journal – Slovak Academy of Sciences)*, 60(1), 3–14.

Morse, R. (2008) *Environmental Justice through the Eye of Hurricane Katrina*. Joint Center for Political and Economic Studies, Health Policy Institute, Washington, DC.

Muis, S., Verlaan, M., Winsemius, H. et al. (2016) A global reanalysis of storm surges and extreme sea levels. *Nature Communications*, 7, 11969. https://doi.org/10.1038/ncomms11969

Nakamura, S. and Oki, T. (2018) Paradigm shifts on flood risk management in Japan: Detecting triggers of design flood revisions in the modern era. *Water Resources Research*, 54, 5504–5515. https://doi.org/10.1029/2017WR022509

Newson, M. (1992) *Land, Water and Development: River Basin Systems and Their Sustainable Management*. Routledge, London.

Paprotny, D., Sebastian, A., Morales-Nápoles, O. et al. (2018) Trends in flood losses in Europe over the past 150 years. *Nature Communications*, 9, 1985. https://doi.org/10.1038/s41467-018-04253-1

Pei, S., Dahl, K.A., Yamana, T.K., Licker, R. and Shaman, J. (2020) Compound risks of hurricaneevacuation amid the COVID-19 pandemic in the United States. *GeoHealth*, *4*, e2020GH000319. https://doi.org/10.1029/2020GH000319

Pielke, R.A. and Downton, M.W. (2000) Precipitation and damaging floods: Trends in the United States, 1932–97. *Journal of Climate*, 13, 3625–3637.

Plate, E.J. (2002) Flood risk and flood management. *Journal of Hydrology*, 267, 2–11.

Rahimi, R., Tavakol-Davani, H., Graves, C., Gomez, A. and Fazel Valipour, M. (2020) Compound inundation impacts of coastal climate change: Sea-level rise, groundwater rise, and coastal watershed precipitation. *WATER*, 12, 2776. https://doi.org/10.3390/w12102776

Ran, J. and Nedovic-Budic, Z. (2016) Integrating spatial planning and flood risk management: A new conceptual framework for the spatially integrated policy infrastructure. *Computers, Environment and Urban Systems*, 57, 68–79. https://doi-org.ezproxy.uwe.ac.uk/10.1016/j.compenvurbsys.2016.01.008

Riley, D. (2007) *Improving public safety – from federal protection to shared risk reduction*. Pre-reading. Gilbert F. White National Flood Policy Forum *2007 Assembly* Floodplain Management 2050, pp90-92.

Rotkin-Ellman, M., Solomon, G., Gonzal, C.R., Agwaramgbo, L. and Mielke, H.W. (2010) Arsenic contamination in New Orleans soil: Temporal changes associated with flooding. *Environmental Research*, 110(1), 19–25. https://doi.org/10.1016/j.envres.2009.09.004

Sayers, P., Li, Y., Galloway, G., Penning-Rowsell, E., Shen, F., Wen, K., Chen, Y. and Le Quesne, T. (2013) *Flood Risk Management: A Strategic Approach*. UNESCO, Paris.

Sadegh, M., Moftakhari, H., Gupta, H. V., Ragno, E., Mazdiyasni, O., Sanders, B., Matthew, R. and AghaKouchak, A. (2018) Multihazard scenarios for analysis of compound extreme events. *Geophysical Research Letters*, *45*, 5470–5480. https://doi.org/10.1029/2018GL077317

Schlosberg, D. (2007) *Defining Environmental Justice: Theories, Movements, and Nature*. Oxford University Press, Oxford.

Scottish Executive (2005) *Final Report of the National Technical Advisory Group on Flooding*. Scottish Executive, Edinburgh, UK,

Scott, M., White, I., Kuhlicke, C., Steinführer, A., Sultana, P., Thompson, P., Minnery, J., O'Neill, E., Cooper, J., Adamson, M. and Russell, E. (2013) Living with flood risk. *Planning Theory & Practice*, 14(1), 103–140. https://doi.org/10.1080/14649357.2012.761904

Serra-Llobet, A., Llobet, A.S., Kondolf, G.M., Schaefer, K. and Nicholson, S. (2018) *Managing Flood Risk: Innovative Approaches from Big Floodplain Rivers and Urban Streams*. Palgrave Macmillan, London, 155pp.

Sigurmundsson, F.S., Gísladóttir, G. and Erlendsson, E. (2014) The Impacts of Advancing Glaciers and Jökulhlaups on the 19th Century Farming Community in the Suðursveit District South of Vatnajökull Glacier, Iceland. *American Geophysical Union, Fall Meeting*, abstract-id. GC21E-0601 https://ui.adsabs.harvard.edu/abs/2014AGUFMGC21E0601S/abstract

Smith, K. and Tobin, G.A. (1979) *Human Adjustment to the Flood Hazard*. Longman, Harlow, UK, 130pp.

Svarstad, H., Sletten, A., Paloniemi, R., Barton, D.N. and Grieg-Gran, M. (2011) Three types of environmental justice – From concepts to empirical studies of social impacts of policy instruments for conservation of biodiversity. PolicyMix Report Issue No 1.

Tariq, M.A.U.R., Farooq, R. and van de Giesen, N. (2020) A critical review of flood risk management and the selection of suitable measures. *Applied Sciences (Switzerland)*, 10(23), 1–18. [8752]. https://doi.org/10.3390/app10238752

Thomson, S. (2015) Flooding of first nations and environmental justice in Manitoba. Case studies of the impacts of the 2011 flood and hydro development in Manitoba, 38-2 *Manitoba Law Journal* 220, 2015 CanLIIDocs 254. https://canlii.ca/t/7cm

Tunstall, S.M., Johnson, C.L. and Penning-Rowsell, E.C. (2004) Flood hazard management in England and Wales: from land drainage to flood risk management. Paper presented at World Congress on Natural Disaster Mitigation, 19–22 February, Conference Proceedings, 2, pp447–454. Vlgyan Bhawan, New Delhi. http://www.fhrc.mdx.ac.uk/resources/publications.html

UK Cabinet Office (2008) *The Pitt review: lessons learned from the 2007 floods*. http://www.cabinetoffice.gov.uk/thepittreview.aspx.

van Ruiten, L. and Hartmann, T. (2016) The spatial turn and the scenario approach in flood risk management—Implementing the European floods directive in the Netherlands. *AIMS Environmental Science*, 3(4), 697–713. https://doi.org/10.3934/environsci.2016.4.697

Wahl, T., Jain, S., Bender, J., Meyers, S.D. and Luther, M.E. (2015) Increasing risk of compound flooding from storm surge and rainfall for major US cities. *Nature Climate Change*, 5. https://doi.org/10.1038/NCLIMATE2736

Walker, G. and Burningham, K. (2011) Flood risk, vulnerability and environmental justice: Evidence and evaluation of inequality in a UK context. *Critical Social Policy*, 31(2), 216–240. https://doi.org/10.1177/0261018310396149

Waller, R.M. (1994) *Ground water and the rural home owner*. USGS. https://pubs.usgs.gov/gip/gw_ruralhomeowner/

Weitkamp, E., McEwen, L.J. and Ramirez, P. (2020) Communicating the hidden: Towards a framework for drought risk communication in maritime climates. *Climatic Change*, 163, 831–850. https://doi.org/10.1007/s10584-020-02906-z

Werritty, A. (2006) Sustainable flood management: Oxymoron or new paradigm? *Area*, 38, 16–23. https://doi.org/10.1111/j.1475-4762.2006.00658.x

Zhang, W. and Villarini, G. (2020) Deadly compound heat stress-flooding hazard across the central United States. *Geophysical Research Letters*, 47, e2020GL089185. https://doi.org/10.1029/2020GL089185

Zscheischler, J., Westra, S., van den Hurk, B.J.J.M. et al. (2018) Future climate risk from compound events. *Nature Climate Change*, 8, 469–477. https://doi-org.ezproxy.uwe.ac.uk/10.1038/s41558-018-0156-3

3

FLOOD HAZARD PERCEPTION, AWARENESS, AND ACTION

From individual to community?

3.1 Introduction

Human perception can be defined in different ways: "the ability to see, hear or become aware of something through the senses" and "the way in which something is regarded, understood, or interpreted" (Lexico Oxford Dictionary[1]). Both meanings are relevant within community-focused, flood risk management (FRM). At-risk communities may have to deal with the sensory experience of actual flood events in their locale, and the psychology of living in exposed at-risk areas – on floodplains, deltas, coasts, and within urban catchments. 'Risk tolerance' refers to an individual's capacity to accept living with a certain amount of risk. The need to 'live with risk' is exacerbated by changing flood risk and increased likelihood of damage due to interactions between climate change and urban development (Chapter 2). Adequate flood preparedness – both individual and collective – are related to flood risk perception.

A key question is why, when faced with evidence of risk, some citizens and communities take precautionary or preparedness actions to mitigate flood risk while others do not. Exploring this research territory requires contributions from different academic disciplines including psychology, sociology, risk communication, education, and management, and interfacing with science and technology. This chapter asks:

- How have the different paradigms – or ways of thinking – within flood management influenced research needs relating to risk perception?
- What factors influence individual and collective flood hazard perception?
- What are the implications of risk perception for the policy and practice of communication strategies to promote action and behaviour change in community-focused FRM?

DOI: 10.4324/9781315666914-3

3.2   Flooding as a risk: Physical factors influencing perception

Water conveys multiple values and meanings (Strang, 2004). As excess water in the wrong place at the wrong time, flooding tends to be highly visual, sensory, and potentially emotive, in comparison to more hidden diffuse risks like drought, although groundwater flooding can be concealed in its early stages (Chapter 2). In river and coastal flooding, it is important to distinguish between public perceptions of routine flooding and its rhythms (i.e., indirect 'seasonal nuisance flooding'; Moftakhari et al., 2018), from those of flood extremes. The former can involve a strong embedded sense of 'living with water' in everyday lives; the latter involves perceptions of rarer, potentially catastrophic flood events, like the hurricane-induced deluges and storm surges in New Orleans, Louisiana, USA (Gotham et al., 2018). Different types of floods also have different psychological impacts. While much attention has been given to perception of river and coastal flooding, perception of other risks is becoming important, with increasing frequency of high-intensity rainfall and surface water flooding. For example, what factors affect motorists' decisions about whether to drive through flash floods (Coles and Hirschboeck, 2020)?

3.2.1   Why risk perception is important

People's risk perceptions are important factors in their risk-related decisions and in making their knowledge actionable in flood preparedness. Community-focused FRM requires empowered citizen agency (Chapters 2 and 8). Within this paradigm, 'adequate' flood risk perception is "one of the premises to lower the vulnerability of society and is an important element in non-structural measures for flood risk reduction" (Krasovskaia et al., 2007, p387). As Bang and Burton (2021) emphasised, limited community perception of their own risks can be reflected in lack of preparedness for contemporary, let alone future, floods. To reduce personal losses, citizens – individually and collectively – need to acknowledge and act to mitigate their residual risk (Chapter 1). In doing so, they must navigate an increasing range of preparedness and adaptive options that require individual and collective agency in their adoption and longer-term meaningful participation in their maintenance. Building community resilience is influenced by perceptions of both the risk itself *and* the efficacy of different measures. Multiple stakeholders – citizens, NGOs, statutory FRM agencies, insurance companies, and the news media – may hold competing views and have different priorities.

It is well recognised that risk communication – scientific and operational – is inadequately informed by understandings of how at-risk communities perceive that risk (Birkholz et al., 2014; Cologna et al., 2017; Chapter 6). More knowledge is required of behavioural elements within FRM strategies, such as how members of a community respond to warnings issued by community flood wardens, or whether households act to put in place property-level resilience measures. FRM professionals need to understand these factors to inform (co-)development of adaptive strategies for local FRM. They also need to consider how to promote more accurate risk perception among members of the public and promote pro-adaptive attitudes and behaviours. This engagement work may require skills development and training among local FRM actors. Such exchanges are critical in seeking consensus between the public/communities and statutory authorities about what might be considered 'tolerable levels of risk' at-a-place. This sensitive negotiation has major implications for citizens and communities who are undergoing post-flood recovery and rebuilding within the wider disaster risk management (DRM) spiral. Here the 'Risk Perception Paradox' requires attention.

3.2.2 The 'Risk Perception Paradox'

As Baker (2007, p164) reflected "one of many well-documented fallacies in flood-hazard science is that knowledge leads to wise action". In navigating the complex territory of perception and preparedness, a major paradox is an assumption that heightened risk awareness leads to personal preparedness and then risk mitigation behaviours (Wachinger et al., 2013).

> However, this is not necessarily true. In fact, the opposite can occur if individuals with high risk perception still choose not to personally prepare themselves in the face of a natural hazard.
>
> *p1049*

Paek and Hove (2017, p2) further observed that "a common assumption in risk perception research is that people's knowledge and certainty about a risk determines how they will perceive it" (i.e., rational choice model of decision-making). It is, however, well established that public (lay) perception of flood risk (subjective risk) can be very different from reality (objective risk) (e.g., Burningham et al., 2008). The emotional dimension is important – how people feel (Baan and Klijn, 2004) and what they value. Risk underestimation and passive attitudes towards flood problems, as well as risk denial, are common. One of Pielke's (1999) nine fallacies about floods is that 'knowledge leads to action', with Knowledge-attitude-behaviour theory long refuted in relation to pro-environmental and risk behaviours (e.g., Bell and Tobin, 2007). It is also widely recognised that a spectrum exists: from lack of individual risk awareness; to awareness; to action/adaptation: to behaviour change. Importantly, it is clear that promoting awareness – and hence simply addressing risk perception 'deficits' through filling a 'knowledge gap' – is unlikely to lead directly to changes in behaviour.

Lack of flood experience may be one explanation for the gap between objective and subjective risk appraisals. However, strong evidence exists that even despite repeated flood experiences, many people negate or underestimate future flood risk for varied reasons (Lechowska, 2018). This influences whether their knowledge becomes actionable and manifested in developing resilience (Chapter 4).

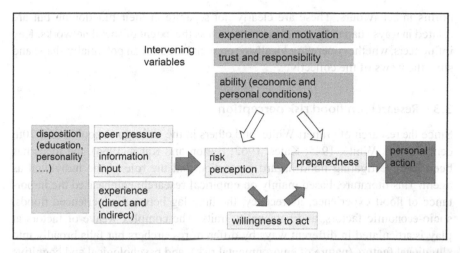

FIGURE 3.1   Visualisation of the "hazard to action-chain" (Wachinger et al., 2013, p1054; Figure 1; Original figure was in colour; redrawn here in B/W)

Wachinger et al. (2013) gave three possible explanations for a weak relationship between risk perception and personal actions, linked to "intervening variables": experience and motivation, trust and responsibility, and personal ability through economic and personal conditions (Figure 3.1; Table 3.1 explains). This assumes awareness of risk, which may itself be changing.

Distinctions also need to be made between perception – or subjective assessment and social construction of risk by the individual as decision-maker, and the collective perceptions of communities that drive dominant narratives and social

TABLE 3.1   Explanations for the Risk Perception Paradox (Figure 3.1) (edited from Wachinger et al., 2013, p1054; permission Ortwin Renn)

Intervening variables	Reasons	References[a]
Experience and motivation	"Individuals understand the risk but choose to accept it due to the fact that the perceived benefits of living close to the river appear to outweigh the potential negative impacts."	Hung et al., (2007); Gough (2000)
Trust and responsibility	"Individuals understand the risk but the responsibility for action is transferred to someone else."	Terpstra (2009)
Personal ability (economic and personal conditions)	"Individuals understand the risk but have little resources to affect the situation."	Njome et al. (2010)

[a] Original references cited.

norms in behaviours. These are clearly not separate in their functioning but are related in ways shaped by social context, such as the extent of social networks. Key influencers, whether operating locally or on social media, can potentially shape and shift the views of the collective.

3.3 Research on flood risk perception

Since the research of Gilbert White[2] and others in the Chicago School in mid-20th century (e.g., White, 1945; Kates, 1962; Burton and Kates, 1964), the focus has been on investigating flood hazard perception and the role of the individual as agent. This literature, based mainly on empirical research, emphasised the importance of flood experience, its recency, the time lag between experienced floods, socio-economic factors, and personality traits. The complex weave of factors at play is articulated in different ways by different researchers but falls broadly into situational factors (nature of environmental risk), and psychological and cognitive factors (Orr et al., 2015; Montz et al., 2017). For example, Carlton and Jacobson (2013) identified risk-specific factors, personal cognitive and affective variables and socio-demographic variables.

Useful reviews of research on risk perception include Wachinger et al. (2013)'s synthesis of European studies (post-2000) on risk perception of natural hazards, and Lechowska's (2018) analysis of empirical research on flood risk perception, focusing on interrelationships between preparedness, worry, and awareness. Wachinger et al. (2013, p1049) emphasised principal causal connections between experience, trust, perception, and preparedness to take protective actions. They distinguished between "personal experience of a natural hazard and trust – or lack of trust – in authorities and experts" as most impactful on risk perception and "mediators and amplifiers of main causal connections" that play a lesser role ("cultural and individual factors such as media coverage, age, gender, education, income, social status"). Lechowska (2018) differentiated factors as primary, secondary, and intervening (describing context), depending on the clarity of their influence on risk perception.

It is useful to classify theories of risk perception into two main categories (Birkholz et al., 2014). Rationalist theory posits that 'objective' reality is fixed and 'to be discovered' through empirical investigation. In contrast, constructivist theory proposes that people make meaning in their own lives, socially constructing knowledge into their mental models of the world (Table 3.2).

Historically, interviews and questionnaires predominate as investigative methods used to explore risk perception under the psychometric paradigm (Lechowska, 2018). This involves assigning numeric values to responses to survey questions given to a sample of individuals affected and then data analysis using statistical methods. An example is Gotham et al.'s (2018) use of regression techniques to analyse experiential, geophysical, and socio-demographic variables in a survey of residents' perceived risk in flooded and unflooded neighbourhoods in New Orleans during post-disaster recovery after Hurricane Katrina. O'Neill et al. (2016) summarised the diverse range of quantitative variables or "construct measures" that have

TABLE 3.2 Some Key Theories of Risk Perception and Their Disciplinary Setting

Theory	Summary	Theoretical approach	Key researcher (discipline)
Psychometric paradigm	Risk understood as a function of general properties of risk object	Rationalist	Slovic (1987) (psychologist)
Cultural Theory of Risk	Risk seen as joint product of knowledge of future and consent about most desired prospects	Constructivist	Douglas (1966) (anthropologist)
Social Amplification of Risk Framework	Concerns about hazards are amplified or attenuated by social, cultural, and institutional processes	Constructivist	Kasperson et al. (1988) (geographer)

(Modified from Marsden. https://risk-engineering.org/risk-perception/, CC BY-SA 4.0).

been used to explain risk perception concerning various flood types. But qualitative approaches such as focus groups (facilitated group discussions) and participatory appraisal methods have also been used to assess community risk perception. Participatory appraisal[3] aims for inclusive participation in data gathering, using flexible and visual tools tailored to local groups. As Birkholz et al. (2014, p19) reflected:

Research has already highlighted that different stakeholders can have different and competing priorities around flood risk management. Constructivist-oriented flood risk perception research could deepen our understanding of these differences and deliver a more nuanced recognition of how flood risk cultures are fashioned by individual and collective experiences, histories, beliefs, relationships, and understandings.

Potential also exists to learn from qualitative research methods into community perception of wider water-risk relationships (e.g., drought; pollution). For example, interest is increasing in the use of community-based, ethnographic methods – in-depth study of interactions, perceptions, and behaviours within groups – in other risk settings (e.g., oil spills; Mundorf and Lichtveld, 2018). More recently, research has highlighted the value of unstructured narrative interviews in making sense of an experience (e.g., Andersen et al., 2020; see sensemaking below). Individual and community narratives have an important role in capturing insights into cultural understandings of risk from purposively (expert) selected community representatives. This allows identification of dominant and counter narratives (Section 4.4 of Chapter 4) that circulate in communities so providing insights into people's different values and worldviews.

In appraising research methods, it is important to establish how individual perceptions have been assessed and aggregated into community risk perceptions. For example, Atmadja and Sills (2016) used a multi-method approach, undertaking both

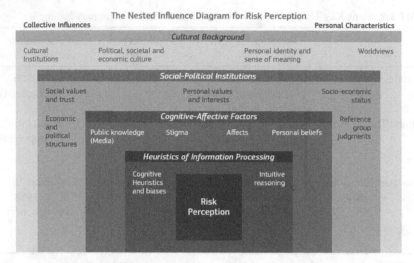

FIGURE 3.2   Multiple influences that interact to form risk perception (modified from Renn and Rohrmann, 2000; source: Science Communication Unit, 2014, Figure 2, reprinted with permission)

probability sampling of local populations (quantitative) and purposive sampling of community representatives for interviews (qualitative). Within a community, complex interactions between individual and collective factors may impact perception of risk. Figure 3.2 encapsulates the nesting of factors that interact to determine people's risk perception (Renn and Rohrmann, 2000; Renn, 2008; SCU, 2014). These combine personal characteristics and collective influences, and both innate and learned aspects.

The sections below explore factors affecting individual, then collective, flood risk perception.

3.4   Individual factors and personal characteristics

Historically, emphasis has been given to individual factors influencing risk perception: floods experienced, relationships to climate change perception, worry and emotion, and sense of place and place attachment.

3.4.1   Floods experienced

Direct personal flood experience has the most significant impact on public flood perception (Gotham et al., 2018), in contrast to learning from science-based probabilistic knowledge of specialists. As Motoyoshi (2006, p122) noted

> People tend to perceive flood disasters as periodic phenomena instead of as probable and random phenomena. Furthermore, people tend to believe that if a major flood disaster occurs in a certain year, no major flood disasters will occur for some time after.

'Risk salience' refers risk being considered important to an individual or connected with. It has two components: relevant prior experience and proximity to the risk (Carlton and Jacobson, 2013). Recency of experience and the physical time-lag between severe floods can be important, with flood acquiescent periods and time-lapses in perception based on low experience, leading to fading memories and inaccuracies in recall. Physical factors like location of dwelling to risk source (e.g., closeness to river or sea), length of residence, and number of floods experienced also impact perception (Heitz et al., 2009). Gotham et al. (2018) found that residents of neighbourhoods that flooded during Hurricane Katrina had higher levels of perceived risk than those in unflooded neighbourhoods, even though the latter were still at risk. Inevitable problems in risk salience exist with more diffuse risks, like surface water flooding, where flood sources may be less obvious in the everyday.

The accuracy of people's factual knowledge of historic flooding can be deconstructed. For example, Pagneux et al. (2011), investigating flood risk perception in an Icelandic town exposed to ice jam floods, found that personal experience was the most critical variable in predicting the accuracy of individuals' knowledge. Incomers' lack of flood experience led to lack of knowledge of risk. It was also found that extreme floods with specific generating conditions (here ice jams) were remembered more accurately than more common routine flood causes (here precipitation).

'Misperception' – or misalignment of lay knowledge with specialist knowledge – can occur in several ways (Chapter 4). One issue relates to flood mapping – of perceived personal location relative to actual flood zones – thinking 'out' of a flood risk area but actually 'in'. O'Neill et al. (2016) found that physical distance to the perceived flood zone (perceived flood exposure) was a key factor in cognitive and affective elements of flood risk perception, rather than reality. 'Cognitive' involves processes of thinking and reasoning while 'affective' relates to emotions and attitudes.

3.4.2 Relationships to climate change perception

Factors affecting flood risk perception nest with those influencing climate change perception. As Whitmarsh and Capstick (2018, np) observed, the latter process is complex involving "a range of psychological constructs, including knowledge, beliefs, attitudes, concern, affect and risk perception". In terms of local knowledge, memory of weather can be heavily conditioned by recollection of recent years (short-term memory bias), with implications for perception of changing weather patterns, particularly when recent patterns counter longer-term climate trends. A challenge of changing flood risk due to climate change is that people's past experiences may not equate with future probabilities of flood occurrence. A further issue impacting public risk perception is diversification of settings at flood risk, for example, with increased high-intensity rainfall leading to more severe surface water flooding in urban areas (e.g., Nanni et al., 2021). There is also likely to be limited perception of possible compound risks yet to be experienced (Chapter 2).

Psychological distance is the "cognitive separation between the self and other instances such as persons, events, or times" (Baltatescu, 2014, p5145), and its application to climate change is well recognised (Spence et al., 2011). Temporal, hypothetical, spatial, and social distance from an event are considered the most critical dimensions to cognitive separation (Trope and Liberman, 2010). This (in)ability for 'mental travel' has implications for uptake of preparedness measures in response to changing risk, an issue that statutory FRM organisations need to address. Maiella et al. (2020), in their systematic literature review, found conflicting results for the impacts of psychological distance on preparedness for climate resilience.

> In general, ... individuals have a higher propensity to perform pro-environmental and resilient behaviors against climate change when it is perceived as more proximal and concrete within the construct of psychological distance. However, not all studies show this result. Some studies showed that, despite people considering climate change as real and tangible, they do not perform mitigation and adaptation behaviors. Other studies showed that people implement these behaviors despite perceiving climate changes as distal [remote] and abstract.
>
> *p1*

A key question is how flood perception and preparedness for floods are affected by the complex relationships between psychological distance, extreme weather events, and preparedness for climate change. The possible impact of emotional reaction (worry) is a further consideration in both risk perception and actions to prepare.

3.4.3 Worry, emotion, and risk perception

Worry is defined as "the state of being anxious and troubled over actual or potential problems" (Oxford Dictionaries). Perceived level of risk involves intellectual judgement while worry refers to emotional reactions. Research into the relationships between worry, risk perception, and preparedness gives ambiguous results (Lechowska, 2018). A key issue is whether worry increases risk preparedness – an "unclear" relation.

Some argue that worry can connect with knowledge of event frequency and risk exposure (Raaijmakers et al., 2008; Bradford et al., 2012). However, Pagneux et al. (2011) found no correlation between risk estimation and levels of worry in the perception of ice dam floods in Iceland. While certain studies have found worry can be a driver for current preparedness levels, others differ. Miceli et al. (2008) found feelings of worry were associated with disaster preparedness,

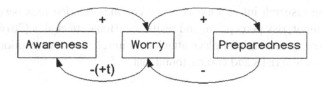

Relationship between flood risk characteristics (Raaijmakers et al., 2008, Figure 1, reproduced by permission of Springer Nature)

but no significant relationship between appraisal of probabilities of risk and take-up of adaptive behaviours to prevent individual loss. Raaijmakers et al. (2008) summarised the relationships between worry, awareness, and preparedness, in terms of positive (increasing) and negative (decreasing) feedback loops (Figure 3.3). "Increased awareness can increase levels of worry, but this may then have a positive effect on level of preparation" (SCU, 2014, p9). However, positive effects on preparation cannot be assumed. "Improved preparation in turn reduces worry. Over a longer timescale (+t), this may reduce awareness" (Raaijmakers et al., 2008).

Worry and trauma can also occur from a mismatch between lay knowledge and the reality of flood risk (McEwen et al., 2016; Chapter 4). For example, after the UK Summer floods 2007, some residents whose properties had flooded reported worry and anxiety every time subsequently that heavy rain occurred. Such anxiety has been linked to mental health impacts of flooding (Chapter 10).

Widening out in focus, Nanni et al. (2021) researched levels of worry about climate change in relation to increased extreme storms and urban surface water flood risk, and preparedness and willingness to act. They found significant disconnects – that increased worry about climate change experienced by people living in exposed settings had no impact on their behaviour and willingness to pay for sustainable drainage measures on their property and locally.

3.4.4 Sense of place, place attachment, and risk perception

Emotional connections can also influence risk construction in other ways, for example, linking attachment (strength of feeling) for a place and belongings, and flood preparedness. Place is defined by both physical location and personal and social or cultural meanings. Hence place-based attachment and "local rationalities" are complex, linking social, cultural, religious, and economic factors (e.g., Miceli et al., 2008). Societal change is also important; Quinn et al. (2018) identified that mobility patterns can generate diverse place attachments and place meanings, important in understanding the dynamics of flood risk perception. Awareness of cultural elements of place attachment is likely to be significant in understanding the diversity of risk perception within multi-cultural geographical communities (Chapter 1).

Increasing research interest exists in the complex relationships between wider place attachment, risk perception, and resilience. Bonaiuto et al.'s (2016, p33) systematic review of research on place attachment in relation to perception of natural and environmental risks and coping found that

(a) strongly attached individuals perceive natural environmental risks but underestimate their potential effects; (b) strongly attached individuals are unwilling to relocate when facing natural environmental risks and are more likely to return to risky areas after a natural environmental disaster; (c) place attachment acts both as a mediating and moderating variable between risk perception and coping.

Quinn et al. (2019) emphasised the importance of sense of place in management of changing risks due to climate change. A key imperative is to understand how place-based risk is perceived by local communities, and to establish how such communities would like these risks to be managed. A predicament at individual and community levels involves people's potential attachment to *both* home and the risk source (e.g., a river or coast).

The well noted restorative aspects of rivers and waterbodies are coupled with risks; 'wild' and 'natural' can have both positive and negative connotations depending on whether one is referring to aesthetic or flood related dynamics of a river.

Quinn et al. (2019, p567)

Place attachment can also encourage the development of local social capital and mobilisation of community in response to local risk (Chapter 8). For example, Clarke et al.'s (2018) Irish study showed place attachment to be strongest in individuals who perceive governance processes as inadequate. They found that neither flood experience nor flood risk affected people's strength of place attachment and their support for implementation of flood defences when these negatively affect place, impacting its local aesthetics, recreational usage, and cultural heritage. This has consequences for any strategy for transformative adaptation "that disrupts place and threatens place attachment" (p81).

3.4.5 Sensemaking, cultural meaning, and risk perception

Risk can also be perceived as 'trouble', with a need for 'sensemaking' of unknown, uncertain, ambiguous, and complex situations to be able to act in them. 'Sensemaking' (Weick, 1995) is used to describe the "processes through which people interpret and give meaning to their experiences" (Urquhart et al., 2019, p1). Sensemaking is both an individual and collective activity, occurring when individuals come to an understanding of the meaning of an experience they have had (Kramer, 2016). Weick (1995, pp17–63) identified seven factors influencing sensemaking: social context, identity formulation, retrospective interpretation, meaningful cues, ongoing events, plausible stories, and effortful enactment (acronym SIR-COPE). Understanding these as leverage points can inform leadership strategies in times of crisis.[4]

This thinking has been applied to understanding relationships between perceptions and behaviours in FRM. For example, Harries et al. (2018) applied a sensemaking approach to understanding small business owners' responses to flood risk. They explored how business owners process their flood experiences and why these do not consistently lead to resilient adaptation of their premises. They found that more significant flood experiences that challenge belief systems can overcome the personal defences that maintain existing interpretative frameworks. This can then lead to preparedness actions such as property-level resilience. They concluded (p712):

> … some of the explanation for low levels of adaptation relates to a desire to defend existing sensemaking structures and associated identities. Sensemaking structures are only revised if these structures are not critical to business identity, or if a flood constitutes an 'ontological shock' and renders untenable existing assumptions regarding long-term business continuity.

3.5 Collective perception of flood risk

Turning from individual to collective perception, geographical communities will vary in their direct and indirect flood experiences, their heterogeneity of demographic factors, complexity of social microclimate (Westcott et al., 2017), and their social capital (Chapter 8). The concept of community perception "links individual experiences together into a collective experience". Atmadja and Sills (2016, p3) described it as

> The collective experience or beliefs of communities [have been] summarized as the fraction of the population that ascribes to particular opinions, the heterogeneity of perceptions that exist in the community, or reasons behind various perceptions.

However, factors affecting individual and collective risk perception within neighbourhoods and communities are difficult to untangle (e.g., socio-environmental characteristics; Miceli et al., 2008). Collective risk perception nests concern for cognitive heuristics and biases, roles of different types of media in public knowledge, and the dynamics of social influence and opinion by culture and media (Renn and Rohrmann, 2000; SCU, 2014). This is impacted by size and density of social networks for sharing knowledge and opinion, and whether and how active remembering and active forgetting are embedded and negotiated in communities (McEwen et al., 2016 in the UK; Ullberg, 2018 in Argentina; Chapter 4). Study of collective perception also needs to consider influencers of groups, social values and trust (of scientists, government), social norms and behaviours (e.g., senses of citizenship), and propensity for collaborative actions like campaigning and activism (Chapter 8; see Orr et al., 2015).

Collective perception, linked to societal norms, has implications for who is considered vulnerable in communities (e.g., children and older people). Such perception also affects diffusion of adaptive knowledge, e.g., local uptake of property-level resilience measures by observation or recommendation. Social identity theory proposes that individuals define their own identities with regard to social groups; this has implications for individual perceptions and group behaviour. At a macro-scale, socio-economic, cultural, and political drivers and perception of the role of institutions (e.g., FRM agencies; Mashi et al., 2020) also exert influence.

3.5.1 Cognitive heuristics and biases

A long-standing psychological literature exists on the influences of heuristics and cognitive biases on risk perception. Heuristics are mental shortcuts in problem-solving, resulting in rapid judgements that lay people are commonly assumed to use to evaluate risks (Paek and Hove, 2017). Cognitive bias is defined as:

> the way a particular person understands events, facts, and other people, which is based on their own particular set of beliefs and experiences and may not be reasonable or accurate.
>
> *Cambridge Dictionary*

Such systematic errors in thinking lead to judgements and decision-making that deviate from rational objectivity, meaning individual risk awareness cannot necessarily be equated with willingness or preparedness to act (e.g., Grothmann and Reusswig, 2006). Oft-cited early papers include Kahneman and Tversky (1973), Kahneman et al. (1982), and Slovic (1987, 2000) on biases that lead to underestimation or negation of risk in general. These lead to the pervasive "myth of flood immunity" (Cook, 2018). For example, optimism bias or "unrealistic optimism" of invulnerability is the difference between a person's subjective expectation and the resulting objective outcome (Weinstein, 1980; Sharot, 2011). Such biases are evidenced in frequently cited narratives reflecting personal flood risk perception (Box 3.1).

These biases may be learned and transferred within families and exchanged within communities. A common perception articulated by at-risk older people, having recently experienced an extreme event, is that "a flood is not going to happen again within my lifespan" (McEwen et al., 2016). This links to long-standing problems in using return periods in flood risk communication (Chapter 6). Weitkamp et al. (2020) outlined some cognitive biases and heuristics reported to affect perceptions of water risks (Table 3.3).

Cognitive dissonance (Cd; Festinger, 1957) is another important concept in understanding people's risk perception. Cd is defined as the feeling of mental

---

BOX 3.1   BIASES THAT ACT AS BARRIERS TO PUBLIC FLOOD RISK PERCEPTION (EDITED FROM MCEWEN, 2011; ADAPTED FROM HALEY, 2008)

- Risk perception – 'It won't happen'
- Threat denial/dissonant perception – 'It will never happen again'
- Unrealistic optimism bias – 'It won't happen in my lifetime'; 'It won't happen to me'
- Determinate perception – 'It won't happen again for [x] years'
- Response efficacy – 'I cannot be bothered – this is more important'
- Outcome expectancy – 'It does not matter what I do'
- Normalisation bias – 'It happens all the time...I'll be alright'
- External locus of control - 'I can't do anything about it'
- Probabilistic perception - 'It was an Act of God'
- Transfer of responsibility – 'the government regulator will protect me'

---

discomfort (stress, anxiety) when an individual holds conflicting beliefs, attitudes, or values as people tend to seek out consistency. Festinger's theory proposes that

> we have an inner drive to hold all our attitudes and behavior in harmony and avoid disharmony (or dissonance). ... When there is an inconsistency between attitudes or behaviors (dissonance), something must change to eliminate the dissonance.
>
> *McLeod (2018 np)*

Such dissonance can be important in how people think, make decisions, and behave in relation to personal flood risk. Examples include avoiding learning new information that conflicts with their existing beliefs, with implications for risk communication strategies (Wood and Miller, 2021; Chapter 6). Cologna et al. (2017, p6) cited the example of flooding still occurring despite new investment in structural flood defences. This can conflict with a person's beliefs of total protection (i.e., "reduced flood risk perception gained from new found security"), and here led to mental stress released by blaming government organisations.

3.5.2   Perception and influence of media narratives

The role of regional and national news media in collective risk perceptions is a focus of ongoing debate. Kasperson and Kasperson (1996) highlighted that most people learn about floods through news media, rather than from personal flood experience. SCU (2014) asked whether media coverage influences or alters public awareness and risk perceptions, or whether it mirrors existing dominant, publicly held views

TABLE 3.3 Cognitive Biases and Heuristics Reported to Affect Perceptions of Water Risks (Weitkamp et al., 2020, Table 1, with author permission)

Cognitive bias	Detail	Reference	Water risk context
Availability heuristic	Person evaluates probability of events by how readily relevant instances come to mind. Memory biased towards recent, unusual, and personally experienced events	Keller et al. (2006); Kahneman and Tversky (1973); Siegrist and Gutscher (2006)	Interpretation of flood risk influenced by people's own flood experience; role of experiential processing; impact of time-lag between experienced events
Affect heuristic	Predicts that concerns about risk will be stronger if concept triggers negative affective experiences	Keller et al. (2006); Lechowska (2018)	Emotional responses to flooding can influence willingness to take action
Collective atrophy	Emphasises importance of social norms in behaviour	Bernedo et al. (2014); Benartzi et al. (2017)	Messaging to prompt water conservation/ efficiency behaviours
Normalcy bias	Refusal to plan for, or react to, a disaster which has never happened before	Omer and Alon (1994); Bryan et al. (2019)	Causes people to underestimate both likelihood of a disaster and expected impacts
(Social and technological) optimism bias	Overestimation of likelihood of positive outcomes	Costa-Font et al. (2009)	Reduced risk perceptions when risks receive significant media attention
Cultural bias	Culturally induced biases that influence risk perception	Overdevest and Christiansen (2013)	People selectively credit or dismiss risks to support cultural predispositions

(e.g., cultural narratives of blame or resilience). Such media outlets are not neutral or value-free, and hence reporting may be weighted and framed in particular ways to sell (Cook, 2018). Bohensky and Leitch (2014) explored the media's role in understanding and internalising the 2011 Brisbane floods as an extreme event in relation to resilience building. Finding that coverage gave inadequate focus to longer-term aspects of regional resilience, they concluded that media discourse could impact resilience by promoting certain blame narratives at the expense of others. In media coverage, the events depicted, and voices and knowledges recognised and represented can strongly influence whether and how flooding is incorporated into policy (Henrique and Tschakert, 2019).

Simultaneously, the nature of media, its channels, and genres is diversifying and fragmenting rapidly – from newspapers and TV to entertainment and social media (Rutsaert et al., 2013). Paek and Hove (2017, p1) identified several different aspects of media as impacting public risk perception, including:

(1) amount of media coverage; (2) frames used for describing risks; (3) valence and tone of media coverage; (4) media sources and their perceived trustworthiness; (5) formats in which risks are presented; and (6) media channels and types. For all of these media factors, albeit to varying degrees, there is theoretical and empirical support for their relevance to risk perceptions.

This complex territory of media as an influencer of flood risk perception requires further research to understand better its role in public discourse and in reinforcing or changing community perceptions (Chapter 10). For example, Cologna et al. (2017), building on Wachinger et al. (2013), proposed a conceptual framework for considering the media's influence on risk perception of flood events (Figure 3.4).

This framework recognises the media's important role as either "an amplifier or attenuator of risks" (p1). They and others argued the need for a critical shift in the dominant media narrative from "once in a lifetime" and blame narratives, to

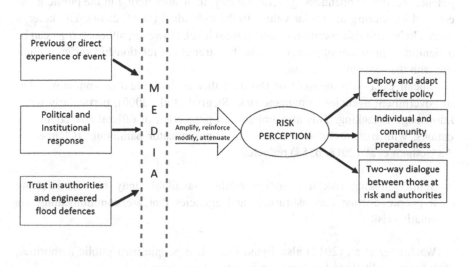

FIGURE 3.4   Framework on the media's influence on risk perception (Cologna et al., 2017, Figure 1, adapted from Wachinger et al., 2013, reproduced by permission of Elsevier)

"be prepared" flood reporting of resilience narratives that promote proactivity and agency in community identity within the wider DRM spiral.

Strong potential exists for working with the news media positively in this context, recognising conflicts between ensuring press independence, and their potential role in entrenching dependency, misconceptions, and deferral of responsibility within flood-affected communities. Opportunities to shift dominant narratives also exist through co-working with new citizen actors (e.g., influencers on social media – trusted and popular – with particular groups of citizens) to help promote preparedness measures. However, social media can also pose potential threats to risk perception "when outdated, false, or misleading information is spread", particularly when there is not time for checking (Alexander, 2014; Du et al., 2017, p9165).

3.5.3 Trust in authority and the perception of responsibility

Trust is a complicated concept in relation to FRM and can act as a primary influence on collective risk perception. It acts in several domains: overarching societal trust in scientists and engineers; in newly constructed flood defence infrastructure (Motoyoshi, 2006); in official flood warning schemes; and climate change predictions. For example, what message is given by defence structures – "we are all protected" or "there is no flood risk here"? It is well cited that the "levee effect" leads to increased investment, false sense of security among at-risk groups (e.g., Ludy and Kondolf, 2012; Hutton et al., 2018), and issues of cognitive dissonance when structures fail (see above). Importantly, Touili et al. (2014) argued that mitigation options, such as engineering structures, are both physically and socially constructed.

Research suggests that several factors influence trust in an organisation: its competence; record of openness, general honesty, its history acting in the public interest; and its sharing of similar values to the individual (SCU, 2014). The latter is more likely with risk communication at local level. However, although trust can be a significant predictor of perceived risk, the strength of relationship varies, including with the scale of actual risk.

'Social trust' is a measure of the trust that an individual or community has in government agencies to manage risk (Siegrist et al., 2000), particularly when knowledge is lacking. This also impacts on perception of official risk communications, with trust and confidence in the communicator paramount (Chapter 6). As Gotham et al. (2018, p354) reflected

> Simply making risk information publicly available may be self-defeating if people distrust the institutions and agencies that are communicating the putative risk.

Wachinger et al. (2013) also found that when people trust public authorities, they are more likely to heed and act on official warnings.

Trust can be hard won and easily lost as floods play out. There is growing recognition by FRM agencies that sometimes the most trusted risk messengers in risk communication may be fellow community members (e.g., community flood wardens). Trust and perception of whether responsibility to take action in mitigating risk rests with the citizen, or is deferred to statutory FRM agencies, politicians, and the state, can also impact in linking awareness and personal action to mitigate losses. Misperceptions about 'who is responsible' among communities can lead to significant flood losses. As Carlton and Jacobson (2013) noted from their study of coastal environmental risk perceptions in Florida, individuals with higher levels of social trust may perceive less risk than those possessing lower levels of social trust. Papagiannaki et al. (2018 in Greece) also found higher levels of trust were associated with lower levels of worry, leading to lower preparedness levels.

3.5.4 Social and cultural factors

Risk can be viewed as socially constructed, framed mostly through values and norms (e.g., Johnson and Covello, 1987; Vanderlinden et al., 2017). Social norms, or informal rules which govern people's behaviour in specific situations, are the normative bases of group behaviour and attitudes. They determine how people situate themselves within their social circles or society. The family tends to be important in primary socialisation of norms and values, while secondary socialisation occurs through educational systems, media, religion, and the workplace (Haralambos and Holborn, 2013). Perceived social norms can have a mediating role between risk perception and adaptive behaviours, such as uptake of insurance cover (Lo, 2013 in Australia).

A tipping point (cf. Gladwell, 2002) may be achieved if enough people or key social influencers in a group or wider society adapt norms. Indeed, governmental policies can be most effective if they stimulate longer-term changes in beliefs and norms. Kinzig et al. (2013, p164) argued for more social science research on emergence of social norms and use of policy instruments in "intractable environmental challenges". At a local level, this might, for example, involve adoption of adaptive measures like sustainable drainage at household-level to help mitigate local pluvial flooding.

At societal level, risk perception also relates to the dominant paradigm of human-environment interactions and world views within different societies and cultures (Palm, 1990). Such relational thinking considers whether humans are subordinate to, or can control, nature.

> Environmental determinism occupies one end of a continuum, cultural determinism occupies the other; each argues that the human condition is determined simply by nature or simply by culture. Between these two extreme positions lies a broad spectrum of positions described variously as 'environmental possibilism' or 'environmental probabilism'.
>
> *Brooke (2016, np)*

Dominance of this thinking can lead to 'act of god' and divine retribution explanations of flooding (Box 3.1). For example, such interpretation of catastrophic floods was prevalent in 19th-century Scotland (McEwen and Werritty, 2007), within Lauder's (1830) account of the catastrophic 1829 floods:

We remembered that the calamity came from the hand of God.

p100

This awful admonition to a sinful land.

p82

This also has implications for shifts in flood management paradigms, whether dominion over nature ('control') or humans as part of nature ('risk management'; Chapter 2).

Demographic and cultural factors (Wachinger et al., 2013) can also be determinants of risk perception. These characteristics and their intersectionality can influence dominant perceptions of risk and vulnerability in communities. It is well established that some groups have a relatively low-risk perception (e.g., the White Male Effect (WME)) in comparison to women (Flynn et al., 1994; Olofsson and Rashid, 2011). Studies have also indicated links between risk perception and socio-economic factors: for example, low-income levels and female gender have been positively related to higher flood risk perception (cf. male gender and higher income levels; Gotham et al., 2018). Macias (2016) studied the nature of, and influences on, environmental risk perception among nine different race and ethnic groups in the USA. He found greater perceived risks (e.g., due to climate change) among BAME communities (cf. white) across generations. Hence, marginalised groups may have distinct risk perceptions leading to social and spatial justice issues.

What is considered an extreme flood is also culturally determined. Such insight may be harnessed, with potential for cross-cultural learning in multi-cultural communities that possess different flood experiences, perceptions, knowledge of effective adaptive strategies, and experiences of community co-working (Chapter 9).

3.5.5 *Wider organisational perceptions of risk*

Community-focused FRM needs to recognise differences in risk perceptions of different stakeholders, for instance between lay and professional stakeholders (e.g., Shen, 2010). This applies to participation and transparency in all stages of decision-making and implementation around risk management strategies (Dixit, 2003). Scientific experts, such as hydrological modellers, rely on scientific information and more objective assessments in evaluating risks. However, there are other FRM actors beyond citizens and scientists; 'professional partners' are likely to be heterogeneous groups, with different interests and emotional connections to flooding and place (Kashefi and Walker, 2009; Buchecker et al., 2013).

Extending out from at-risk settings, Birkholz et al. (2014, p12) emphasised the importance of diversifying away from a historic prioritised focus on "the cognitive perceptions of those at risk' to understanding a more diverse range of flood risk perceptions" (e.g., of policymakers or tax-payers living outside flood risk areas). This is important in investigating relationships between their perceptions and adaptive measures that require collective input and response.

3.6   Perception and behaviour change

As well as its relevance to risk appraisal, perception influences behavioural response – that is, the decision on whether or not to take action (Harries, 2007; Figure 3.5).

Returning to the 'Risk Perception Paradox' (Wachinger et al., 2013; Section 3.2), a key question in FRM is why some people at flood risk act to prepare and mitigate personal losses while others do not (Andrasko, 2021). Both individual and collective perceptions interact in the efficacy of precautionary and reactive responses in private adaptation (cf. Grothmann and Reusswig, 2006). These perceptions impact both adoption of property-level resilience measures and behavioural FRM strategies, such as the public's response to official flood warnings (Chapter 7).

Self-efficacy is "the belief in one's capabilities to organize and execute the courses of action required to manage prospective situations" (Bandura, 1995, p2). This is critical for personal resilience behaviours but also in the role of key individuals within collective resilience actions. Social capital in communities (Chapter 8) also has a complex role in risk perception. Babcicky and Seebauer (2017, 1030/1) found that social capital "cuts both ways":

> while it lowers flood risk perceptions of private households, it increases perceived self-efficacy and positively contributes to coping capacity.

Moreover, expectation of social support could downplay local risk, with the consequence that preparedness actions by households became less likely.

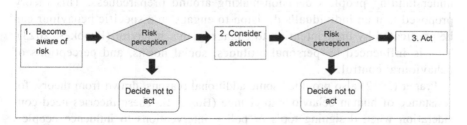

FIGURE 3.5   Representation of the process of risk perception and response (Harries, 2007 unpublished PhD thesis, Middlesex University with permission; redrawn by author)

3.6.1 Theories of behaviour change

Human behaviour is highly resistant to change for various reasons, and behaviour change is a complicated process – not resulting from one cognitive decision. A key concern within the "behavioural turn" in FRM (Kuhlicke et al., 2020) is how to motivate residents in at-risk settings to act in response to residual risk for disaster preparedness. Theories that link perception, action, and behaviour change are therefore important for FRM; these include 'Protection Motivation Theory', 'Theory of Planned Behaviour', and 'Theory of Reasoned Action'.

Protection Motivation Theory (PMT; Rogers, 1975) has been used to describe how individual people and families are motivated to react in a self-protective way towards a perceived health threat. Grothmann and Reusswig (2006), in exploring 'private precautionary action', developed a socio-psychological model based on PMT. This explained such action in terms of several factors:

> residents' perceptions of previous flood experience, risk of future floods, reliability of public flood protection, the efficacy and costs of self-protective behavior, their perceived ability to perform these actions, and non-protective responses like wishful thinking.

They argued (p1) that motivating is not just about communicating the risk itself effectively, but also the communication of the "possibility, effectiveness and cost" of potential measures. In extending the work of Rogers (1975) and Grothmann and Reusswig (2006) concerning Australian bushfire risk, Westcott et al., (2017, p7) emphasised the importance of trust – now added in the "coping appraisal" part of the PMT model (Figure 3.6). They also inserted "understanding complexity of the social microclimate" as an additional factor in determining behavioural responses (Section 3.5).

Other relevant theories include the Theory of Reasoned Action, which emphasises the role of beliefs, attitudes, and intentions in predicting human behaviours (Fishbein and Ajzen, 1975). Often considered together, the cognitive Theory of Planned Behaviour (Ajzen, 1985) also has relevance to understanding people's decision-making around preparedness. This theory proposed that an individual's decision to engage in a specific behaviour can be predicated by their intention to engage in that behavior (Brookes, 2023). This is influenced by personal attitudes, social norms, and perceptions of behavioural control.

Prager (2012) also provided some additional reasons, drawn from theory, for resistance of human behaviour to change (Box 3.2). These theories need consideration when designing tools or policy interventions to influence people's behaviour.

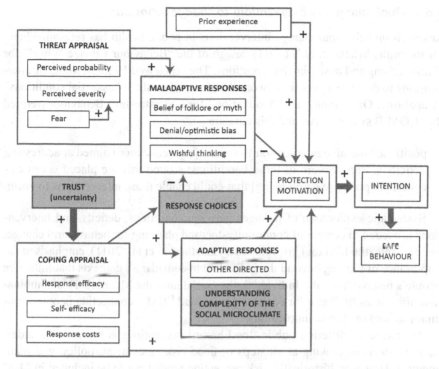

FIGURE 3.6   Westcott et al.'s (2017) expansion of Protection Motivation Theory
(Figure 1 after Rogers, 1975 and adapted from Grothmann and Reusswig,
2006; permission: Rachel Westcott). Grey shading indicates elements of
their proposed PMT expansion

BOX 3.2   REASONS FOR RESISTANCE OF BEHAVIOUR TO CHANGE (PRAGER, 2012, P6, AFTER MEARNS 2012, WITH PERMISSION)

- People are creatures of habit, and we try to achieve maximum gain with minimum effort.
- Information and exhortation are among the least effective ways of influencing behaviour (Campbell, 1963; Bandura, 1977) [...].
- We are focused on the short-term not the long-term (immediate survival was critical for our ancestors); in contrast, the very gradual, slow pace of environmental change represents a cognitive barrier (Kollmuss and Agyeman, 2002).
- We do not necessarily respond well to *being told* what to do (related to ideas of personal freedom/rights especially in Western societies) (Branson et al., 2012).

3.6.2 Tools and policy interventions to change behaviour

Research on behaviour change interventions in public health has relevance here, for example, Michie et al.'s (2011) design of the "Behaviour Change Wheel" for characterising and designing interventions. They reviewed 19 frameworks of interventions to develop a new framework that focused on three "essential conditions" "Capability, Opportunity, and Motivation" for understanding Behaviour (termed the 'COM-B system'). Around this core are

> positioned the nine intervention functions [e.g., education] aimed at addressing deficits in one or more of these conditions; around this are placed seven categories of policy [e.g., planning] that could enable those interventions to occur.

Systematic exploration of linkages between conditions, deficits, and interventions is needed in context of community-focused planning for behavioural change. For example, the ISM tool,[5] developed by Southerton et al. (2011), emphasised the importance of moving beyond the individual to consider all contexts that influence people's behaviours – the Individual, the Social, and the Material. This combination offers a useful frame for community-focused FRM – connecting people, communities, and resilience innovations.

Awareness of differing public flood hazard perceptions is essential in understanding decision-making at all steps of flood risk assessment, policy, and management. However, historically, risk perception tended not to be included in FRM assessments as predominantly scientific. To address this, Buchecker et al. (2013), working in Zurich, Switzerland, argued (p3014) for a "socio- and economic oriented approach" that takes account of the "social dimension of flood risk" and "stakeholders' risk perception". They proposed and investigated a participatory procedure with strong attention to risk communication that allowed inclusion of stakeholders' perceptions of alternative preventative measures within risk assessments. They found that this procedure worked well in flood risk assessment, but with the need for adaption for wider use.

3.7 Conclusions

Risk perception is a complex but core territory for investigation in community-focused FRM. Understanding perceptions gained from personal experience or inherited knowledge, and those of likely future exposure, is important for local resilience-building, and in the design and assessment of effective, community-focused FRM strategies. Risk perception has implications for community participation in adaptive measures that require collective agency like flood warden schemes.

Many factors affect the public's acceptance of their residual risk, and the likelihood that they will take action to mitigate or reduce it. Recognition of likely heterogeneity of perception in geographical communities is also critical. Historically, much attention has been given to theorising individual perceptions and behaviours. Key areas for investigation now focus on relationships and interactions between individual and collective perceptions, how perception and behaviours are influenced by social norms and influence of others, how perception links to social and environmental justice issues (Chapter 2), and how perception links to collective efficacy and behaviour change for adaptation and transformation (Chapter 10).

Strategies for behaviour change need to embrace both the individual (household resilience measures) and the collective (strategies implemented by communities). More research is needed to understand the dynamics between perception, action, and behaviour change within 'the material' – an increasingly diverse suite of possible precautionary measures, innovations, and activities for direct adaptation and wider community resilience building. Such research should identify the different and new social influencers that have potentially transformative roles beyond statutory FRM agencies. Awareness of key factors like social trust affecting perception, and willingness to act to mitigate personal and collective risk, is fundamental underpinning to effective community-focused, flood risk communication and education strategies (Chapters 6 and 9).

Notes

1 https://www.lexico.com/definition/perception.
2 http://www.aag.org/cs/membership/tributes_memorials/sz/white_gilbert_f.
3 https://involve.org.uk/resources/methods/participatory-appraisal.
4 https://www.bus.umich.edu/FacultyResearch/Research/TryingTimes/Rules.htm.
5 https://www.gov.scot/publications/influencing-behaviours-moving-beyond-individual-user-guide-ism-tool/.

References

Ajzen, I. (1985) From intentions to actions: A theory of planned behavior. In J. Kuhl and J. Beckmann (eds.) *Action Control*. SSSP Springer Series in Social Psychology. Springer, Berlin, Heidelberg. https://doi.org/10.1007/978-3-642-69746-3_2

Alexander, D.E. (2014) Social media in disaster risk reduction and crisis management. *Science and Engineering Ethics*, 20(3), 717–733. https://doi.org/10.1007/s11948-013-9502-z

Andersen, D., Ravn, S. and Thomson, R. (2020) Narrative sensemaking and prospective social action: Methodological challenges and new directions. *International Journal of Social Research Methodology*, 23(4), 367–375. https://doi.org/10.1080/13645579.2020.1723204

Andráško, I. (2021) Why people (do not) adopt the private precautionary and mitigation measures: A review of the issue from the perspective of recent flood risk research. *Water*, 13, 140. https://doi.org/10.3390/w13020140

Atmadja, S.S. and Sills, E.O. (2016) What is a "Community perception" of REDD+? A systematic review of how perceptions of REDD+ have been elicited and reported in the literature. *PLoS ONE*, 11(11), e0155636. https://doi.org/10.1371/journal.pone.0155636

Baan, P.J.A. and Klijn, F. (2004) Flood risk perception and implications for flood risk management in the Netherlands. *International Journal of River Basin Management*, 2(2), 113–122. https://doi.org/10.1080/15715124.2004.9635226

Babcicky, P. and Seebauer, S. (2017) The two faces of social capital in private flood mitigation: Opposing effects on risk perception, self-efficacy and coping capacity. *Journal of Risk Research*, 20(8), 1017–1037. https://doi.org/10.1080/13669877.2016.1147489

Baker, V. (2007) Flood hazard science, policy, and values: A pragmatist stance. *Technology in Society*, 29, 161–168.

Baltatescu, S. (2014) Psychological distance. In A.C. Michalos (eds.) *Encyclopedia of Quality of Life and Well-Being Research*. Springer, Dordrecht. https://doi.org/10.1007/978-94-007-0753-5_2306

Bandura, A. (1977) *Social Learning Theory*. Prentice Hall, Englewood Cliffs.

Bandura, A. (1995) *Self Efficacy in Changing Societies*. Cambridge University Press, New York.

Bang, H.N. and Burton, N.C. (2021) Contemporary flood risk perceptions in England: Implications for flood risk management foresight. *Climate Risk Management*, 32, 100317. https://doi.org/10.1016/j.crm.2021.100317

Bell, H.M. and Tobin, G.A. (2007) Efficient and effective? The 100-year flood in the communication and perception of flood risk. *Environmental Hazards*, 7(4), 302–311. https://doi.org/10.1016/j.envhaz.2007.08.004

Benartzi, S., Beshears, J., Milkman, K.L., Sunstein, C.R., Thaler, R.H., Shankar, M., Tucker-Ray, W., Congdon, W.J. and Galing, S. (2017) Should governments invest more in nudging? *Psychological Science*, 28(8), 1041–1055. https://doi.org/10.1177/0956797617702501

Bernedo, M., Ferraro, P.J. and Price, M. (2014) The persistent impacts of norm-based messaging and their implications for water conservation. *Journal of Consumer Policy*, 37, 437–452. https://doi.org/10.1007/s10603-014-9266-0

Birkholz, S., Muro, M., Jeffrey, P. and Smith, H.M. (2014) Rethinking the relationship between flood risk perception and flood management. *Science of the Total Environment*, 478, 12–20. https://doi.org/10.1016/j.scitotenv.2014.01.061

Bohensky, E.L. and Leitch, A.M. (2014) Framing the flood: A media analysis of themes of resilience in the 2011 Brisbane flood. *Regional Environmental Change*, 14, 475–488. https://doi.org/10.1007/s10113-013-0438-2

Bonaiuto, M., Alves, S., De Dominicis, S. and Petruccelli, I. (2016) Place attachment and natural hazard risk: Research review and agenda. *Journal of Environmental Psychology*, 48, 33–53.

Bradford, R.A., O'Sullivan, J.J., van der Craats, I.M., Krywkow, J., Rotko, P., Aaltonen, J., Bonaiuto, M., De Dominici, S., Waylen, K. and Schelfaut, K. (2012) Risk perception—Issues for flood management in Europe. *Natural Hazards and Earth System Sciences*, 12, 2299–2309.

Branson, C., Duffy, B., Perry, C. and Wellings, D. (2012) Acceptable Behaviour? Public Opinion on Behaviour Change Policy. Ipsos MORI, Social Research Institute, London.

Brooke, J.L. (2016) Environmental determinism. *Oxford bibliographies*. https://doi.org/10.1093/OBO/9780199363445-0045

Brookes, E. (2023) The theory of planned behavior: Behavioral intention. *Simply Psychology*. https://www.simplypsychology.org/theory-of-planned-behavior.html

Bryan, K., Ward, S., Barr, S. and Butler, D. (2019) Coping with drought: Perceptions, intentions and decision-stages of SouthWest England households. *Water Resources Management*, 33(3), 1185–1202. https://doi.org/10.1007/s11269-018-2175-2

Buchecker, M., Salvini, G., Di Baldassarre, G., Semenzin, E., Maidl, E. and Marcomini, A. (2013) The role of risk perception in making flood risk management more effective. *Natural Hazards and Earth System Sciences*, 13, 3013–3030. https://doi.org/10.5194/nhess-13-3013-2013

Burningham, K., Fielding, J. and Thrush, D. (2008) 'It'll never happen to me': Understanding public awareness of local flood risk. *Disasters*, 32(2), 216–238.

Burton, I. and Kates, R. (1964) The perception of natural hazards in resource management. *Natural Resources Journal*, 3, 412–441.

Campbell, D. (1963) Social attitudes and other acquired behavioural dispositions. In S. Koch (ed.) *Psychology: A Study of a Science*. McGraw Hill, New York, pp94–172.

Carlton, S.J. and Jacobson, S.K. (2013) Climate change and coastal environmental risk perceptions in Florida. *Journal of Environmental Management*, 130, 32–39.

Clarke, D., Murphy, C. and Lorenzoni, I. (2018) Place attachment, disruption and transformative adaptation. *Journal of Environmental Psychology*, 55, 81–89. https://doi.org/10.1016/j.jenvp.2017.12.006

Coles, A.R. and Hirschboeck, K.K. (2020) Driving into danger: Perception and communication of flash-flood risk. *Weather, Climate, and Society*, 12, 387–404. https://doi.org/10.1175/WCAS-D-19-0082.1

Cologna, V., Bark, R.H. and Paavola, J. (2017) Flood risk perceptions and the UK media: Moving beyond "once in a lifetime" to "Be Prepared" reporting. *Climate Risk Management*, 17, 1–10.

Cook, M. (2018) "It will Never Happen Again": The myth of flood immunity in Brisbane. *Journal of Australian Studies*, 42(3), 328–342. https://doi.org/10.1080/14443058.2018.1487871

Costa-Font, J., Mossialos, E. and Rudisill, C. (2009) Optimism and the perceptions of new risks. *Journal of Risk Research*, 12(1), 27–41. https://doi.org/10.1080/13669870802445800

Dixit, A. (2003) Floods and vulnerability: Need to rethink flood management. *Natural Hazards*, 28(1), 155–179. https://doi.org/10.1023/A:1021134218121

Douglas, M. (1966) *Purity and Danger: An Analysis of Concepts of Pollution and Taboo*. Vol. 2. Routledge,, London, 188pp.

Du, E., Cai, X., Sun, Z. and Minsker, B. (2017) Exploring the role of social media and individual behaviors in flood evacuation processes: An agent based modeling approach. *Water Resources Research*, 53, 9164–9180. https://doi.org/10.1002/2017WR021192

Festinger, L. (1957). *A Theory of Cognitive Dissonance*. Stanford University Press, Stanford, CA.

Fishbein, M. and Ajzen, I. (1975) *Belief, Attitude, Intention and Behavior: An Introduction to Theory and Research*. Addison-Wesley, Reading, MA.

Flynn, J., Slovic, P. and Mertz, C.K. (1994) Gender, race, and perception of environmental health risks. *Risk Analysis*, 14, 1101–1108. https://doi.org/10.1111/j.1539-6924.1994.tb00082.x

Gladwell, M. (2002) *The Tipping Point: How Little Things Can Make a Big Difference*. Abacus; New Edition.

Gotham, K.F., Campanella, R., Lauve-Moon, K. and Powers, B. (2018) Hazard experience, geophysical vulnerability, and flood risk perceptions in a postdisaster city, the case of New Orleans. *Risk Analysis*, 38(2). https://doi.org/10.1111/risa.12830

Gough, J. (2000) Perceptions of risk from natural hazards in two remote New Zealand communities. *Australiasian Journal of Disaster and Trauma Studies*, 2000(2). https://trauma.massey.ac.nz/issues/2000-2/gough.htm

Grothmann, T. and Reusswig, F. (2006) People at risk of flooding: Why some residents take precautionary action while others do not. *Natural Hazards*, 38(1–2), 101–120.

Haley, J. (2008) *Improving flood education for the community*. Presentation for Victoria State Emergency Service's FLOODSMART programme.

Haralambos, M. and Holborn, M. (2013) *Sociology Themes and Perspectives*. Collins, New York.

Harries, T. (2007) Householder responses to flood risk; the consequences of the search for ontological security. *Unpublished PhD thesis. FHRC Middlesex University*.

Harries, T., McEwen, L. and Wragg, A. (2018) Why it takes an 'ontological shock' to prompt increases in small firm resilience: Sensemaking, emotions and flood risk. *International Small Business Journal*, 36(6), 712–733.

Heitz, C., Spaeter, S., Auzet, A.V. and Glatron, S. (2009) Local stakeholders' perception of muddy flood risk and implications for management approaches: A case study in Alsace (France). *Land Use Policy*, 26(2), 443–451.

Henrique, K.P. and Tschakert, P. (2019) Taming São Paulo's floods: Dominant discourses, exclusionary practices, and the complicity of the media. *Global Environmental Change*, 58. https://doi.org/10.1016/j.gloenvcha.2019.101940

Hung, H.V., Shaw, H. and Kobayashi, M. (2007) Flood risk management for the RUA of Hanoi: Importance of community perception of catastrophic flood risk in disaster risk planning. *Disaster Prevention and Management*, 16(2), 245–258.

Hutton, N.S., Tobin, G.A. and Montz, B.E. (2018) The levee effect revisited: Processes and policies enabling development in Yuba County, California. *Journal of Flood Risk Management*, 12, e12469. https://doi.org/10.1111/jfr3.12469

Johnson, B.B. and Covello, V.T. (eds.) (1987) *The Social and Cultural Construction of Risk: Essays on Risk Selection and Perception*. Reidel, New York.

Kahneman, D., Slovic, P. and Tversky, A. (1982) *Judgment Under Uncertainty: Heuristics and Biases*. Cambridge University Press, Cambridge.

Kahneman, D. and Tversky, A. (1973) On the psychology of prediction. *Psychological Review*, 80(4), 237–51.

Kashefi, E. and Walker, G.P. (2009) How the public and professional partners make sense of information about risk and uncertainty. Literature Review – Science Project. DEFRA/ Environment Agency Flood and Coastal Erosion Risk Management R&D Programme (Great Britain).

Kasperson, R.E. and Kasperson, J.X. (1996) Social amplification and attenuation of risk. *Annals of the American Academy of Political and Social Science*, 545(1), 95–105.

Kasperson, R.E., Renn, O., Slovic, P., Brown, H.S., Emel, J., Goble, R., Kasperson, J.X. and Ratick, S. (1988) The social amplification of risk: A conceptual framework. *Risk Analysis*, 8, 177–187. https://doi.org/10.1111/j.1539-6924.1988.tb01168.x

Kates, R.W. (1962) *Hazard and Choice Perception in Flood Plain Management*. University of Chicago, Department of Geography Research, Paper No. 78. University of Chicago.

Keller, C., Siegrist, M. and Gutscher, H. (2006) The role of the affect and availability heuristics in risk communication. *Risk Analysis*, 26(3), 631–639. https://doi.org/10.1111/ j.1539-6924.2006.00773.x

Kinzig, A.P., Ehrlich, P.R., Alston, L.J. et al. (2013) Social norms and global environmental challenges: The complex interaction of behaviors, values, and policy. *BioScience* 63: 164–175. https://doi.org/10.1525/bio.2013.63.3.5

Kollmuss, A. and Agyeman, J. (2002) Mind the gap: Why do people act environmentally and what are the barriers to pro-environmental behavior? *Environmental Education Research*, 8, 239–260.

Kramer, M.W. (2016). Sensemaking. In C.R. Scott, J.R. Barker, T. Kuhn, J. Keyton, P.K. Turner and L.K. Lewis (eds.) *The International Encyclopedia of Organizational Communication*. https://doi.org/10.1002/9781118955567.wbieoc185

Krasovskaia, I., Gottschalk, L., Skiple Ibrekk, A. and Berg, H. (2007) Perception of flood hazard in countries of the North Sea region of Europe. *Hydrology Research*, 38(4–5), 387–399. https://doi.org/10.2166/nh.2007.019

Kuhlicke, C., Seebauer, S., Hudson, P., Begg, C., Bubeck, P., Dittmer, C. et al. (2020) The behavioral turn in flood risk management, its assumptions and potential implications. *WIREs Water*, 7(3). https://doi.org/10.1002/wat2.1418

Lauder, T.D. (1830) *An Account of the Great Floods of August 1829, in the Province of Moray, and Adjoining Districts*. Adam Black, Edinburgh.

Lechowska, E. (2018) What determines flood risk perception? A review of factors of flood risk perception and relations between its basic elements. *Natural Hazards*, 94, 1341–1366. https://doi.org/10.1007/s11069-018-3480-z(012

Lo, A.Y. (2013) The role of social norms in climate adaptation: Mediating risk perception and flood insurance purchase. *Global Environmental Change*, 23(5), 1249–1257. https://doi.org/10.1016/j.gloenvcha.2013.07.019

Ludy, J. and Kondolf, G.M. (2012) Flood risk perception in lands "protected" by 100-year levees. *Natural Hazards*, 61, 829–842. https://doi.org/10.1007/s11069-011-0072-6

Macias, T. (2016) Environmental risk perception among race and ethnic groups in the United States. *Ethnicities*, 16(1), 111–129. https://doi.org/10.1177/1468796815575382

Maiella, R., La Malva, P., Marchetti, D., Pomarico, E., Di Crosta, A., Palumbo, R., Cetara, L., Di Domenico, A. and Verrocchio, M.C. (2020) The psychological distance and climate change: A systematic review on the mitigation and adaptation behaviors. *Frontiers in Psychology*, 11, 2459. https://doi.org/10.3389/fpsyg.2020.568899

Mashi, S.A., Inkani, A.I., Obaro, O. et al. (2020) Community perception, response and adaptation strategies towards flood risk in a traditional African city. *Natural Hazards* 103, 1727–1759. https://doi.org/10.1007/s11069-020-04052-2

McEwen, L.J. (2011) Approaches to community flood science engagement: The lower River Severn catchment, UK as case-study. *International Journal of Science in Society*, 2(4), 159–179.

McEwen, L.J., Garde-Hansen, J., Holmes, A., Jones, O. and Krause, F. (2016) Sustainable flood memories, lay knowledges and the development of community resilience to future flood risk. *Transactions of the Institute of British Geographers*, 42(1), 14–28.

McEwen, L.J. and Werritty, A. (2007) 'the Muckle Spate of 1829': The physical and human impact of a catastrophic nineteenth century flood on the River Findhorn, Scottish Highlands. *Transactions of the Institute of British Geographers* NS, 32, 66–89.

McLeod, S.A. (2018, February 05). *Cognitive dissonance*. Simply Psychology. https://www.simplypsychology.org/cognitive-dissonance.html

Mearns, K. (2012) Behaviour Change. Presentation at the Climate X Change meeting 21–22 February 2012, Edinburgh.

Miceli, R., Sotgiu, I. and Settanni, M. (2008) Disaster preparedness and perception of flood risk: A study in an alpine valley in Italy. *Journal of Environmental Psychology*, 28, 164–173.

Michie, M., van Stralen, M.M. and West, R. (2011) The behaviour change wheel: A new method for characterising and designing behaviour change interventions. *Implementation Science*, 6, 42. https://doi.org/10.1186/1748-5908-6-42

Moftakhari, H.R., AghaKouchak, A., Sanders, B.F., Allaire, M. and Matthew, R.A. (2018) What is nuisance flooding? Defining and monitoring an emerging challenge. *Water Resources Research*, 54, 4218–4227. https://doi-org.ezproxy.uwe.ac.uk/10.1029/2018WR022828

Montz, B.E., Tobin, G.A. and Hagelman, R.R. (2017) *Natural Hazards: Explanation and Integration* (2nd edn). Guilford Publications, New York..

Motoyoshi, T. (2006) Public perception of flood risk and community-based disaster preparedness. In S. Ikeda, T. Fukuzono and T. Sato (eds.) *A Better Integrated Management of Disaster Risks: Toward Resilient Society to Emerging Disaster Risks in Mega-Cities.* TERRAPUB and NIED, pp121–134.

Mundorf, C.A. and Lichtveld, M.Y. (2018) Using community-based, ethnographic methods to examine risk perceptions and actions of low-income, first-time mothers in a post-spill environment. *Journal of Risk Research*, 21(3), 308–322. https://doi.org/10.1080/13669 877.2016.1200656

Nanni, P., Peres, D.J., Musumeci, R.E. and Cancelliere, A. (2021) Worry about climate change and urban flooding risk preparedness in southern Italy: A survey in the Simeto River valley (Sicily, Italy). *Resources*, 10, 25. https://doi.org/10.3390/resources 10030025

Njome, M.S., Suh, C.E., Chuyong, G. and deWit, M.J. (2010) Volcanic risk perception in rural communities along the slopes of Mount Cameroon, West-Central Africa. *Journal of African Earth Sciences*, 58, 608–622.

O'Neill, E., Brereton, F., Shahumyan, H. and Clinch, P.J. (2016) The impact of perceived flood exposure on flood-risk perception: The role of distance. *Risk Analysis*, 36(11), 2158–2186. https://doi.org//10.1111/risa.12597

Olofsson, A. and Rashid, S. (2011) The white (male) effect and risk perception: Can equality make a difference? *Risk Analysis: An Official Publication of the Society for Risk Analysis*, 31(6), 1016–1032.

Omer, H. and Alon, N. (1994) The continuity principle: A unified approach to disaster and trauma. *American Journal of Community Psychology*, 22(2), 273–287.

Orr, P., Forrest, S., Brooks, K. and Twigger-Ross, C. (2015) *Public dialogues on flood risk communication, Literature review*. SC120010/R3 Technical Report.

Overdevest, C. and Christiansen, L. (2013) Using "cultural cognition" to predict environmental risk perceptions in Florida water-supply planning process. *Society and Natural Resources*, 26(9), 987–1007. https://doi.org/10.1080/08941920.2012.724152

Paek, H.-J. and Hove, T. (2017) Risk perceptions and risk characteristics. *Oxford Research Encyclopedias, Communication.* https://doi.org/10.1093/acrefore/9780190228613. 013.283

Pagneux, E., Gísladóttir, G. and Jónsdóttir, S. (2011) Public perception of flood hazard and flood risk in Iceland: A case study in a watershed prone to ice-jam floods. *Natural Hazards*, 58(1), 1269–1287.

Palm, R.I. (1990) *Natural Hazards: An Integrative Framework for Research and Planning.* John Hopkins University Press, Baltimore.

Papagiannaki, K., Kotroni, V., Lagouvardos, K. and Papagiannakis, G. (2018) How awareness and confidence affect flood-risk precautionary behavior of Greek citizens: The role of perceptual and emotional mechanisms. *Natural Hazards and Earth System Sciences*, 19, 1329–1346. https://doi.org/10.5194/nhess-19-1329-2019

Pielke, R.A. (1999) Nine fallacies of floods. *Climate Change*, 42, 413–438.

Prager, K. (2012) Understanding behaviour change. How to apply theories of behaviour change to SEWeb and related public engagement activities. *Report for SEWeb LIFE10 ENV-UK-000182.*

Quinn, T., Bousquet, F., Guerbois, C., Sougrati, E. and Tabutaud, M. (2018) The dynamic relationship between sense of place and risk perception in landscapes of mobility. *Ecology and Society*, 23(2), 39. https://doi.org/10.5751/ES-10004-230239

Quinn, T., Bousquet, F., Guerbois, C. et al. (2019) How local water and waterbody meanings shape flood risk perception and risk management preferences. *Sustainability Science*, 14, 565–578. https://doi.org/10.1007/s11625-019-00665-0

Raaijmakers, R., Krywkow, J. and Veen, A. (2008) Flood risk perceptions And spatial multi-criteria Analysis: An exploratory research for hazard mitigation. *Natural Hazards*, 46(3), 307–322.

Renn, O. (2008) *Risk Governance: Coping with Uncertainty in a Complex World*. Earthscan, London.

Renn, O. and Rohrmann, B. (2000) Cross-cultural risk perception research: State and challenges. In O. Renn and B. Rohrmann (eds.) *Cross-Cultural Risk Perception. A Survey of Empirical Studies*. Springer, Dordrecht and Boston, pp211–233.

Rogers, R.W. (1975) A protection motivation theory of fear appeals and attitude change. *The Journal of Psychology*, 91, 93–114.

Rutsaert, P., Regan, Á, Pieniak, Z., McConnon, Á, Moss, A., Wall, P. and Verbeke, W. (2013) The use of social media in food risk and benefit communication. *Trends in Food Science & Technology*, 30(1), 84–91. https://doi.org/10.1016/j.tifs.2012.10.006

Science Communication Unit, University of the West of England, Bristol (2014) *Science for Environment Policy Future Brief: Public risk perception and environmental policy*. Report produced for the European Commission DG Environment, September 2014. Available at: http://ec.europa.eu/science-environment-policy

Sharot, T. (2011) The optimism bias. *Current Biology*, 21(23), R941–R945. https://doi.org/10.1016/j.cub.2011.10.030

Shen, X. (2010) Flood risk perception and communication within risk management in different cultural contexts: a comparative study between Wuhan, China, and Cologne, Germany. PhD thesis, UN University Institute for Environment and Human Security (UNU-EHS), Bonn, Germany.

Siegrist, M., Cvetkovich, G. and Roth, C. (2000) Salient value similarity, social trust, and risk/benefit perception. *Risk Analysis*, 20, 353–362.

Siegrist, M. and Gutscher, H. (2006) Flooding risks: A comparison of lay people's perceptions and expert assessments in Switzerland. *Risk Analysis*, 26(4), 971–979. https://doi.org/10.1111/j.1539-6924.2006.00792.x

Slovic, P. (1987) Perception of risk. *Science*, 236, 280–285. https://doi.org/10.1126/science.3563507

Slovic, P. (2000) *The Perception of Risk*. Earthscan Publications, London.

Southerton, D., McMeekin, A. and Evans, D. (2011) *International Review of Behaviour Change Initiatives*. Scottish Government Report, The Scottish Government. https://webarchive.nrscotland.gov.uk/3000/https://www.gov.scot/Resource/Doc/340440/0112767.pdf

Spence, A., Poortinga, W., Butler, C. and Pidgeon, N.F. (2011) Perceptions of climate change and willingness to save energy related to flood experience. *Nature Climate Change*, 1, https://doi.org/10.1038/NCLIMATE1059

Strang, V. (2004) *The Meaning of Water*. Berg, Oxford and New York, 274pp.

Terpstra, T. (2009) *Flood Preparedness: Thoughts, Feelings and Intentions of the Dutch Public*. Thesis, University of Twente, Twente.

Touili, N., Baztan, J., Vanderlinden, J.-P., Kane, I.O., Diaz-Simal, P. and Pietrantoni, L. (2014) Public perception of engineering-based coastal flooding and erosion risk mitigation options: Lessons from three European coastal settings. *Coastal Engineering*, 87, 205–209.

Trope, Y. and Liberman, N. (2010) Construal-level theory of psychological distance. *Psychological Review*, 117(2), 440–463. https://doi.org/10.1037/a0018963

Ullberg, S.B. (2018) Forgetting flooding?: Post-disaster livelihood and embedded remembrance in Suburban Santa Fe, Argentina. *Nature and Culture*, 13, 27–45.

Urquhart, C.L., Lam, M.C., Cheuk, B. and Dervin, B.L. (2019) Sense-Making/Sensemaking. In P. Moy (ed.) *Oxford Bibliographies in Communication*. Oxford University Press. https://doi.org/10.1093/obo/9780199756841-0112

Vanderlinden, J.-P., Baztan, J., Touili, N., Kane, I.O., Rulleau, B., Diaz Simal, P., Pietrantoni, L., Prati, G. and Zagonari, F. (2017) Coastal flooding, uncertainty and climate change: Science as a solution to (mis) perceptions? A qualitative enquiry in three coastal European settings. In M.L. Martinez, A.Taramelli and R. Silva (eds.) *Coastal Resilience: Exploring the Many Challenges from Different Viewpoints. Journal of Coastal Research*, 77, 127–133. Coconut Creek (Florida).

Wachinger, G., Renn, O., Begg, C. and Kuhlcke, C. (2013) The risk perception paradox— Implications for governance and communication of natural hazards. *Risk Analysis*, 33, 1049–1065. https://doi.org/10.1111/j.1539-6924.2012.01942.x

Weick, K.E. (1995) *Sensemaking in Organizations*. Sage Publications, Thousand Oaks.

Weinstein, N.D. (1980) Unrealistic optimism about future life events. *Journal of Personality and Social Psychology*, 39(5), 806–820.

Weitkamp, E., McEwen, L.J. and Ramirez, P. (2020) Communicating the hidden: Towards a framework for drought risk communication in maritime climates. *Climatic Change*, 163, 831–850. https://doi.org/10.1007/s10584-020-02906-z

Westcott, R., Ronan, K., Bambrick, H. and Taylor, M. (2017) Expanding protection motivation theory: Investigating an application to animal owners and emergency responders in bushfire emergencies. *BMC Psychology*, 5(1), 13. https://doi.org/10.1186/s40359-017-0182-3

White, G.F. (1945) *Human Adjustment to Floods*. Department of Geography Research Paper No. 29. University of Chicago Press, Chicago.

Whitmarsh, L. and Capstick, S. (2018) Perceptions of climate change. In S. Clayton and C. Manning (eds.) *Psychology and Climate Change: Human Perceptions, Impacts, and Responses*. Academic Press, Cambridge. pp13–33. https://doi.org/0.1016/B978-0-12-813130-5.00002-3

Wood, E. and Miller, S.K. (2021) Cognitive dissonance and disaster risk communication. *Journal of Emergency Management and Disaster Communications*, 2(1), 39–56. https://doi.org/10.1142/S2689980920500062

4
DIFFERENT FLOOD KNOWLEDGES
Conflict or integration?

4.1 Introduction

Up-to-date, attuned, and actionable knowledge is critical for citizens and communities so they can contribute to effective community-focused, flood risk management (FRM). This chapter explores different types of knowledge (e.g., scientific, technical, organisational, lay, local, indigenous, vernacular, entrepreneurial, archival) that can be brought to decision-making by communities and other stakeholders in local FRM. It asks:

- What are the different kinds of knowledge and their relationships? This includes perceived knowledge hierarchies, what knowledge is valued by whom, and who is considered expert in local FRM.
- What are the characteristics and roles of brokers and translators of different knowledges? What can be learnt by FRM from other disciplines and professions?
- What is the potential for knowledge integration, bringing together specialist and lay perspectives, in local FRM?
- What are the issues in dialogue, knowledge exchange, and co-production when working with different kinds of knowledge?

The chapter reflects on where issues of 'paradigm lock' or 'inability to shift beyond a given mindset' or 'know-do gaps' occur in FRM. This is where domains of academic research and professional practice sit in unconnected realms, underused for addressing challenges beyond their own domain. Exploring the changing role of different knowledges in FRM requires contribution from various academic disciplines including geography, knowledge management, risk and health sciences, and sociology, attending to their interfaces.

DOI: 10.4324/9781315666914-4

4.2 Background context: The nature of knowledge and its systems

Before exploring specific knowledge systems and flood knowledges from a community perspective, a basic understanding of the relationships between knowledge and understanding is useful. Ackoff (1989), the pioneering systems thinker, usefully classified the 'content of the mind' into categories of increasing value (data, information, knowledge, and wisdom). The DIKW knowledge pyramid or hierarchy (Figure 4.1A) is a useful way of modelling their relationships although Spiekermann et al. (2015, p101) noted that the categories are interwoven:

> with interweaving bi-directional threads so as to show that data, information, and knowledge are intertwined more closely than a pyramid can illustrate. In accordance with perception and context, data is collected to improve understanding of a certain reality. Knowing the context in which data is collected and transformed into information (contextualised, categorised, calculated, connected and/or condensed) is highly significant.

Regarding the 'value' of the categories, Ackoff (1999, np) observed that still "educational systems and most managers" tend to allocate more time to the acquisition of data rather than to wisdom, which is "inversely proportional to their importance" (Chapter 9).

Knowledge systems are defined as "a body of propositions that are adhered to, whether formally or informally, and are routinely used to claim truth" (Intergovernmental Science-Policy Platform on Biodiversity and Ecosystem Services,[1] np).

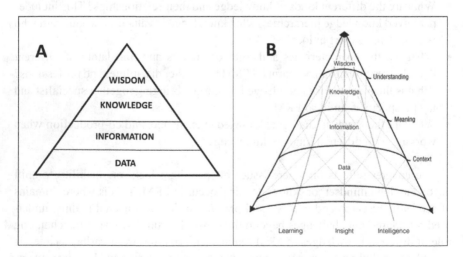

FIGURE 4.1 (A) DIKW pyramid or knowledge hierarchy (based on Ackoff, 1989); (B) the DIKW web (Spiekermann et al., 2015, Figure 2, reproduced by permission of Elsevier)

Calls pervade for an opening up and transformation of knowledge systems to inform better responses to environmental challenges – scaling up from local to global (e.g., Cornell et al., 2013). This involves acknowledging the value of, and integrating, different knowledge systems (e.g., as Assemblages of Local Knowledge) which requires "communication, translation and mediation across boundaries" (Cash et al., 2003, p8086)

Western science draws strongly from philosophies of Ancient Greece and the Renaissance and can be conceived as defining knowledge as objective and rational. However, Kuhn's (1962) philosophy of science highlighted socially constructed paradigm shifts – or historical changes in worldview – that have characterised 'science' in practice. Moreover, traditional lay knowledge systems have developed radically different strategies to create and transmit knowledge. It is important to understand the nature of, and connections between, different types of knowledge to establish what is, and is not being, analysed within any theory of knowledge. Increasingly, it is recognised that different forms of knowledge can, and need to, learn from each other (Mazzocchi, 2006). An important skill for resilience is the ability to operate outside familiar knowledge systems in the awareness of different paradigms – or socially and culturally constructed ways of thinking – through which the world is viewed in sense-making (Chapter 3). For example, awareness and deployment of both explicit, codified knowledge and tacit, implicit knowledge (see Table 4.1) are fundamental to community-focused FRM.

Other conceptual framings of knowledge help in understanding the potential of different knowledge contributions in FRM. These include the concepts of Post-Normal Science (PNS) and Sustainability Science. The concept of PNS was developed by Funtowicz and Ravetz (1993), who recognised that increases in both systems uncertainties and stakes in decision-making necessitated new problem-solving strategies. PNS is a socially aware approach to knowledge production at the "contested interfaces of science and policy" (Ravetz, 2005, p12). It is applied to "issue-driven science relating to environmental debates, [where] typically facts are uncertain, values in dispute, stakes high, and decisions urgent" (Ravetz, 1999, p479). The case of climate science exemplifies this – where the risks to be communicated frequently shift as available knowledge and understanding increase (Hulme, 2007).

PNS emphasises the importance of social, cultural, and historical influences, and changing values on science. It also values the public contributing to scientific thinking and practice. In doing so, it aims to increase the quality of both science and its underpinned policies. As Ravetz and Funtowicz (1999, p641) observed

> Public participation in decision-making on science-related issues is now fashionable. Post-Normal Science shows why it is necessary, not merely for political justice, but also for the quality of the decisions themselves.

TABLE 4.1 Characteristics of Explicit and Tacit Knowledge (Adapted from Virkus, 2014 and Talisayon, 2008[a], cclfi.international)

Type of knowledge	Explicit knowledge (already codified)	Tacit (implicit) knowledge (embedded in the mind)
Character	• Objective, rational, technical, structured	• Personal, subjective, cognitive, experiential learning
Examples[a]	• Manuals	• Experience of what works
	• Blue print	• Expertise
	• Scientific formula	• Wisdom
	• Work template	• Indigenous knowledge
	(e.g., organisational flood manuals)	• Informal business knowledge *(e.g., local flood knowledge)*
Documentation	• Easily documented; easy to codify	• Hard to document; difficult to capture and codify
Transfer and sharing	• Easily transferred/ taught/learned	• Hard to transfer/teach/learn
Advantages[a]	• Very easy to multiply and share	• Sticky and difficult to steal
	• More visible to others	• Grows and evolves with practice
	• Easy to manage	• Rich in nuances
	• Often measurable	• Linked to personality of owner
	• Easy to facilitate with IT	• Sharing is personal and contextual
Disadvantages[a]	• Must be adapted to new contexts	• Less visible or recognisable
	• Leaky and easily stolen	• Difficult to 'manage'
	• Utility depends on skill of user	• Difficult to track or measure
	• Utility changes if context changes	• Very difficult to facilitate with IT
	• Must be practised to be owned as tacit	• Lost if the person leaves or dies

[a] https://apintalisayon.wordpress.com/2008/11/28/d3-tacit-knowledge-versus-explicit-knowledge/

Another useful framing of knowledge that can be drawn on in FRM is that of 'Sustainability Science', bringing sustainability thinking into multi-stakeholder problem-solving. Sustainability Science is:

> ... an emerging field of research dealing with the interactions between natural and social systems, and with how those interactions affect the challenge of sustainability: meeting the needs of present and future generations while substantially reducing poverty and conserving the planet's life support systems.
>
> *Proceedings of National Academy of Sciences of United States of America*[2]

This is a research field that is "designed by the problems it addresses rather than by the disciplines it employs" (Clark, 2007, p1737). Kates (2011) recognised the challenge for Sustainability Science as

> a different kind of science that is primarily use-inspired ... with significant fundamental and applied knowledge components, and commitment to moving

such knowledge into societal action. .... However, its real test of success will be in implementing its knowledge to meet the great environment and development challenges of this century.

Kates (2011, p19450)

This thinking explicitly deals with interactions between natural and social systems that underpin sustainable water management and UN Sustainable Development Goals.

4.3   Specialist scientific and technical knowledge of flooding

Against this background, traditionally FRM has depended strongly on the development of Western framings of 'expert science' (Mazzocchi, 2006). This involves ever-improved flood risk modelling for prediction and forecast, and development of technological solutions for flood mitigation. But what are its strengths and limitations, and implications for knowledge-power relations among local stakeholders including communities? Specialist science is traditionally produced and validated as expert knowledge (McEwen et al., 2014b), but even scientific knowledge can be considered socially constructed (Chapter 6).

In terms of community engagement, it is also useful to distinguish between 'cutting edge' specialist scientific knowledge, generated and validated by professional scientists, and the basic scientific building blocks needed for public understanding. Defining a floodplain is a good example of where specialist science and school-level education occupy the two ends of a spectrum. The basic idea of a floodplain, where a river overflows overbank and 'takes its rent', is traditional fare in formal geographical learning in schools. However, temporal and spatial issues and uncertainties in defining floodplains are rarely mentioned. In practice, modelling floods and zoning levels of risk are complex exercises involving iterative improvement, with increasingly sophisticated modelling, technological and computing advances, and more accurate and extensive data garnered (e.g., through Lidar) during successive extreme floods (Porter, 2009).

Strong evidence also exists that science is not always used effectively, with barriers at the science-policy-practice interface in mobilising 'Research to Action'. A science-based approach to integrated FRM needs iterative knowledge exchanges between research, policy, and practice. In analysing the issue of 'paradigm lock' in integrated river catchment management, Falkenmark (2004) emphasised the importance of ongoing dialogue between scientists, stakeholders, and policymakers. Paradigm lock occurs because "scientists do not grasp what managers require, and managers and stakeholders do not appreciate the scientific alternatives available" (Gregory et al., 2008, p201). Hence there is a need for translation of knowledge to ensure its effective application in practice (Section 4.6) and the development of knowledge co-production strategies (see below).

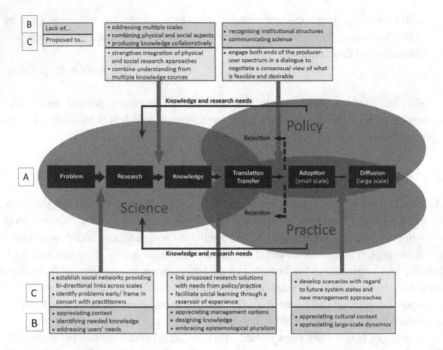

FIGURE 4.2 Pitfalls of, and propositions for, research to increase its effectiveness (Source: Spiekermann et al., 2015 adapted from Weichselgartner and Marandino, 2012, p327; reproduced by permission of Elsevier). Original figure was in colour; adapted here to B/W with letter codes

From a concern about knowledge fragmentation within different phases of Disaster Risk Reduction (DRR), Spiekermann et al. (2015, p100) critiqued the interfaces between science, policy, and practice, identifying challenges and opportunities for research to increase its effectiveness (Figure 4.2).

They outlined how in the "traditional pipeline model" [A], the research problem is identified and investigated by the scientists in isolation, with results only transferred to potential 'users' at the end of the process. This transference of research-based knowledge tends to have pitfalls [B] "assuming that [results] diffuse automatically through the policy and practice communities". They then proposed ways [C] to increase the effectiveness of research-based knowledge that would result in

more socially robust and context-sensitive knowledge within the framework of knowledge-brokering, where the research and knowledge needs of end-users are explicitly expressed.

Such thinking requires sensitive negotiation and application within community-focused, disaster risk management (DRM), with its diverse range of community and organisational stakeholders.

4.4   Local or lay knowledge of flooding

Awareness of the value of unorganised, lay, place-based knowledge, in complementing Western knowledge systems, spans many decades in some disciplines and contexts. In the context of knowledge in economics, Hayek (1945, p521) reflected on the value of informal knowledge:

> Today it is almost heresy to suggest that scientific knowledge is not the sum of all knowledge. But a little reflection will show that there is beyond question a body of very important but unorganized knowledge which cannot possibly be called scientific in the sense of knowledge of general rules: the knowledge of the particular circumstances of time and place.

From a Western science perspective, 'lay' implies being non-professional: knowledge not embedded in formal scientific networks. However, such place-based knowledge includes what Raygorodetsky (2017, np) described, in context of resource management, as "indigenous knowledge rooted in a millennia of meticulous on-the-land observations". He and others observed, in building resilience to change, it is critical to draw on the best available knowledge, which may derive from different knowledge systems. Nakashima (2015) and Nakashima et al. (2018) emphasised the status and value of such indigenous or traditional ecological knowledge, and a "holistic view of community and environment" in climate change adaptation (IPCC, 2014, p19):

> Local and indigenous knowledge is not something new. Indeed, it is as old as humanity itself. What is new, however, is its growing recognition by scientists and policy-makers around the world, on all scales and in a rapidly growing number of domains.
>
> *Nakashima (2015, p15)*

The Intergovernmental Panel on Climate Change (IPCC, 2022, p9) also emphasised the value of diverse forms of knowledge

> such as scientific, as well as Indigenous knowledge and local knowledge in understanding and evaluating climate adaptation processes and actions to reduce risks from human-induced climate change.

Indigenous or local knowledge contributes an important cultural dimension to human security to climate change impacts (Adger et al., 2014). There are increasing calls for inclusion of local (environmental) knowledges as forms of expertise in processes of equitable decision-making within climate change adaptation to flooding (e.g., Alba et al., 2020). However, there can be tensions and contestation between indigenous and scientific knowledge systems that reflect long histories of being overlooked or disapproved of (Whyte, 2013). A major concern is for knowledge justice – recognition and respect for different ways of knowing – in deliberative systems (Fan, 2021).

Within FRM, indigenous or local knowledge includes experiential or intergenerational knowledge of weather, flood patterns, and what works locally in coping and adapting to floods – knowledge gained individually and collectively through living with water and its associated risks (Chapter 5). This results from observing and experiencing interactions of flooding with people, buildings, and the landscape at-a-place over time (McEwen and Jones, 2012). Such knowledge can be of different types, forms, and sources, including intergenerational, experiential, vernacular, hobbyist knowledge, and citizen science (Table 4.2; McEwen, 2023).

TABLE 4.2 Different Types of Local Flood Knowledge with Their Characteristics (See McEwen, 2023)

Type of knowledge	Definition	Form	Additional detail in flood context	Reference
Archival knowledge	Knowledge contained in archival records (formal and informal) and its management	Historic documents, visuals (drawings, plans, photograph)	Written and visual records in both government and citizen/community curated flood archives	McEwen et al. (2018)
Intergenerational or inherited knowledge	Local knowledge unique to given culture; passed generation to generation	In written archival or oral histories; cultural rituals	Knowledge for resilience, e.g., how to live sustainably; of coping strategies (traditional)	McEwen et al. (2016)
Experiential knowledge	Gained from personal experience of living through events	Oral narratives and anecdotes	Knowledge gained from living through a flood(s)	McEwen et al. (2016)
Vernacular or everyday knowledge	'Integrates expert science and local knowledge with community beliefs and values'	A socially shared belief or way of doing things	Floods as explained within regional daily media Traditional ways of building to mitigate flood risk; improve recovery in vernacular architecture	Simpson et al. (2015, p252) Degen et al. (2003) Le Lay and Rivière-Honegge (2009)

(Continued)

TABLE 4.2 (Continued)

Type of knowledge	Definition	Form	Additional detail in flood context	Reference
Vernacular science knowledge	Scientific facts used in everyday life that are scientifically wrong but still have value in local belief systems and discourses	"Metaphoric and iconic representations of scientific facts"	The efficacy of interventions in flood management	Wagner (2007, p7)
Hobbyist knowledge	Knowledge of local factual information and skills Characterised by: • situated Learning • social enterprise/ communities of practice • Mastery and expertise • Motivation by interest	Writing in local society journals, newspapers Visuals include photography, cine, videos	Can be linked to retired professions (e.g., researching flood histories; historic river management) Involves skills (e.g., work in archives; recording of floods by film/ photographic clubs) Links to archival knowledge (Chapter 5)	Longnecker Lab, University of Otago[a] Liu and Falk (2014)
Citizen science	Quantitative and descriptive recording –sometimes working with academic researchers or motivated by citizens' own drive and interest in 'care of place'	Collection of observational 'data' – scientific, visual (photographic)	Observation of place at different stages in DRM spiral (during and between events); of local rainfall, runoff processes, or flow patterns; impact of structures, blockage of water courses	Citizen observatories (e.g., Mazumdar et al., 2016)

[a] https://www.longneckerlab.net/scom406-2019-blog/2019/9/23/gaining-knowledge-through-having-a-hobby-creating-motivated-learners.

While forms of scientific knowledge comprise measurements and (often big) data, lay knowledge tends to be more narrative and anecdotal in form. Within local knowledge

> knowledge claims are not adjudicated by absolute standards; rather their authority is established through the workings of local negotiations and judgments in particular contexts.
>
> *Turnbull (2008, p1)*

Stories capture and share local knowledge as thick data (Wang, 2016), and community narratives can reflect complex and diverse local contexts. The concepts of dominant, counter, and sticky narratives are important in navigating narrative evidence. A dominant cultural narrative is an overlearned and repeated "explanation or story that is told [with authority] in service of the dominant social group's interests and ideologies"[3] and communicated through "mass media or other large social and cultural institutions and social networks" (Rappaport, 2000, p4). Examples might include narratives about perceived responsibilities in local FRM. In contrast, counter-narratives are "stories which people tell and live which offer resistance, either implicitly or explicitly, to dominant cultural narratives" (Andrews, 2002, p1). Examples might involve narratives of community empowerment and agency in FRM, where deferral of responsibility otherwise pervades. Sticky narratives defy evidence but are difficult to disengage from, and hence have persistence (Heath and Heath, 2007; e.g., narratives attributing extreme floods to impacts of particular interventions).

Direct experience of floods links to different types of knowledge. Mondino et al. (2020), working after a flash flood in Italy, explored their interplay in risk awareness. They found that previous experience influences flood risk awareness not only directly but also indirectly through the acquisition of experiential knowledge. They proffered a typology of experience-knowledge comprising four classes:

Class 1: Inertia (low experience; low knowledge)
Class 2: Tacit/empirical knowledge (high experience, low knowledge)
Class 3: Theoretical knowledge (low experience, high knowledge)
Class 4: Wisdom (high experience, high knowledge)

Higher-level wisdom (Section 4.2) comes from integrating experience and knowledge. They recommend (p11) future research that assesses the role of experience should appraise

> not only the experience in itself (in terms of intensity and frequency), but also, and especially, the knowledge produced as a result of the experience and the knowledge acquired through secondary sources.

The kinds of stories recorded by lay people about floods are gained from observing what floodwater does at a 1:1 scale, as a form of vernacular science

(Wagner, 2007). This detailed knowledge of very local water geographies then becomes a critical form of knowledge that makes specialist knowledge workable. Such knowledge can be invaluable in establishing contributions from different causes of floodwater in complex situations – to ground truth, and sometimes challenge, specialist hydrological modelling (Chapters 8 and 10). This perspective is critical to knowledge as practice, which develops and tests coping and adaptive strategies to flood situations.

Flood awareness and place-based, lay knowledge are interlinked. At its most fundamental, traditional ecological knowledge can be embedded in watery place names[4] (e.g., Jones, 2016). Development of local environmental knowledge, involving cognitive or mental marking of flood levels, can occur when citizens monitor floods relative to known personal points of reference. Materialisation of lay flood knowledge could be visible in the domestic environment (e.g., old floodwater marks or flood protection products; family/community archives as photographs, videos, diaries) or in the landscape (e.g., flood marking or epigraphic records, and stage boards by rivers; Figure 4.3; McEwen et al., 2018). Community flood knowledge can be captured and archived in varied ways as flood 'heritage' (Chapter 5).

FIGURE 4.3   Forms of flood materialisation: collective and personal settings. A/B, epigraphic marking of varying persistence; C, remembering personal flood levels; D, July 2007 'a weekend in history'; E, historic 1947 flood photographs; F/H, sharing 'flood albums'; G, historic community protection measures; I, preserved flood water (McEwen et al., 2016, Figure 1)

Alongside community lay knowledge, small businesses may use their entrepreneurial knowledge (skills and innovative mindset) in dealing with diverse risks, whether economic, political, or environmental (Durst and Zieba, 2020). This can be seen in bricolage – "ability to adapt or create necessary resources and solutions from internal resources" or draw on external network relationships to gain resources – as a strategy during floods (McGuinness and Johnston, 2014, p447). They cited an example of staff in a factory making wire products utilising their wire and fixing devices to create a safety rope across a flooded road to help stranded workers getting to safety. Relationships between entrepreneurial knowledge and lay knowledge will depend on business proximity to, and degree of connection with, local place-based communities.

However, as Šakić Trogrlić et al. (2019, pp1,4) found from their work in Malawi, local knowledge is not "homogenous" or "equally available" within a particular community; it is also dynamic and evolving. While local knowledge of adaptive strategies may carry over time, local experiential knowledge of physical flood character may be less applicable and challenged in some settings, e.g., due to climate or land use changes altering local flood characteristics.

4.4.1 Knowledge, memory, and sense of place

So how can local knowledge lead to a deeper understanding of flooding and place? The *Sustainable Flood Memories*[5] project found complex relationships between flood memory and lay knowledge in the development of community resilience, after the extreme summer 2007 floods in Gloucestershire, UK (McEwen et al., 2016; Chapter 5). This involves the will – individually and collectively – to actively remember or forget floods, and how these tensions are negotiated in communities (see 'knowledge loss' below). Laudan et al. (2020) found that frequency of remembering an event is positively connected to preparedness intentions. This emphasises the importance of keeping flood risk in the consciousness of at-risk communities and ensuring positive reinforcement whilst recognising the sensitivities.

People living in flood-risk settings that routinely experience floods, and where risk of more extreme flooding forms part of everyday life, may possess what has been termed a "watery sense of place" – a strong sense of how to live with water (McEwen and Jones, 2012; Chapter 3). Communities that might be described as having this sense of place in terms of their knowledge, identity, and resilience include inhabitants of low-lying areas (e.g., river deltas, estuaries, and coasts) at long-standing flood risk. A watery sense of place may be more frequently present in rural communities with livelihoods that have long-term connections to interactions between land and water. However, it can also be found in historic urban community settings that experience routine flooding as well as extremes, and where the community is established over several generations. This watery sense of place can also apply to coastal identities where communities have to live with extreme weather in an exposed open landscape. As Ratter (cited in Helmholtz-Zentrum, 2014[6]) observed

> What is new, is the realisation that there is a sense of coastal identity which reaches across the borders. Whether Danes, Dutch or Germans – for these coastal inhabitants the Wadden Sea is their homeland. They all feel at home and at ease in an area in which they are surrounded by the forces of nature.

Such sense of place is important in collective risk perception (Chapter 3). Ideas of place, place attachment, and sense of place have been linked to the German concept of 'Heimat'. This is influenced by social norms and values, and constructed individually (Ratter and Gee, 2012), combining social, emotional, and spatial components. In England, such inextricably connected communities and watery places include villages on the Somerset Levels, with their long history of river and coastal inundation and land drainage. This sense of place is captured in prose and photographs within Sutherland and Nicholson's (1987) book *Wetland: Life in the Somerset Levels*. The same can apply to those living in historic river-side towns with frequent experience of routine floods, and long histories of extremes, such as Tewkesbury, UK (McEwen et al., 2014a).

4.4.2   Applying local knowledge

A growing research and policy literature exists on the value of lay flood knowledge in local resilience planning, and its influence on the power dynamics of FRM (McEwen and Jones, 2012). This emphasises the importance of taking lay knowledge seriously and considering how it meets specialist scientific and technical knowledge. Indeed, lay knowledge can be considered authoritative and 'expert' in terms of subject and locality by communities themselves and other FRM actors such as politicians. Collins and Evans (2002) reinforced this by distinguishing two types of "experts" – "certified" and "non-certified". In engaging such knowledge controversies, opportunities and methodologies are being sought for integrating specialist scientific and lay knowledges.

Lay knowledge is being applied in FRM in various ways. These include co-developing within local communities and FRM agencies an understanding of how place-based, land-water-people interactions build up into whole catchment, hydrosocial systems (Chapter 7). Local knowledge is reinforced by care for place, and the desire to act and prevent flooding of home and its environs. For example, such knowledge is being used in towns affected by the Summer 2007 floods, UK, as evidence to support promotion of locally implemented, creative flood risk solutions (e.g., sustainable drainage systems), and in campaigning to statutory FRM agencies for maintenance of small water courses (McEwen and Jones, 2012). Its value can be reinforced when local knowledge is linked to funding for community agency (Chapter 8).

Local knowledge (as photographs, videos, and narrative evidence) is also being used to highlight and fill gaps in specialist scientific knowledge. For example, evidence gathered by citizens during floods caused by intense rainfall is helping validate special-ist scientific modelling (e.g., in SINATRA project[7]; Chapters 8 and 10).

4.4.3 Loss of place-based local knowledge

Risks of loss or fragmentation of knowledge, and disconnects in knowledge networks (Durst and Zieba, 2020), are manifest in varied ways among communities, small businesses, and other FRM stakeholders. Baubion (2015) expressed concern about the loss of collective memory due to time elapsed since the last major flood on the River Seine, Paris. Loss of local knowledge may also be attributed to the 'disin-tegration' of communities, for example, through increased transience of urban popu-lations and disconnects between older and younger residents. Relationships between memory and local knowledge are complex, with varied reasons for forgetting about extreme floods in communities (Figure 4.4; McEwen et al., 2016; Ullberg, 2018).

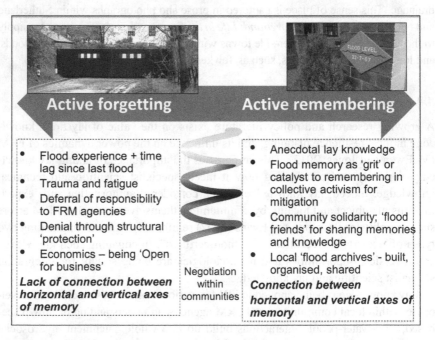

Active forgetting   **Active remembering**

- Flood experience + time lag since last flood
- Trauma and fatigue
- Deferral of responsibility to FRM agencies
- Denial through structural 'protection'
- Economics – being 'Open for business'

Lack of connection between horizontal and vertical axes of memory

Negotiation within communities

- Anecdotal lay knowledge
- Flood memory as 'grit' or catalyst to remembering in collective activism for mitigation
- Community solidarity; 'flood friends' for sharing memories and knowledge
- Local 'flood archives' - built, organised, shared

Connection between horizontal and vertical axes of memory

FIGURE 4.4 Negotiation between active remembering and active forgetting within communities

The relationships between memory, knowledge, and emotion and affect are critical. This negotiated interaction has implications for local resilience building.

Tensions can also exist between sustainability of community flood memory and that of statutory FRM agencies (see 'knowledge risks'). Agencies might not main-tain significant archives, and there is a loss of organisational flood knowledge as individuals are seconded to other jobs or retire. In UK government, for example, staff move to new roles every 2–3 years as a normalised element of staffing strategies. Local governments also increasingly outsource their management of key flood risk functions to cheaper operators in budgetary cuts, so losing important local organisational knowledge. A good example is in maintenance of urban drains in Gloucester city, UK – losing knowledge of seasonal pinch points that get blocked by autumnal leaf fall, and hence need clearing to avoid local surface water flooding issues (McEwen and Jones, 2012). Knowledge can also be lost in other ways during attempts to connect and integrate different knowledge systems.

4.5 Knowledge brokers and translators: Implications for FRM

In working with, and integrating, different types of knowledge systems, a key consideration is 'knowledge brokering'. This concept involves transfer and exchange of knowledge to improve processes or aid innovation (Meyer, 2010). Much learning about knowledge brokering has come from health and organisational research, with the concept applied in associated policy settings. Knowledge brokers are increasingly advocated as a solution for bridging the gap between specialist science and decision-making (Cvitanovic et al., 2017). As either people or organisations, they act as knowledge intermediaries, providing links between producers (traditionally researchers) and users of knowledge (traditionally organisations). Hargadon (1998), working in business management, observed that knowledge brokers are typically involved in assessing barriers in how to access knowledge; working out how to link separate knowledge pools; facilitating individual and organisational capacity development for using knowledge; and implementing knowledge in new settings (e.g., combining existing knowledge in new ways). Meyer (2010) argued that knowledge brokers do not simply move knowledge but also produce new types of knowledge as 'brokered knowledge'.

Interactive knowledge brokering approaches – with their focus on mechanisms, relationships, and trust – can be considered under the banner of science communication practices (Chapter 6).

> Communicating science, therefore, has expanded to include knowledge translation in which scientific information is packaged to the preferences, channels, and timescales of particular audiences, and knowledge brokering in which intermediaries (knowledge brokers) link the producers and users of knowledge to strengthen the generation, dissemination and eventual use of that knowledge.
>
> *Bielak et al. (2008, p203)*

Among the skills of a successful knowledge broker are those of synthesis and ability to adapt information for application in specific local contexts. A key skill in FRM is in identifying what knowledge needs to be translated and communicated, and then establishing who the knowledge brokers are, or could be, within a specific geographical or policy/practice setting. Knowledge brokers, as influential individuals and their developing networks as collective learning spaces, can be important in FRM governance (Witting et al., 2021). Within community-focused FRM, this brokering role can be undertaken by a range of key individuals and community-facing organisations. In the UK, examples of knowledge brokers include Long-Dhonau[8] and her 'Know your flood risk™' campaign, and UK NGO Groundwork's 'Communities Prepared' project,[9] funded through diverse income streams.

4.5.1 Knowledge translation: The knowledge-to-action process

Knowledge translation, as a metaphor, involves "the synthesis, exchange, and application of knowledge to accelerate benefits" (PAHO/WHO, ud[10]). The need for 'knowledge translation' is a term applied to long-standing problems of underused evidence-based research, linking to 'paradigm lock' (Section 4.3). This is "often described as a gap between 'what is known' and 'what is currently done' in practice settings" (National Centre for the Dissemination of Disability Research, 2005, p1). Similar issues can occur in FRM research (e.g., around what works in community engagement and participatory methods). Translation of research can be required for diverse reasons including awareness-raising, to inform decision-making within policy and practice, or to facilitate public action. In one model in health, Graham et al. (2006) provided a conceptual framework for a dynamic, integrated Knowledge-to-Action (KTA) process that emphasised collaboration and tailoring of knowledge from its creation to application (Figure 4.5).

Importantly, as Sudsawad (2007, p8) emphasised, KTA processes can accommodate different forms of knowledge. While knowledge is

> mainly conceptualized as empirically derived (research-based) knowledge; however, it encompasses other forms of knowing, such as experiential knowledge. The KTA framework also emphasizes the collaboration between the knowledge producers and knowledge users throughout the KTA process.

The Canadian Institutes of Health Research (CIHR, 2005) conceptual model of the generic research cycle identified six opportunities (KT1-6) at which "the interactions, communications, and partnerships that will help facilitate KT could occur" (Figure 4.6). Such opportunities (KT1-6) can be transferred into a community-focused, flood research context (Table 4.3; Section 4.6).

While knowledge brokering tends to be construed in terms of sharing specialist science with stakeholders to support their decision-making, equally brokers may be needed to support sharing of lay knowledge as thick data within FRM. This involves communicating how community narratives, vignettes, and anecdotes might be brought into the evidence base for multi-stakeholder local decision-making. The arts can act as a valuable knowledge broker and translator within communities, negotiating between different types of knowledge (Chapter 6; Table 6.7). Examples

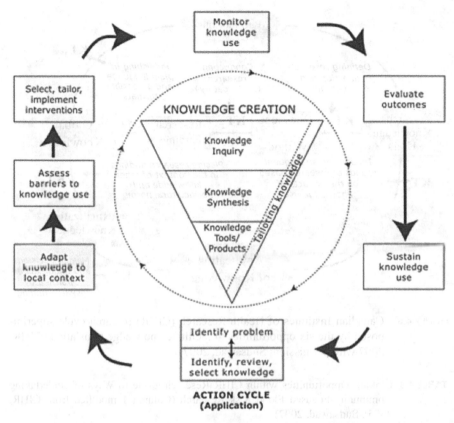

FIGURE 4.5 A visual representation of the KTA process (Graham et al., 2006; Sud-sawad, 2007, p9; permission Wolters Kluwers Health)

include visual artist Simon Read's maps[11] that explore the concept of change and possible futures of vulnerable coastlines – used to support decision-making with local communities (Jones et al., 2012); and *Highwaterline's* art collaboration with communities in coastal cities to mark flood lines from flood risk maps within the cityscape (Chapter 6).

The dominance of the 'translation' metaphor in health research and practice has, however, been challenged, recognising the term as value-laden with a power dynamic. For example, Greenhalgh and Wieringa (2011, p501) argued for applying a wider range of metaphors, models, and conceptualisations of knowledge, as "'created', 'constructed', 'embodied', 'performed' and 'collectively negotiated'". They proposed that this would allow more creative and critical ways of researching links between knowledge and practice. Additionally, they acknowledged the importance of research into situation-specific practical wisdom and tacit knowledge. Such proposals for other knowledge transmission models have led to more recent work on how to co-produce and integrate knowledges.

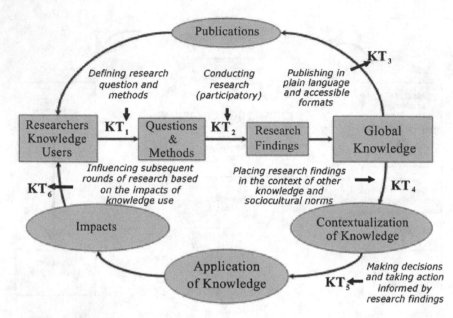

FIGURE 4.6 Canadian Institutes of Health Research (CIHR) research cycle superimposed by the six opportunities to facilitate knowledge translation (CIHR, 2007 with permission; Sudsawad, 2007)

TABLE 4.3 Linking Opportunities within CIHR Research Cycle to Ways of Undertaking Community-Focused Flood Risk Research (Column 1 modified from CIHR, 2005[a]; Sudsawad, 2007)

[a]CIHR Research Cycle opportunities	Diversifying the translation metaphor and linking to research processes in community-focused FRM
KT1: Defining research questions and methodologies	Using co-production approaches to problem definition
KT2: Conducting research (as in the case of participatory research)	Using participatory methods involving communities
KT3: Publishing research findings in plain language and accessible formats	Recognising different publication formats tailored to differing community capital
KT4: Placing research findings in the context of other knowledge and sociocultural norms	Connecting specialist knowledge with lay knowledge (meeting, overlaying, braiding)
KT5: Making decisions and taking action informed by research findings	Developing co-production processes that deliver knowledge exchange throughout a research process
KT6: Influencing subsequent rounds of research based on the impacts of knowledge use	Feeding back from knowledge use in practice into co-design of research questions

4.6   Knowledge controversies and need for knowledge integration

Work on knowledge integration emphasises the importance of avoiding quick judgements about the value and applicability of different forms of knowledge. Tangihaere and Twiname (2011), in the context of Maori culture, argued for the importance of making space for different types of knowledge, indigenous language, and cultural values in organisational practice (e.g., public health). Alongside challenges, opportunities exist for FRM to draw on, and integrate, local/lay experiential knowledges – with implications for their mapping, sharing, and redistribution (Lane et al., 2011; McEwen and Jones, 2012; Donaldson et al., 2013). Indeed, the focus on public participation in environmental risk and resource management is reconfiguring notions of expertise (Landström, 2020; Chapter 8).

"Knowledge controversies" (Whatmore, 2009) reflect mismatches between local knowledge and scientific flood risk assessments. Knowledge conflicts can occur within different phases of the DRM spiral (Chapter 1), and issues of disparity in the sway of different knowledges, power, and who is considered "expert" can regularly be witnessed as flood situations play out. Significant conflicts between scientific and lay knowledge can occur for varied reasons in emotionally charged community settings during floods. Recent examples include the debates about impacts of dredging as a river management strategy (Chapter 2). During the winter 2013/2014 floods, communities in the Somerset levels (SW UK) petitioned the UK government for a re-adoption of past dredging practices whereas scientific reports on the efficacy of river dredging, previously commissioned by the environmental regulator, had indicated that it was not a sustainable solution to siltation problems. A high-profile public debate ensued, highlighting issues in effective knowledge exchange, and the question of whose knowledge is promulgated and trusted by the public, politicians, and the media (Monbiot, 2014). In the end in a political response, the rivers *were* dredged, despite limited scientific evidence for its efficacy.[12]

Outside the acute event phase, local experiential knowledge can contest specialist knowledge through mismatches between official risk communication and local people's longer-term, assimilated place-based knowledge (Chapter 6). This can occur in government-produced flood risk maps used for local planning decisions in urban development (e.g., Lower Severn and Avon Combined Flood Group, 2008). After Hurricane Katrina devastated New Orleans in 2005, there were tensions between local knowledge of Afro-Americans and the government disaster response in reconstruction of historic building fabric (Allen, 2007). Here, accessing local knowledge would have led to "better practices and better policy recommendations" (p158). It is also important to be aware of how knowledge of different actors interacts in, and through, relationships at hyperlocal scale in designing adaptive strategies for FRM. This is seen, for example, in the implementation of embedded nature-based solutions – spatially distributed within catchments (Chapter 8). Here local knowledges of varied actors, produced through their situated relationships and practices, can usefully inform decision-making processes (Carnelli et al., 2020).

4.6.1 Modes and scales of knowledge integration

Significant opportunities for knowledge integration sit at the intersection of Western and indigenous knowledges. As Nicholas (2018, np) noted

> As ways of knowing, Western and Indigenous Knowledge share several important and fundamental attributes. Both are constantly verified through repetition and verification, inference and prediction, empirical observations and recognition of pattern events.

Awareness of areas of commonality and difference within different knowledge systems is critical in developing strategies for their integration across different risk and resource domains. Arnold et al. (2014, p21) emphasised that in community-driven development "Information is key" with the need to "blend climate science with local knowledge" to avoid maladaptation. Several researchers have recently used the metaphor of "braiding" or "weaving" in bringing science together with indigenous knowledge. These emphasised the need to develop respectful, equitable, and culturally appropriate ways of generating critical knowledge for use in addressing complex environmental problems, and climate adaptation and mitigation responses (e.g., Johnson et al., 2016; Wilcox et al., 2023). A process of braiding – or bringing multiple knowledge systems together:

> creates opportunities to develop a deeper understanding of observed events and their consequences. It facilitates joint assessment of information, leading to new insights and innovations, and results in better-informed actions.
>
> *Raygorodetsky (2017, np)*

However, Mazzocchi (2018, p19) cautioned that

> the possibility of a sound integration depends on the possibility to accommodate different interpretations of reality and knowledge criteria, recognising the value of pluralism and mutual learning.

Processes of integrating – whether 'weaving', 'braiding', or 'bridging' – multiple knowledge systems need to be underpinned by equity and ways of working that build relationships, embed mutual responsibility, and ensure reciprocity of understanding for mutual learning and application (Alexander et al., 2019; Bowles, 2021). This ethos of working is likely to improve the likelihood of expectation matching reality for participants in the co-management of place-based knowledge, as an evidence-base for decision-making.

There are various equally valid and relevant approaches to management and transmission of different knowledges and expertise in FRM. Selection requires attention to scaling, stakeholder roles and values, and identifying collaborative fora. In its application, the river basin or catchment is a valuable and scalable unit for integrated approaches to managing humans and water, but also for thinking and knowledge integration. There is strong value in blurring knowledge producers and users in catchments, working throughout the FRM spiral (Chapter 1). For example, in independent reviews to learn after impactful floods, local and specialist knowledges need to be integrated to develop shared corporate memory. This potentially involves validation of specialist knowledge by local experts and vice versa. Here the notion of vernacular knowledge is helpful (Table 4.2). For example, as Simpson et al. (2015, p352) observed

> collaborative approaches to water problem-solving can provide forums for bringing together diverse, and often competing, interests to produce vernacular knowledge through deliberation and negotiation of solutions.

Extending this further, Haughton et al. (2015, p375) argued for "hybrid knowledge formations" in relation to co-production of flood knowledge. They also cautioned against simplistic ways of thinking about such knowledge that deem 'local' knowledge as positive and 'expert knowledge' [specialist knowledge] as in some way dubious.

Attention is therefore needed in developing strategies that value and protect different knowledges while promoting their integration. Echoing the importance of process as well as outcome, Haughton et al., (2015, p375) proposed that:

> Experiments in the co-production of flood risk knowledge need to be seen as part of a spectrum of ways for producing shared knowledge.

Before 'experimenting', however, it is important to be aware of the principles of co-production in interdisciplinary and transdisciplinary research and practice (e.g., Polk, 2015).

4.6.2   Co-production and knowledge co-generation: Implications for FRM

Co-production of knowledge is defined with differing emphases in research and practice. It is

> A process where people intentionally try to collaborate on equal terms to develop a more collective wisdom, which can become a basis for making the quality of life "better."
>
> *Romm (2017, p22)*

> An approach in which researchers, practitioners, and the public work together, sharing power and responsibility from the start to the end of the project, including the generation of knowledge.
>
> *Hickey et al. (2021, p1)*

Co-production has been described as "a rather heterogeneous umbrella concept" (Verschuere et al., 2012, p1094), involving input of end users working together throughout any project that affects them (McEwen, 2023). This collaborative working in co-generation of knowledge integrates thinking about research initiation, process, and outputs and can be construed as an aspiration or ideal. Co-production has been celebrated as a "magic concept" (Sorrentino et al., 2018, p284) in research, policymaking, and governance (e.g., in climate change; health and social care). However, there is no "co-production panacea" (Jam and Justice[13]), with well-rehearsed challenges in putting co-production principles into practice in organisations with "embedded hierarchies and structural inequalities" (Farr et al., 2021, p1; Reddel and Ball, 2021). The author's personal experience indicates that use of the term can be considered tokenistic by communities and NGOs, based on sub-optimal prior experience of work delivered under that banner. Meaningful co-production requires strong attention to power relations and effective inclusive participation in knowledge co-generation throughout a process. Such processes need to create safe spaces for ongoing dialogue between different styles and ways of knowing. As Filipe et al. (2017, p1) observed

Co-production can be understood as an exploratory space and a generative process that leads to different, and sometimes unexpected, forms of knowledge, values, and social relations.

Potential barriers include expectations, language and terminology, and the ability and willingness to integrate different knowledges in decision-making processes (Briley et al., 2015). Various investigative domains (e.g., public health) have proposed principles for co-production as important in underpinning their research practice. National Institute of Health and Care Research (Hickey et al., 2021) provided an accessible and transferable articulation of principles and features – valuable for sharing during early dialogue in multi-stakeholder working (Box 4.1). Formative and summative co-evaluation of co-production processes is essential.

Different disciplinary clusters (e.g., social sciences) are recognising the importance of this space and their potential contributions within it. Banks et al. (2019), in exploring university-community relations, proposed a community development approach to co-production that privileges community agency in the ways a research process is co-designed and co-delivered. The move to co-production has also been driven by arts-based research as practice, and interest in the 'participatory turn' in producing knowledge with communities as co-researchers (e.g., Facer and Enright, 2016). As Liguori et al. (2023, p14) emphasised, working with people's stories of their memories, experiences, and visions is a way

of dismantling knowledge hierarchies and blurring the boundaries between knowledge producers, knowledge implementers, and knowledge consumers – we can all partake of any of these roles and real dialogue between participants becomes possible.

BOX 4.1   KEY PRINCIPLES AND FEATURES OF CO-PRODUCTION (HICKEY ET AL., 2021, PERMISSION FROM NIHR)

Key principles

- *Sharing of power* – the research is jointly owned and people work together to achieve a joint understanding
- *Including all perspectives and skills* – make sure the research team includes all those who can make a contribution
- *Respecting and valuing the knowledge of all those working together on the research* – everyone is of equal importance
- *Reciprocity* – everybody benefits from working together
- *Building and maintaining relationships* – an emphasis on relationships is key to sharing power

Key features

- Establishing ground rules
- Continuing dialogue
- Joint ownership of key decisions
- A commitment to relationship building
- Opportunities for personal growth and development
- Flexibility
- Continuous reflection
- Valuing and evaluating the impact of co-producing research

4.6.3   Examples of co-production in flood and water management

There are calls for knowledge co-production in FRM research and practice (e.g., Chrs, 2017), and much to be learnt from experiments (cf. Haughton et al., 2015) involving different disciplinary collaborations. In co-production of environmental knowledge, co-development of processes and agreed ways of working are as important as the modelled outcome and impact its success. A valuable case study is work by researchers Lane (hydrological modelling), Whatmore (social geography), Landström (technology management), and others within an extended research collective involving communities in a small rural town, Pickering, UK, where flooding presented a controversial issue. Here specialist flood science came out from its normal institutional setting, entering a dialogic process that both empowered local communities and led to a new computer model that embodied the co-produced flood risk knowledge (Landström et al., 2011). This collaborative process – termed a "Stakeholder Competency Group" (SCG) way of working – recognised "modelling as a scientific practice shaped in complex relationships, and hence changeable" (p1621). The ongoing collaboration brought together expertise and concerns from community and researchers to shape more locally acceptable FRM measures. This approach recognised that much is learnt from exploring the parallels, controversies, and frictions between different forms of competency and knowledge, and dismantling understandings of expertise. Ward et al. (2014) summarised this dialogic process – a slowing down of reasoning – and its impacts (e.g., confidence building; creating new knowledge and knowledge communities) as a way to support decision-making in FRM at community level.[14]

Participatory Action Research (PAR) is one model of co-production that conceives research as not just for knowledge but also for social change (Kindon et al., 2010). This can involve co-working with diverse and marginalised communities throughout a research process, as exemplified by a university-community PAR collaboration[15] between Carleton College and community members and institutions in Faribault, Minnesota, USA. Here PAR is defined as:

> A framework for conducting research and generating knowledge centered on the belief that those who are most impacted by research should be the ones taking the lead in framing the questions, the design, methods, and the modes of analysis of such research projects. The framework is rooted in the belief that there is value in both traditionally recognized knowledge, such as scholarship generated by university-based researchers, and historically de-legitimized knowledge, such as knowledge generated within marginalized communities.

Longitudinal PAR involving local education institutions and their proximal place-based, at-risk communities can have transformative potential in problem-solving for local FRM (Chapter 10).

Finally, co-production is also increasingly an integral part of research design within research projects focused on improving communication of flood risk knowledge between scientists and FRM practitioners for application in community settings. For example, Alexander et al. (2014) considered how complex specialist flood risk science could be translated for professionals in emergency response organisations. They found that a paradigm shift from knowledge transfer to co-production and participatory approaches for knowledge exchange was important in developing a Geographical Information System (GIS) flood risk decision support tool – KEEPER (Knowledge Exchange Exploratory tool for Professionals in Emergency Response) (Chapter 6).

4.7 Applying flood knowledge in practice

For any kind of knowledge to be utilised by individuals or communities with practical impact, it needs to become actionable knowledge (Antonacopolou, 2008). As Argyris (1993, p1) reflected, its implications are wider: "actionable knowledge is not only relevant to the world of practice, it is the knowledge that people use to create that world". In actionable knowledge, 'data' and 'information' need to be structured, contextualised, and made 'sense of' (Weick, 1995; Chapter 3). Actions and practices can be seen as manifestations of that knowledge – attuned or otherwise.

In the context of Community-based Disaster Risk Reduction (CBDRR), it is important to understand the contexts to applied knowledge (e.g., issues with Knowledge-Action-Behaviour theory; Chapter 3). While logic suggests knowledge might be applied, there can be many reasons (perception, values, worldview, priorities, resources) why it may not be. As Norton et al. (2015, p135) stated

> No matter how useful, knowledge itself is not a panacea for DRR. Decision-making is invariably influenced by conflicting priorities, objectives, and constraints, and not necessarily in all stakeholders' interests or even reflecting their objectives.

This contextualisation of whether and how knowledge is used can be critical at a local level when developing and delivering meaningful equitable participation and learning strategies. While FRM stakeholders may appreciate the importance of local knowledge in their rhetoric, its inclusion in DRR practice can remain more limited (Šakić Trogrlić et al., 2021).

4.8 Conclusions

Building actionable flood knowledge at-a-place is a critical concern in community-focused FRM. This involves being aware of the nature and value of different knowledges (specialist; lay) and forms of expertise while avoiding value judgements. Caution is required in what knowledge is privileged in different decision-making fora, with need for awareness about how different flood knowledge cascades and scales. This includes a strong attention to knowledge system functioning and the value of building and connecting local flood knowledge networks up to a catchment scale. Importantly, thinking beyond linearity has value in establishing how knowledge moves. It is also critical to understand the possible roles of specialist and lay (local) flood knowledge throughout the phases in the DRM spiral, and how these are generated, transmitted, meet, and interact locally. This foregrounds knowledge justice and how different knowledges are viewed, integrated, and applied by different stakeholders, including local communities in learning and decision-making for resilience.

Knowledge systems management is emerging as an important research area, with metaphors like brokering, translation, braiding, and co-production of knowledge benefitting from inter-professional sharing and critique. This involves increased recognition of the values and power dynamics of distinct co-production models. Different academic disciplines and professions contribute valuable perspectives, with a strong potential for FRM to learn from other risk engagement settings like public health, social care, and business. Co-production of knowledge in participatory research and governance needs to be meaningful, with a strong aspiration for equity. Experiments in different ways of integrating knowledge, working with multi-stakeholders, have high value. Emphasis is needed on mutual validation and co-evaluation *with* communities to establish the key enhancers and inhibitors of co-production processes.

Notes

1 Intergovernmental Science-Policy Platform on Biodiversity and Ecosystem Services https://ipbes.net/glossary/knowledge-systems.
2 https://www.pnas.org/sustainability-science.
3 "Dominant narratives", *Inclusive Teaching at U-M*, https://sites.lsa.umich.edu/inclusive-teaching/wp-content/uploads/sites/853/2021/09/Dominant-Narratives-Draft.pdf. Accessed 30 June 2023.
4 https://waternames.wordpress.com/.
5 The SESAME project was funded through the UK Engineering and Physical Sciences Research Council Fund.
6 https://www.waddenacademie.nl/en/news/news-archive-article/in-the-land-of-the-frisians-coastal-identity-in-germany-denmark-and-the-netherlands.
7 https://www.ncl.ac.uk/engineering/research/research-case-studies/water/sinatra/.
8 https://marydhonau.com/about-us/.
9 https://www.communitiesprepared.org.uk/.
10 https://www.paho.org/en/evidence-and-intelligence-action-health/knowledge-translation-and-evidence.
11 https://debensoundings.wordpress.com/.
12 https://www.therrc.co.uk/news/floods-and-dredging-reality-check.
13 https://jamandjustice-rjc.org/blog/theres-no-co-production-panacea.
14 http://www.humanitarianfutures.org/wp-content/uploads/2014/05/CS5-Competency.pdf.
15 https://participatoryactionresearch.sites.carleton.edu/about-par/.

References

Ackoff, R.L. (1989) From data to wisdom. *Journal of Applied Systems Analysis*, 16, 3–9.
Ackoff, R.L. (1999) A lifetime of systems thinking. *The Systems Thinker*, 10, 1–4. https://thesystemsthinker.com/a-lifetime-of-systems-thinking/
Adger, W.N., Pulhin, J.M., Barnett, J., Dabelko, G.D., Hovelsrud, G.K., Levy, M., Oswald Spring, Ú and Vogel, C.H. (2014) Human security. In *Climate Change 2014: Impacts, Adaptation, and Vulnerability. Part A: Global and Sectoral Aspects. Contribution of Working Group II to Fifth Assessment Report of Intergovernmental Panel on Climate Change*. Cambridge University Press, Cambridge and New York, pp755–791.
Alba, R., Klepp, S. and Bruns, A. (2020) Environmental justice and the politics of climate change adaptation – the case of Venice. *Geographica Helvetica*, 75, 363–368. https://doi.org/10.5194/gh-75-363-2020
Alexander, S.M., Provencher, J.F., Henri, D.A., Taylor, J.J., Lloren, J.I., Nanayakkara, L., Johnson, J.T. and Cooke, S.J. (2019) Bridging indigenous and science-based knowledge in coastal and marine research monitoring, and management in Canada. *Environmental Evidence*, 8, 36. https://doi.org/10.1186/s13750-019-0181-3
Alexander, M., Viavattene, C., Faulkner, H. and Priest, S. (2014) Translating the complexities of flood risk science using KEEPER – a knowledge exchange exploratory tool for professionals in emergency response. *Journal of Flood Risk Management*, 7(3), 205–216. https://doi.org/10.1111/jfr3.12042

Allen, B.L. (2007) Environmental justice, local knowledge, and after-disaster planning in New Orleans. *Technology in Society*, 29(2), 153–159. https://doi.org/10.1016/j.techsoc.2007.01.003

Andrews, M. (2002) Counter-narratives and the power to oppose. *Narrative Inquiry*, 12(1), 1–6. https://doi.org/10.1075/ni.12.1.02and

Antonacopolou, E.P. (2008) Actionable knowledge. In R. Clegg and J. Bailey (eds.) *International Encyclopaedia of Organization Studies*. Sage, London.

Argyris, C. (1993) *Knowledge for Action. A Guide to Overcoming Barriers to Organizational Change*. Jossey-Bass Publishers, San Francisco.

Arnold, M., Mearns, R., Oshima, K. and Prasad, V. (2014) *Climate and Disaster Resilience: The Role for Community-Driven Development*. Social Development Department. World Bank, Washington, DC.

Banks, S., Hart, A., Pahl, K. and Ward, P. (2019) *Co-Producing Research: A Community Development Approach*. Policy Press, Bristol, United Kingdom.

Baubion, C. (2015) Losing memory – the risk of a major flood in the Paris region: Improving prevention policies. *Water Policy*, 17, 156–179.

Bielak, A.T., Campbell, A., Popec, S., Schaefera, K. and Shaxso, L. (2008) From science communication to knowledge brokering: The shift from 'Science Push' to 'Policy Pull'. In D. Cheng et al. (eds.) *Communicating Science in Social Context*. Springer Science+Business Media B.V, Berlin.

Bowles, E. (2021) *Practices for braiding Indigenous knowledge systems and Western science for research and monitoring of terrestrial biodiversity in Canada*. Canadian Government Evidence Brief. https://www.sshrc-crsh.gc.ca/society-societe/community-communite/ifca-iac/evidence_briefs-donnees_probantes/earth_carrying_capacity-capacite_limite_terre/bowles-eng.aspx

Briley, L., Brown, D. and Kalafatis, S.E. (2015) Overcoming barriers during the co-production of climate information for decision-making. *Climate Risk Management*, 9, 41–49. https://doi.org/10.1016/j.crm.2015.04.004

Canadian Institutes of Health Research CIHR (2005) *About knowledge translation*. http://www.cihr-irsc.gc.ca/e/29418.html

Canadian Institutes of Health Research (2007) *Knowledge Translation [KT] within the Research Cycle Chart*. Canadian Institutes of Health Research, Ottawa.

Carnelli, F., Mugnano, S. and Short, C. (2020) Local knowledge as key factor for implementing nature-based solutions for flood risk mitigation. *Rassegna Italiana di Sociologia*, 2, 381–406. https://doi.org/10.1423/97838

Cash, D.W., Clark, W.C., Alcock, F., Dickson, N.M., Eckley, N., Guston, D.H., Jäger, J. and Mitchell, R.B. (2003) Knowledge systems for sustainable development. *Proceedings of the National Academy of Sciences of the United States of America*, 100(14), 8086–8091. https://doi.org/10.1073/pnas.1231332100

Chrs, C. (2017) Models, the establishment, and the real world: Why do so many flood problems remain in the UK? *Journal of Geoscience and Environment Protection*, 5, 44–59. https://doi.org/10.4236/gep.2017.52004

Clark, W.C. (2007) Sustainability science: A room of its own. *Proceedings of National Academy of Science Editorial*, 104(6), 1737–1738.

Collins, H.M. and Evans, R. (2002) The third wave of science studies: Studies of expertise and experience. *Social Studies of Science*, 32(2), 235–296.

Cornell, S., Berkhout, F., Tuinstra, W. et al. (2013) Opening up knowledge systems for better responses to global environmental change. *Environmental Science & Policy*, 28, 60–70. https://doi.org/10.1016/j.envsci.2012.11.008

Cvitanovic, C., Cunningham, R., Dowd, A.-M., Howden, S.M. and van Putten, E.I. (2017) Using social network analysis to monitor and assess the effectiveness of knowledge brokers at connecting scientists and decision-makers: An Australian case study. *Environmental Policy and Governance*, 27, 256–269.

Degen, M., Whatmore, S., Hinchcliffe, S. and Kearnes, M. (2003) Re-inhabiting the city: Making vernacular knowledge count in urban wildlife policy and practice. www.open. ac.uk/socialsciences/habitable-cities/published-papers-and-papers-in-progress.php. Accessed 5 May 2012.

Donaldson, A., Lane, Ward, R. and Whatmore, S. (2013) Overflowing with issues: Following the political trajectories of flooding. *Environment and Planning C: Government and Policy*, 31, 603–618. https://doi.org/10.1068/c11230

Durst, S. and Zieba, M. (2020) Knowledge risks inherent in business sustainability. *Journal of Cleaner Production*, 251, 119670.

Facer, K. and Enright, B. (2016) *Creating Living Knowledge: The Connected Communities Programme, Community University Relationships and the Participatory Turn in the Production of Knowledge*. University of Bristol/AHRC Connected Communities, Bristol.

Falkenmark, M. (2004) Towards integrated catchment management: Opening the paradigm locks between hydrology, ecology and policy-making. *International Journal of Water Resources Development*, 20, 275–281.

Fan, M.-F. (2021) Indigenous participation and knowledge justice in deliberative systems: Flooding and wild creek remediation controversies in Taiwan. *Environment and Planning C: Politics and Space*, 39(7), 1492–1510. https://doi.org/10.1177/23996544211044505

Farr, M., Davies, P., Andrews, H., Bagnall, D., Brangan, E. and Davies, R. (2021) Co-producing knowledge in health and social care research: Reflections on the challenges and ways to enable more equal relationships. *Humanities and Social Sciences Communications*, 8, 105. https://doi.org/10.1057/s41599-021-00782-1

Filipe, A., Renedo, A. and Marston, C. (2017) The co-production of what? Knowledge, values, and social relations in health care. *PLoS Biology*, 15(5), e2001403. https://doi.org/10.1371/journal.pbio.2001403

Funtowicz, S. and Ravetz, J. (1993) Science for the post-normal age. *Futures*, 25, 735–755. https://doi.org/10.1016/0016-3287(93)90022-L

Graham, I.D., Logan, J., Harrison, M.B., Straus, S.E., Tetroe, J., Caswell, W. and Robinson, N. (2006) Lost in knowledge translation: Time for a map? *Journal of Continuing Education in the Health Professions*, 26(1), 13–24. https://doi.org/10.1002/chp.47

Greenhalgh, T. and Wieringa, S. (2011) Is it time to drop the 'knowledge translation' metaphor? A critical literature review. *Journal of the Royal Society of Medicine*, 104(12), 501–509. https://doi.org/10.1258/jrsm.2011.110285

Gregory, K.J., Benito, G. and Downs, P.W. (2008) Applying fluvial geomorphology to river channel management: Background for progress towards a palaeohydrology protocol. *Geomorphology*, 98(1–2), 153–172.

Hargadon, A.B. (1998) Firms as knowledge brokers: Lessons in pursuing continuous innovation. *California Management Review*, 40(3), 209–227. https://doi.org/10.2307/41165951

Haughton, G., Bankoff, G. and Coulthard, T.J. (2015) In search of 'lost' knowledge and outsourced expertise in flood risk management. *Transactions of Institute of British Geographers*, 40, 375–386.

Hayek, F.A. (1945) The use of knowledge in society. *The American Economic Review*, XXXV, 519–530.

Heath, C. and Heath, D. (2007) Made to stick: Why some ideas survive and others die. *Journal of Product Innovation Management*, 25(1), 103–105. https://doi.org/10.1111/j.1540-5885.2007.00285.x

Hickey, G., Brearley, S., Coldham, T., Denegri, S., Green, G., Staniszewska, S., Tembo, D., Torok, K. and Turner, K. (2021) *Guidance on coproducing a research project.* INVOLVE; NIHR Update.

Helmholtz-Zentrum Geesthacht (2014) *In the land of the Frisians – coastal identity in Germany, Denmark and the Netherlands.* https://www.waddenacademie.nl/en/news/news-archive-article/in-the-land-of-the-frisians-coastal-identity-in-germany-denmark-and-the-netherlands

Hulme, M. (2007) Understanding climate change – the power and the limit of science. *Weather*, 62, 243–244. https://doi.org/10.1002/wea.108

Intergovernmental Panel on Climate Change (IPCC; 2014) *Climate Change 2014: Synthesis Report Summary for Policy-makers.* https://www.ipcc.ch/pdf/assessment-report/ar5/syr/AR5_SYR_FINAL_SPM.pdf

Intergovernmental Panel on Climate Change (IPCC; 2022) *Impacts, Adaptation and Vulnerability Summary for Policymakers.* https://www.ipcc.ch/report/ar6/wg2/downloads/report/IPCC_AR6_WGII_SummaryForPolicymakers.pdf

Johnson, J.T., Howitt, R., Cajete, G., Berkes, F., Louis, R.P. and Kliskey, A. (2016) Weaving indigenous and sustainability sciences to diversify our methods. *Sustainability Science*, 11, 1–11. https://doi.org/10.1007/s11625-015-0349-x

Jones, R. (2016) Responding to modern flooding: Old English place-names as a repository of traditional ecological knowledge. *Journal of Ecological Anthropology* 18(1). http://dx.doi.org/10.5038/2162-4593.18.1.9

Jones, O., Read, S. and Wylie, J. (2012) Unsettled and unsettling landscapes: Exchanges by Jones, Read and Wylie about living with rivers and flooding, watery landscapes in an era of climate change. *Journal of Arts & Communities*, 4(1), 76–99. https://doi.org/10.1386/jaac.4.1-2.76_1

Kates, R.W. (2011) What kind of a science is sustainability science? *Proceedings of the National Academy of Sciences of the United States of America* (PNAS), 108(49), 19449–19450. https://doi.org/10.1073/pnas.1116097108

Kindon, S. , Pain, R. and Kesby, M. (eds.) (2010) *Participatory Action Research Approaches and Methods.* Routledge, London.

Kuhn, T.S. (1962) *The Structure of Scientific Revolutions.* University of Chicago Press. Chicago.

Landström, C. (2020) *Environmental Participation: Practices Engaging the Public with Science and Governance.* Palgrave Macmillan, Switzerland.

Landström, C., Whatmore, S.J., Lane, S.N., Odoni, N.A., Ward, N. and Bradley, S. (2011) Coproducing flood risk knowledge: Redistributing expertise in critical 'participatory modelling'. *Environment and Planning A*, 43, 1617–1633. https://doi.org/10.1068/a43482

Lane, S.N., Odoni, N., Landström, C., Whatmore, S.J., Ward, N. and Bradley, S. (2011) Doing flood risk science differently: An experiment in radical scientific method. *Transactions of the Institute of British Geographers*, 36, 15–36. https://doi.org/10.1111/j.1475-5661.2010.00410.x

Laudan, J., Zöller, G. and Thieken, A.H. (2020) Flash floods versus river floods – A comparison of psychological impacts and implications for precautionary behaviour. *Natural Hazards and Earth System Sciences*, 20, 999–1023. https://doi.org/10.5194/nhess-20-999-2020

Le Lay, Y.F. and Rivière-Honegge, A. (2009) Expliquer l'inondation: la presse quotidienne régionale dans les Alpes et leur piedmont (1882–2005). *Geocarrefour*, 84(4), 259–270. https://doi.org/10.4000/geocarrefour.7555

Liguori, A., Le Rossignol, K., Kraus, S., McEwen, L. and Wilson, M. (2023) Exploring the uses of arts-led community spaces to build resilience: Applied storytelling for successful co-creative work. *Journal of Extreme Events* https://doi.org/10.1142/S2345737622500075

Liu, C.C. and Falk, J.H. (2014) Serious fun: Viewing hobbyist activities through a learning lens. *International Journal of Science Education*, 4(4), 343–355. https://doi.org/10.1080/21548455.2013.824130

Lower Severn and Avon Combined Flood Group (2008) *Building on the floodplain is misguided (a proof of evidence)*. https://taagroup.co.uk/wp-content/uploads/2021/09/Flood-Report-Revised-2008.pdf Flood-Report-Revised-2008.pdf

Mazumdar, S., Lanfranchi, V., Ireson, N. et al. (2016) Citizens observatories for effective Earth observations: The WeSenseIt approach. *Environmental Scientist*, 25(2), 56–61.

Mazzocchi, F. (2006) Western Science and traditional knowledge. *EMBO Reports (European Molecular Biology Organisation)*, 7, 463–466.

Mazzocchi, F. (2018) Why "Integrating" Western science and indigenous knowledge is not an easy task: What lessons could be learned for the future of knowledge? *Journal of Futures Studies*, 22(3), 19–34.

McEwen, L.J. (2023) Co-production and the role of lay knowledge in community resilience: Learnings for local flood risk management. In J. Lamond, D. Proverbs, and N. Bhattacharya-Mis, *Research Handbook on Flood Risk Management*. Edward Elgar Publishing, Cheltenham, UK

McEwen, L.J., Garde-Hansen, J., Holmes, A., Jones, O. and Krause, F. (2016) Sustainable flood memories, lay knowledges and the development of community resilience to future flood risk. *Transactions of the Institute of British Geographers*, 42(1), 14–28.

McEwen, L.J., Garde-Hansen, J., Robertson, I. and Holmes, A. (2018) Exploring the changing nature of flood archives: Community capital for flood resilience. In A. Metzger and J. Linton (eds.) *La Crue, l'inondation: un patrimoine*. L'Harmattan Publishing House, France.

McEwen, L.J. and Jones, O. (2012) Building local/lay flood knowledges into community flood resilience planning after the July 2007 floods, Gloucestershire, UK. *Hydrology Research, Special Issue*, 43, 675–688.

McEwen, L.J., Jones, O. and Robertson, I. (2014a) "A glorious time?" Reflections on flooding in the Somerset Levels. *The Geographical Journal*, 180, 326–337.

McEwen, L.J., Stokes, A., Crawley, K. and Roberts, C.R. (2014b) Using role-play for expert science communication with professional stakeholders in flood risk management. *Journal of Geography in Higher Education*, 38(2), 277–300. https://doi.org/10.1080/03098265.2014.911827

McGuinness, M. and Johnston, N. (2014) Exploiting social capital and path-dependent resources for organisational resilience: Preliminary findings from a study on flooding. *Procedia Economics and Finance*, 18, 447–455. https://doi.org/10.1016/S2212-5671(14)00962-9

Meyer, M. (2010) The rise of the knowledge broker. *Science Communication*, 32, 118–127.

Monbiot, G. (2014) Dredging rivers won't stop floods. It will make them worse. *The Guardian*. https://www.theguardian.com/commentisfree/2014/jan/30/dredging-rivers-floods-somerset-levels-david-cameron-farmers

Mondino, E., Scolobig, A., Borga, M. and Di Baldassarre, G. (2020) The role of experience and different sources of knowledge in shaping flood risk awareness. *Water*, 12(8), 2130. https://doi.org/10.3390/w12082130

Nakashima, D.J. (2015) *Local and Indigenous Knowledge at the Science–Policy Interface*. UNESCO.

Nakashima, D., Krupnik, I. and Rubis, J. (eds.) (2018) *Indigenous Knowledge for Climate Change Assessment and Adaptation*. Cambridge University Press, Cambridge. https://doi. org/10.1017/9781316481066

National Centre for the Dissemination of Disability Research NCDRR (2005) What is Knowledge Translation? Focus, Technical Brief 10. https://ktdrr.org/ktlibrary/articles_pubs/ncddrwork/focus/focus10/Focus10.pdf

Nicholas, G. (2018) It's taken thousands of years, but Western science is finally catching up to Traditional Knowledge. *The Conversation.* https://theconversation.com/its-taken-thousands-of-years-but-western-science-is-finally-catching-up-to-traditional-knowledge-90291

Norton, J., Atun, F. and Dandoulaki, M. (2015) Exploring issues limiting the use of knowledge in disaster risk reduction. *TeMA – Journal of Land Use, Mobility and Environment*, 8, 135–154. http://www.serena.unina.it/index.php/tema/article/view/3032

Polk, M. (2015) Transdisciplinary co-production: Designing and testing a transdisciplinary research framework for societal problem solving. *Futures*, 65, 110–122. https://doi.org/10.1016/j.futures.2014.11.001

Porter, J. (2009) *Extreme flood outline: Co-producing flood risk mapping and spatial planning in England and Wales*. Unpublished PhD thesis, King's College, University of London.

Rappaport, J. (2000) Community narratives: Tales of terror and joy. *American Journal of Community Psychology*, 28(1), 1–24.

Ratter, B.M.W. and Gee, K. (2012) Heimat – A German concept of regional perception and identity as a basis for coastal management in the Wadden Sea. *Ocean and Coastal Management*, 68, 127–137. https://doi.org/10.1016/j.ocecoaman.2012.04.013

Ravetz, J.R. (1999) What is post-normal science? *Futures*, 31(7), 647–654.

Ravetz, J.R. (2005) The post-normal sciences of precaution. *Water Science and Technology*, 52(6), 11–7.

Ravetz, J. and Funtowicz, S. (1999) Post-Normal Science—an insight now maturing. *Futures* 31, 641–646.

Raygorodetsky, G. (2017) Braiding science together with indigenous knowledge. *Scientific American.* https://blogs.scientificamerican.com/observations/braiding-science-together-with-indigenous-knowledge/

Reddel, T. and Ball, S. (2021) Knowledge coproduction: Panacea or placebo? Lessons from an emerging policy partnership. *Policy Design and Practice.* https://doi.org/10.1080/25741292.2021.1992106

Romm, N.R.A. (2017) Researching indigenous ways of knowing-and-being: Revitalizing relational quality of living. In P. Ngulube (ed.) *Handbook of Research on Theoretical Perspectives on Indigenous Knowledge Systems in Developing Countries*. pp 22–49. https://doi.org/10.4018/978-1-5225-0833-5.ch00

Šakić Trogrlić, R., Duncan, M., Wright, G., van den Homberg, M., Adeloye, A., Mwale, F. and McQuistan, C. (2021) External stakeholders' attitudes towards and engagement with local knowledge in disaster risk reduction: Are we only paying lip service? *International Journal of Disaster Risk Reduction*, 58, 102196. https://doi.org/10.1016/j.ijdrr.2021.102196

Šakić Trogrlić, R., Wright, G., Duncan, M.J., van den Homberg, M.J., Adeloye, A.J., Mwale, F.D. and Mwafulirwa, J. (2019) Characterising local knowledge across the flood risk management cycle: A case study of Southern Malawi. *Sustainability*, 11, 1681. https://doi.org/10.3390/su11061681

Simpson, H., de Löe, R. and Andrey, J. (2015) Vernacular knowledge and water management – Towards the integration of expert science and local knowledge in Ontario, Canada. *Water Alternatives*, 8(3), 352–372.

Sorrentino, M., Sicilia, M. and Howlett, M. (2018) Understanding co-production as a new public governance tool. *Policy and Society*, 37(3), 277–293. https://doi.org/10.1080/14494035.2018.1521676

Spiekermann, R., Kienberger, S., Norton, J., Briones, F. and Weichenselgarter, J. (2015) The disaster-knowledge matrix: Reframing and evaluating the knowledge challenges in disaster risk reduction. *International Journal of Disaster Risk Reduction*, 13, 96–108. https://doi.org/10.1016/j.ijdrr.2015.05.002

Sudsawad, P. (2007) *Knowledge Translation: Introduction to Models, Strategies, and Measures*. Southwest Educational Development Laboratory, National Center for the Dissemination of Disability Research, Austin, Texas. https://ktdrr.org/ktlibrary/articles_pubs/ktmodels/

Sutherland, P. and Nicholson, A. (1987) *Wetland: Life in the Somerset Levels*. Mermaid Books. London.

Talisayon, A. (2008) Tacit Knowledge versus Explicit Knowledge. https://apintalisayon.wordpress.com/2008/11/28/d3-tacit-knowledge-versus-explicit-knowledge/

Tangihaere, T.M. and Twiname, L. (2011) Providing space for indigenous knowledge. *Journal of Management Education*, 35(1), 102–118.

Turnbull, D. (2008) Knowledge systems: Local knowledge. In H. Selin (ed.) *Encyclopaedia of the History of Science, Technology, and Medicine in Non-Western Cultures*. Springer, Dordrecht. https://doi.org/10.1007/978-1-4020-4425-0_8705

Ullberg, S.B. (2018) Forgetting flooding?: Post-disaster livelihood and embedded remembrance in Suburban Santa Fe, Argentina. *Nature and Culture*, 13, 27–45.

Verschuere, B., Brandsen, T. and Pestoff, V. (2012) Co-production: The state of the art in research and the future agenda. *Voluntas*, 23(4), 1083–1101.

Virkus, S. (2014) *Key Concepts in Information and Knowledge Management: Learning Object*. Tallinn University. https://www.tlu.ee/~sirvir/Information and Knowledge Management/Key_Concepts_of_IKM/tacit_and_explicit_knowledge.html

Wagner, W. (2007) Vernacular science knowledge: Its role in everydaylife communication. *Public Understanding of Science*, 16, 7–22.

Wang, T. (2016) Why Big Data needs Thick Data. *Ethnography Matters*. https://medium.com/ethnography-matters/why-big-data-needs-thick-data-b4b3e75e3d7

Ward, N., Whatmore, S. and Landström, K. (2014) Case Study 5: Competency Groups bringing together scientific and local knowledge to develop collaborative flood risk modelling. http://www.humanitarianfutures.org/wp-content/uploads/2014/05/CS5-Competency.pdf

Weichselgartner, J. and Marandino, C.A. (2012) Priority knowledge for marine environments: Grand challenges at the science-society nexus. *Current Opinion in Environmental Sustainability*, 3, 323–330.

Weick, K.E. (1995) *Sensemaking in Organizations*. Sage Publications, Thousand Oaks.

Whatmore, S. (2009) Mapping knowledge controversies: Science democracy and the redistribution of expertise. *Progress in Human Geography*, 33, 587–598.

Whyte, K.P. (2013) On the role of traditional ecological knowledge as A collaborative concept: A philosophical study. *Ecological Processes*, 2, 7. https://doi.org/10.1186/2192-1709-2-7

Wilcox, A.A.E., Provencher, J.F., Henri, D.A. et al. (2023) Braiding indigenous knowledge systems and Western-based sciences in the Alberta oil sands region: A systematic review. *FACETS*, 8, 1–32. https://doi.org/10.1139/facets-2022-0052

Witting, A., Brandenstein, F. and Kochskämper, E. (2021) Evaluating learning spaces in flood risk management in Germany: Lessons for governance research. *Journal of Flood Risk Management*, 14, e12682. https://doi.org/10.1111/jfr3.12682

5

LINKING FLOOD HERITAGE AND LOCAL COMMUNITY RESILIENCE

5.1 Introduction

This chapter explores 'flood heritage' as a concept, process, and practice, and its relationships with building community resilience in local flood risk management (FRM). This territory intersects several academic discipline areas including environmental history, heritage studies, the new 'environmental humanities', new media practices, (changing) archival practice, information management, technology, and the arts. The chapter explores these questions:

- How can 'flood heritage' be defined for communities within a particular locale? How might these explorations develop our understanding of local resilience?
- What is the nature of the 'flood archive' and how is it changing – in what is captured and shared, how, and with whom? What are the implications of these changes?
- What can be learnt from researching how people in a specific locale dealt with flood risk, its uncertainty, and floods in the past?
- (How) can this archival knowledge, as community capital, be protected, accessed, and applied by communities in FRM in the future?

5.2 Flood heritage contexts

Several disciplinary strands need exploration as background to the conceptual framing of 'flood heritage'. Reconstruction of flood and broader environmental histories over decadal and centurial timespans, using documentary evidence and historical accounts, has been a focus of environmental historians, hydrologists, and professional risk assessors (Section 2.2 of Chapter 2). Examples include MacDonald and

DOI: 10.4324/9781315666914-5

Sangster (2017) for the UK since 1750 and Retsö (2015) for Sweden 1400–1800. This work recognised the importance of understanding local catchment flood histories for better flood risk assessment beyond that achievable using shorter gauged, river flow records. However, embedded in historical accounts of physical flood character and damage can be other evidence about the social, cultural, and religious settings for flood impacts (e.g., Dournel, 2016). These play out as complex interactions of physical and human factors at-a-place over time.

Allan et al. (2016, p164) emphasised the importance of "human interpretations of climate history" and value of multi-disciplinary working beyond the physical sciences.

> Climate change has become a key environmental narrative of the 21st century. However, emphasis on the science of climate change has overshadowed studies focusing on human interpretations of climate history, of adaptation and resilience, and of explorations of the institutions and cultural coping strategies that may have helped people adapt to climate changes in the past.

Flood archives can act as valuable evidence bases for developing detailed insights into both societal impacts of extreme floods at a specific time and changing paradigms for understanding human-environment relations (McEwen and Werritty, 2007). Such shifts have had major implications for dominant environmental hazard perception (Palm, 1990), related knowledge, and contemporary flood management strategies (Chapters 3 and 4). For example, in Scotland, debates took place in the late 18th and early 19th centuries on the role of uniformity and catastrophism in landscape evolution (Davies, 1969). This was linked to strong contemporary interest in biblical explanations of natural phenomena like extreme floods, with attribution of flood damage to a higher power (e.g., McEwen and Werritty, 2007; Chapter 3).

Alongside, the arts and humanities have been reflecting critically on their own contemporary relevance (e.g., Holm et al., 2015), with 'environmental humanities' an emergent field of multi-disciplinary enquiry. Framing of 'blue humanities' has a specific focus on 'knowing water', including oceans and dams,[1] in different ways within modern Western culture through arts and literature (e.g., Gillis, 2013; Campbell and Paye, 2020).

5.3 Framing 'flood heritage'

A close, complex, non-linear relationship exists between memory, heritage, and identity. Framing the concept of 'flood heritage' needs cognisance of a fundamental dichotomy in understanding 'heritage' that has fuelled debate (McEwen et al., 2017). One perspective is that of a conservative, nostalgic project which deploys a romanticised and idealised view of past to reinforce old certainties at times of significant change (Lowenthal 1985, 1988; Hewison, 1998). An alternative more optimistic perspective recognises a more democratic form of heritage – emphasising 'spirit of local places' (Samuel 1994, p158). From this second perspective, it is not memorialisation that is celebrated but 'memorialism' (Dicks 2000), evidenced in attempts by local communities to make and maintain their own heritage. Smith (2006) stressed fluidity of both heritage and identity as an active process or relationship; she argued strongly for understanding of heritage experience as 'something vital and alive … a moment of action' (2006, p83). This moved the debate over the value, ownership, and nature of heritage. In *Heritage from Below*, Robertson (2012) explored uses of the past in the present only minimally related to the economic; such heritage can function as cultural resources for local 'memoryscapes'.

A critical concern within flood heritage is the nature of archival evidence about local flood histories, their human and physical impacts, changing social vulnerability, and adaptive strategies for living with water excess. This takes a longer-term view on civic agency for flood resilience. For example, research has focused on floodwater, water management, and heritage (e.g., Hein, 2020), recognising the value of indigenous (direct experiential) knowledge of how water meadows work, within traditional maintenance of European water meadows (Renes et al., 2020; Chapter 4).

The cumulative record of floods at-a-place over time comprises the local 'flood archive' in all its forms – as evidenced in the physical landscape (geomorphic, sedimentary, ecological), the built environment, and the human record with scientific and narrative resources (archival, oral history, art), artefacts (physical, visual, sensory), and embedded flood knowledges that can be drawn on as resources for social learning by local communities. Such place-based knowledge can support individual and collective resilience to routine and severe events, and in wider living with uncertainty. This is a complex and multi-faceted domain, critical in understanding how past climate and its extremes have been constructed by local communities. As Hulme et al. (2009, p198) reflected

> For people living in particular places and particular cultures, climate is constructed as a function of their experiences and memories of past weather events, and what is socially learned from previous generations. These climates may often be reified through paintings or photographs of physical markers, such as a flood, drought, or a rare snowfall.

The potential exists to draw across different types of archival sources in reconstructing the nature and impacts (physical, human) of extreme floods, but also in establishing longer-term local and regional flood histories. Benito et al. (2015, p3517) noted:

> In the last two decades, there has been growing scientific and public interest in understanding long-term patterns of rare floods, in maintaining the flood heritage and memory of extremes, and developing methods for deterministic and statistical application to different scientific and engineering problems.

This chapter now considers physical flood archives before exploring other archival forms.

5.4 Physical flood archives

5.4.1 Reconstructing local flood histories

Researching local flood histories within river catchments can contribute to both scientific flood reconstruction and longer-term understandings of community resilience (e.g., McEwen, 2006). Such reconstruction, using documentary archival sources but also correlating physical sedimentary archives of rivers and lakes, has been progressed to extend shorter gauged flow records. One reason for detailed reconstruction of flood archives is to gain more realistic estimates of the probabilities of the most extreme floods on record – of particular interest to the insurance industry, flood zoning, and in design of flood protection measures (e.g., Benito et al., 2004; Kjeldsen et al., 2014; Blöschl et al., 2019; Chapter 2). Over the past 20 years, several European projects have worked to extend and analyse flood records at regional, national, and international levels (e.g., EU SPHERE project; Thorndycraft et al., 2002). The British Hydrological Society's *Chronology of British Hydrological Events*, a public repository launched in 1998, is one such crowd-sourced database of archival records on floods and droughts (Black and Law, 2004), contributed by volunteer scientists and enthusiasts. This contrasts with research-led archives like the European Floods Database[2] based on monitored gauged flows (1960–) for use by specialist hydrologists (Hall et al., 2015).

Such archival sources provide quantitative and qualitative information about historic floods (their magnitude, frequency, seasonality, generating characteristics, and intensifying factors). When drawing on archival evidence for scientific or lay investigation of floods, it is important, wherever possible, to triangulate and validate different sources to build as reliable and accurate a picture as possible of individual flood events (see PAGES – Past Global Changes Floods Working Group[4]).

5.4.2 Flood archives in the physical landscape

Extreme floods can leave a significant imprint on the physical landscape when large volumes of high-calibre sediment are transported, and new river channel alignments carved. The scale and character of impact depend on the nature of the river environment (its energy, controls such as channel slope, and volume and accessibility of sediment, etc.). In many cases, however, sedimentary archives of past floods can be hidden buried, with discrete layers of alluvial sediments laid down on floodplains, and within lakes and marine environments (Figure 5.1).

Coring and dating of such sediments (e.g., through radiocarbon dating of organic matter) can reveal detailed flood histories over 100s and 1000s years (Davis et al., 2018), or the timing of catastrophic storm surges and tsunami (e.g., Haslett and Wong, 2021). Such buried evidence can be concealed from community awareness

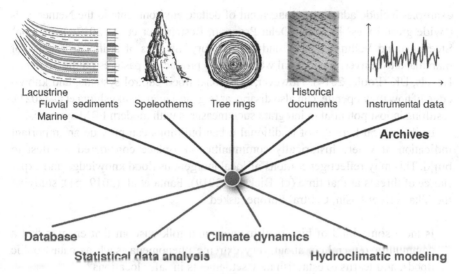

Lacustrine
    Fluvial sediments   Speleothems   Tree rings   Historical documents   Instrumental data
Marine

Archives

Database   Climate dynamics

Statistical data analysis   Hydroclimatic modeling

FIGURE 5.1   Data on past floods collected from a variety of archive types, integrated into a database, statistically analysed, and climatically modelled and interpreted (Permission: PAGES Past Global Changes Floods Working Group http://pastglobalchanges.org/science/wg/floods/intro)

and understanding, although erosion by subsequent extreme events sometimes exposes these sedimentary archives. Visual evidence may need specialist interpretation; an important skill is the ability to 'read' a landscape for signs of changing river processes and occurrence of past floods. This is the territory of fluvial or flood geomorphologists and 'stormy geomorphology' (Naylor et al., 2017). The calibre of boulders transported can affect the persistence and visibility of flood-created landforms, and whether they are reworked in subsequent lesser flood events (e.g., McEwen and Werritty, 1988). For example, huge boulders perched on Icelandic outwash plains evidence the occurrence of episodic jökulhlaups (major glacial ice burst floods; Chapter 2). Of course, flood-deposited boulders may be artificially cleared away off floodplains in recovery to allow the return of agriculture (e.g., in 19th-century Scotland after the 1829 'Muckle Spate'[3]; Lauder, 1830); that record is then erased from the landscape for future communities.

5.4.3   Flood heritage within built environments and human-made landscapes

Much can be learnt about physical flood resilience[4] from past building and land management strategies. For example, built heritage is itself resilient by the fact it has survived. Long histories of managing floods can be captured in physical landscape archaeology on low-lying floodplains, evidenced as relict structures or old management practices (e.g., dikes, rhines, canals and drainage ditches, sluices, etc.). Good examples include adaptive management of deltaic environments in the Netherlands ('wide green dikes' in Dutch Delta Program, Restemeyer et al., 2017; Van Loon-Steensma and Vellinga, 2019), and 19th-century histories of artificial drainage of marshland to convert low coastal wetlands into productive pastures on the Somerset Levels, UK (Poole, 2014). However, traditional flood-control systems, and knowledge of how they operate, can also disappear (e.g., in Japan, Itsukushima et al., 2021), resulting in lost potential to integrate such measures with modern FRM systems.

Built cultural heritage of traditional urban buildings can provide an important indication of where historically communities at-a-place considered it safest to build. This may reflect generations of local indigenous flood knowledge and experience of threats at that time (cf. Bicksler, 2019). Fanta et al. (2019, p1), studying the Vltava river basin, Central Europe, asked

Is there some kind of historical memory and folk wisdom that ensures that a community remembers about very extreme phenomena, such as catastrophic floods, and learns to establish new settlements in safer locations?

Working with an evidence base of 1,293 settlements originating over nine centuries with seven severe floods, they explored relationships between water as risk and resource in selecting locations to settle. They found, however, that while the next generation used knowledge of past extreme floods in choosing that location, flood memory faded after two generations. They concluded that such memory was dependent on human witnesses (Chapter 3), posing questions about how floods are materialised in the landscape and kept alive in collective or communal memory.

Past coping and adaptive strategies can also be reflected in historic structural approaches to building. Traditional low-technology coping measures were incorporated within buildings in preparation for impending floods, for instance, flood boards placed across portals to resist water coming into properties. Such adaptations provide everyday visual indicators of flood risk. Another form of physical evidence is the resilient ways that houses were built to minimise damage with water egress (e.g., with slab stone floors; lime plaster that dried out quickly), as found in medieval properties along the River Severn, UK. However, in the 21st century, physical evidence of property-level flood protection (PLP) can mean different things in the public psyche: a resilient community living with floods or a flood-risk area where it is potentially difficult to secure insurance or sell property. Hence, physical evidence of flood risk to buildings (e.g., flood marking) can be deliberately removed or hidden.

5.5 Cultural flood archives and flood archiving

Sources that reveal human elements of the flood archive vary with locality, region, nation, and timescale, depending on local archiving practices and factors like governance histories, histories of political stability or unrest, demographics, culture, and changing archival and arts practices. Archiving of flood knowledge can embrace written, oral, and visual artefacts, including literature, poetry, painting, and song. McEwen et al. (2012) reflected on the changing nature and persistence of flood archives over time, noting that floods are materialised in increasingly varied ways and settings. This in turn supports more diverse flood archives, and a spectrum of archival practice from "scientific, objective, mechanical and validated, to creative processes and artistic practices" (p48). Such archiving can be considered as:

> process, engagement, and recovery – in the past, present and future - interwoven around the theme of recording changing knowledge, memory and power relationships in relation to floods, communities, and other stakeholders.
>
> *p63*

Flood archiving practices influence sharing of local knowledge. However, what was recorded or not within the flood archive reflects the power dynamics of that time, and whose voices were captured and now sit protected (e.g., in government archives). In the 18th and 19th centuries in the UK, those documenting floods were frequently scientists, engineers, religious leaders, and landowners with an interest in recording unusual events and their impacts on their land and property. In such archives, minority voices – by gender, class, or ethnicity – were more rarely captured. Exceptions, giving perceptions 'from below', might lie within witness statements provided by local people stored in legal archives (e.g., in prosecutions for dike breaking on the Somerset Levels, UK; Poole, pers. comm.; flood damage reports for tax reduction from Norwegian farming communities; Engeland et al., 2018). This poses questions about whose flood histories and heritage are evidenced within flood archives. Carey (2012) discussed the importance of uncovering more diverse climate meanings and narratives (e.g., by race, class, and gender) in understanding climate history. This would allow insights into the social relations and power dynamics that affected past interactions between humans and floods.

Archiving is a dynamic and evolving practice with technological and social-cultural change. Formal archives are now often renamed history or heritage centres. Digitisation of catalogues and artefacts has a "democratising tendency" (Poole, pers. comm), making local heritage resources, like newspaper flood accounts, more accessible to public audiences. More recently, new forms of marking floods and archiving flood knowledge involving technology and social media (Youtube, Flikr, and Twitter) have gained traction, democratising both archives and the archivist so that individual citizens have the power and skills to adopt that role. Ubiquitous technological advances in devices for capturing stories and eye-witness testimonies, recording visual imagery (flood photographs and video resources), alongside the desire of citizens to share experiences through online tools and social media, have led to a vast archive of citizen-generated content or 'flood data'. Such archives with volunteered geographic information and citizen observation have potential for other uses in science and community resilience (Chapters 8 and 10).

This poses questions about the implications of new ways of archiving for flood memory and experience. Derrida (1996, p18) proposed "what is no longer archived in the same way, is no longer lived in the same way". Within new media practices, personal and community archival material can now be shared online locally and globally for solidarity in risk and events. These capitalise on new means of recording and distributing flood information (e.g., through online digital archives and mobile technologies). However, a major concern for social learning of present and future generations is the persistence or transience of early 21st-century flood archives with these technological advances. Will these archives (e.g., on tape, CDs) be accessible in the near and far future like paper-based archives recounting 17th- and 18th-century floods within government archives? Additionally, informal visual flood archives are often created for other reasons, e.g., as personal records for flood insurance claims, and may only be shared within families.

5.5.1 Physical flood marking as an archive

Traditional forms of archiving include acts of flood marking in the landscape (MacDonald, 2007). 'Flood marking', in its most literal sense, is the physical marking (or 'epigraphic marking') of maximum instantaneous flood levels in public or private places like bridges and buildings (Figure 5.2A–G; see also Figure 4.4).

FIGURE 5.2   Examples of flood marks in different cultural settings (composite by author): (A) 'Flommerker' showing 1789 flood ("Storofsen"), Lalm, Norway; (B) 'Wassermerke' (1594–) at Schloss Ort Castle, Lake Traunsee, Gmunden, Austria; (C) Watergate (1770–), Worcester, UK; (D) Upton on Severn (1947 flood marker corrected); (E) La Fleix (1728–), Dordogne, France; (F) Exceptional high tide 1883, Chepstow, UK; (G) Flood marks at the west entrance to King's Lynn Minster, King's Lynn, Norfolk, England

Such markers, when prominently placed or locally known, provide local communities with evidence to differentiate routine from extreme floods at-a-place. The proliferation of such physical markings varies with cultural traditions. For example, in some European countries, practices of marking long flood series on old river bridges date back to the 17th century (e.g., on the Danube and Loire). Norwegians mark extreme floods within and on the landscape with 'Flommerker'; the Austrians with 'Wassermerker' (Figure 5.2). French culture, with its strong sense of Patrimoine or the value of collective heritage (Metzger and Linton, 2017), has traditionally marked formal flood 'lignes' on buildings and bridges in even small villages (e.g., the River Dordogne, France; Figure 5.2). Such marking can also be considered a voice or call for action from the past that warns local communities in, and for, the future. For example, Japanese tsunami stones alert communities of past inundation levels for future building.

Key in building and protecting local knowledge is the relationship between private (informal) and public (official, formal) marking. Physical flood marking can have variable persistence, depending on whether marks are legally protected or are deemed controversial and removed by citizens in active forgetting of local environmental knowledge within communities. A different type of marking is cognitive or perceptual marking, where citizens monitor and rank floods relative to their known local points of reference (Chapter 4.4). Such environmental knowledge also varies in the extent of its sharing.

5.5.2 Oral history and flood stories

Oral history (Vansina, 1985), or recording of accounts from people with personal knowledge of past events, can capture routine 'living with water' and experience of extreme floods to support adaptive learning by communities and other FRM stakeholders. In the former, a watery sense of place can be encapsulated through oral histories and archival accounts of regular floods. For example, this sense of flooding and place is captured in poetry and imagery in *In Time of Flood* by Crowden and Wright (1996). In the latter, oral histories represent a further resource for the local flood archive, documenting severe impactful events and their aftermath, recording rare experiences and meaning making, negotiating conflicting emotions such as fear and awe, and providing insights into how those narrating at the time understood extreme events like floods and the natural world. Academic institutions can have key co-ordinating roles; examples include The Centre for Oral History and Cultural Heritage (University of Southern Mississippi), which documented impacts of Hurricane Katrina (in 2005) within the Mississippi region, USA (Sloan, 2008).

There are accompanying methodological and ethical issues in how and when in the rawness of flood experience, oral histories are collected. Sheftel and Zembrzycki (2016, p338) cited two fears of oral historians: "the fear of failing as researchers and the fear of failing our narrators and doing harm". Oral historians garnering recordings in flood risk settings can be formally trained professional external observers, local amateur historians embedded within their communities, or individuals or organisations simply motivated to capture accounts of an exceptional event while it is still fresh in their community's memory. Examples include: The 2015 South Carolina Flood Oral History Collection[5] and the UK's Women's Institute after *The Gloucestershire Floods 2007* (Thomas and Wilson, 2007). Timing of capturing accounts can affect focus and content; an immediate description of impacts may be the initial emphasis, with critical reflections on flood experiences often requiring elapsed time.

Again, there may be issues about whose voices are captured. Oral historian Sloan (2008, p182), in 'Shouts and Silences', reflects on the 'power' and 'value' of oral history in garnering individual experiences:

> As oral historians, we often argue that the work we do is important because established frameworks of history fall short — at worst they are flawed, and at best they are embedded with significant silences. Why wait for these structures to do our work? Rather, let us be a larger part of the process.
>
> The worth is not in the narratives' power to achieve an immediate level of synthesis or greater meaning, but in their power to glean out individual experience and meaning, even in raw form.

With memory, there can also be emotional trauma. Sloan (2008, p178) also reflected on the oral historian's role in "working between tragedy and memory".

> It is a precarious point in time to conduct oral history. There are ethical issues involved, from discounting loss to compounding grief. It is an invasive exercise, the oral historian stepping in while others are trying to put their lives back together. [...] Working at such a moment requires more of us as professionals.

In some accounts, oral history and reliving events are conceived as part of a cathartic, healing process in recovery. For example, Eugene's (2006) *Bridges of Katrina: Three survivors, one interview* reflected on connecting to a moment through stories. This process required concerns for ethics and responsibility, as Sloan (2008, p185) reflected personally after his process of capturing stories:

> Somebody asked me what I was going to take from this experience, and even though these people have nothing to give, they have given me a lot. And the one thing that I am gonna take from here is their stories. And I am going to tell their stories and I promised them, every one of them, that they will never be forgotten again. And I will not forget them, and I will make sure other people do not forget them.

Oral history accounts have been used as powerful stimuli for performance theatre shared as lived knowledge within and between communities (e.g., The

Katrina Project: 'Hell and High Water'[6]). 'The Caravan', an intimate documentary verbatim theatre performance, shared personal experiences of "ordinary suffering and stoical acceptance" in recovery from the UK 2007 floods (Moran, 2009, np) within and beyond those affected communities.

At differing scale, and to connect flood risk communities to past flood memories and local knowledge of their older residents, the Multi-Story Water project[7] captured oral accounts of the extreme 1968 flood, Bristol, SW UK. These oral histories became stimuli for a sited performance that aimed to engage local at-risk communities, without recent flood experience, with the opportunities and risks of living with their river and environmental change[8] (Bottoms and McEwen, 2014).

Hence, oral histories can provide valuable insights into coping and adaptive strategies in FRM and capture diverse voices on what 'flood recovery' can mean – for example, in terms of property, emotions, and community functioning. Sharing these insights within communities can promote social learning to improve resilience. When shared beyond communities, this can also support learning within FRM agencies to help understand the needs of specific place-based communities.

5.5.3 Newspaper accounts of flood impacts

Over longer periods, accounts of floods and their impacts in newspapers and other historical print media (e.g., small pamphlets like chapbooks and broadsides) can be used to extend local flood archives back into the 18th and 19th centuries. While common issues like syndication exist, newspaper analysis can potentially be used where empirical data on hazard impacts, and individual and community memory, is absent. Indeed, the growing proliferation of local weekly and daily newspapers between 1760 and 1860 was responsible for wide distribution of flood knowledge across geographies in ways that could never have happened previously (Poole, pers. comm.). Newspaper articles can provide insights into dominant perceptions and discourses of the time, and public framings of floods. Timing of these articles can influence dominant media narratives, whether promulgating blame and denial, or promoting community resilience (Chapter 3).

Care needs to be taken in interpreting such potentially subjective sources and their framing of (flood) disaster news. For example, Rashid (2011) in *Tale of two floods in the Red River valley, Manitoba, Canada* (in 1950 and 1997) used content analysis of newspaper discourses to establish how newspapers reported and contextualised flood impacts. He found regional newspapers, through in-depth local reporting, provided rich contexts about flood causes, impacts, and management issues, as distinct from de-contextualised national reporting. There is a need to explore relationships between public flood risk perceptions and such locally 'rich', contextualised media reporting (historical and contemporary) and community risk communication (Chapters 3 and 6).

Newspaper contributions to the flood archive can also encompass photographic evidence (e.g., picture spreads from Somerset in Illustrated London News in

1920s – Poole, pers. comm.). Such visual imagery can act as valuable resources for community engagement, reminiscence, and social learning around historical floods. However, the status of historic 'staged' photographs, captured by official media, can be contested by residents as an equitable and reliable flood record. The professional photograph also dictates what is seen (recorded) and unseen (unrecorded).

Attention also needs to be given to impacts of news media reporting during flood disasters, along with its responsibility in defining and accurately depicting different communities at that time. Ganje and Kenney (2004) researched a regional newspaper's photographic representation of an area's minority community over two months immediately before, during, and after the 1997 flood in North Dakota and Minnesota. The "subtle messaging" was described as "stereotypical" and "misleading" about who was within a community, reinforcing "a prevailing image of minority people as untrustworthy, transient and community outsiders" (p78).

5.5.4 Visual media and flood artefacts

Photographic and film archives provide another form of evidence of cultural flood memory. One example is photographic archives[9] capturing the devastation from the catastrophic St Johnstown flood of 1889, caused by a dam failure in Pennsylvania, USA (Chapter 2). In another, the US American Army Corps of Engineers' photographic archives of the extreme 1927 Mississippi flood – a major disaster within American history – provide evidence of environmental injustices, showing African-American plantation workers forced to work shoring up levees, the huge scale of displacement, and the favouring of white Americans in evacuation and rescue efforts. Figure 5.3 provides examples that capture the scale, impact, and social dynamic of that flood (see also Barry, 1997; Rivera and Miller, 2007).

Floods have also been archived through film in the 20th century, using historic footage that captures perceptions of key issues at the time (e.g., conflicts, community tensions, and inequalities). Examples include the documentary of the 1927 Mississippi floods – "Fatal Flood",[10] drawing on oral history accounts and aerial flood film footage[11] (1927) by International Newsreel and showing disintegrating levees, tent colonies, and Secretary of Commerce Hoover's visit. This captures in 2.39 minutes the disaster as portrayed and broadcast by contemporary media.

> When that river gets to the top of the levee is lapping over, it is guarded. Arkansas people guard the levee against Mississippi people; Mississippi people guard the levee against the Arkansas people. Afraid somebody will blow the levee and turn the water loose on the other side.
>
> *Irene Leblouhs, resident, Transcript, Fatal Flood[12]*

Historic film footage of the same flood in "The Great Betrayal"[13] shows the dynamiting of a levee to try and save the city of New Orleans from flooding but that instead flooded primarily local African American parishes downstream (Rivera and Miller, 2007; see 'flood poetry' below).

FIGURE 5.3   During the 1927 flood: (A) A refugee camp along Mississippi River levee, Greenville, Mississippi; (B) Depicts prisoners brought in to place sandbags on Mississippi River levee (both reproduced by courtesy of USACE)

Film is also used as an important cultural marker in active remembering that supports governmental adaptive strategies. For example, historic video footage of the human impact of the 1953 storm surge in The Netherlands – the trigger for the national Dutch flood infrastructural development of the Rhine Delta Works scheme – is shared in the Watersnoodmuseum[14] (flood museum). The footage sits along with 20,000 photographs to materialise and share national cultural flood memory (Table 5.1).

Preservation of authentic physical artefacts of resilience ('Thing power'; Bennett, 2004) can also capture events, evoke memory, and promote awareness in different ways. These can represent flood heritage in museum collections – in debris indicating flood impacts on livelihoods and the nature of rescue attempts (e.g., a quilt in Johnstown Flood Museum, USA; Table 5.1). In the exhibit 'Flooded House Museum', in a neighbourhood of New Orleans, USA, local artists in 2018 recreated a post-flooded home[15] ("what evacuees came back to") to commemorate Hurricane Katrina (2005) for residents and visitors. Artefacts preserved in museum collections can also capture adaptive strategies for living with water, e.g., traditional flat-bottomed boats and drain-ditching equipment used on UK Somerset Levels.

TABLE 5.1 International Examples of Flood Museums Promoting Flood Heritage

Museum and event commemorated	Learning resources
Watersnoodmuseum,[a] Rhine Delta Scheme – "National Knowledge and Remembrance Centre for the Floods of 1953"	"The extraordinary stories of ordinary people about their experiences with water". Involves cultural messaging – 'Keep our feet dry'. Connects children to 1953 floods through Youth Flood Museum
Johnstown Flood Museum, Pennsylvania, USA[b] commemorating the 1889 dam-burst flood that occurred during extreme storm	Provides "a story of great tragedy, but also of triumphant recovery" Interactive maps; flood artefacts 'recovered'; "a quilt used to drag survivors to safety, a set of keys, a trunk, various household items recovered from the debris, and even a bottle of floodwater"
Lynmouth Flood Memorial Hall, UK (permanent exhibition remembering 1952 flash flood; Chapter 2)	Includes scale model of village pre-flood, along with images of buildings destroyed and how to identify their sites Personal accounts, photos, and material on theory that storm may have been result of military cloud-seeding experiments

[a] https://watersnoodmuseum.nl/en/.
[b] https://www.jaha.org/attractions/johnstown-flood-museum/.

5.5.5 Memorialisation and community memory

Zavar and Schumann (2019, p157) observed: "memorials distilling survivor memories impel community recovery differently than memorials that reconstruct imagined pasts". Flood memorialisation can take place formally as public monuments and informally through community agency. Examples of national memorials in the USA include The Flood Memorial, Grand Forks, North Dakota (with flood levels marked), and Johnstown Flood National Memorial, Pennsylvania. Such memorials can be constructed by governments immediately post events or later to mark anniversaries remembering the loss of life. There can be strong cultural preferences to memorialisation practices; Monument Australia (monumentaustralia.org.au) records 53 Australian examples for floods (e.g., 100th anniversary of 1916 flood, Clermont, Queensland), within a large range of other monuments to disasters (earthquake, drought, fire, storms), culture, and conflicts. Frequently, the physical structure can be linked to extreme flood levels, sometimes with symbolism. In Clermont, the monument is shaped like a tree that, in rapidly rising floodwaters, helped save lives.

But what of less formal memorialisation in, for, and by communities within unsettling landscapes where risk has been realised? Community-led initiatives and petitioning for memorials in active remembering may occur after some time has elapsed. Examples include campaigning for a memorial 44 years on to remember those who died in the 1953 storm surge flooding on Canvey Island, in the Thames estuary, UK, that sits below sea-level with seawall protection. After the Lockyer Valley flood in Australia (2011), the term 'flood memorial' was used both to express a ceremony – a community coming together to remember and look forward – and to represent a physical place and construction of a sculpture.[16] "After the past"[17] – created in local sandstone to commemorate these floods – was purchased from a government-funded community recovery package, and inscribed with this message by the sculptor

> The tragedy shrunk our hearts along with our hopes and dreams but we should never forget, as the past will always be present. Through this adversity we remember the future is yet to be built. Tears will not drown our hope! Motivation and effort will shape our future dreams. After the past, we will keep the spirit alive.
>
> *Jose Carlos Millan*

Additionally, 'guerrilla' acts of flood memorialisation can also occur alongside formal heritage-making or fill its gap in public spaces within flooded communities. These can be disrupting and contested but potentially complement the formal (cf. Muzaini and Minca, 2018). Some later become permanent and formalised. Examples include "Houston Strong" flood graffiti by visual artist Nicky Davis[18] 2017 after Hurricane Harvey, USA; Figure 5.4).

FIGURE 5.4   "Houston Strong" by Houston mural artist Nicky Davis 2017 (reproduced with permission)

5.5.6 Artistic and creative archiving practices

Art, both high and folk art, can capture and mark historic floods as a social archive, or be inspired to record them later. This is seen in painting, literature, poetry, and song, often interwoven in evidence and stimuli. Good examples of visual narratives include Sir Edwin Landseer's 1860 painting 'Flood in the Highlands' (Figure 5.5A) – inspired by the 'Muckle Spate of 1829' in north-east Scotland – a romanticised image 'mythologising the natural'. A contrasting example of a visual essay on the power of nature is Erik Paulsen's painting of Sarpsfossen during Storosfen, the extreme 1789 flood during the Little Ice Age, Norway (Figure 5.5B; Roald, 2013). Unlike the Landseer, this painting says nothing about the human cost of flooding but reflects the ability of the cultivated middle class to rationalise and aestheticise it (Poole, pers. comm.).

A contemporary woodcut in a 17th-century chapbook about the catastrophic 1607 floods[19] in the Severn estuary, UK, emphasises long-standing recognition of the power of visual media in community engagement (Figure 5.6). However, visualisations of catastrophic 17th-century floods sometimes involved repurposing generic images with their local distinctiveness removed, so promoting "homogeneous causes, effects, and interpretations" (Morgan, 2015, p146). For example, this woodcut image was later culturally attuned to represent the 1651 Rhine floods in Germany.

Contemporary literature and poetry can also capture critical local flood narratives in their cultural settings. 'Flood poetry', as a cultural marking of events and issues, is well showcased in the USA, where it was used to capture the environmental injustices badly affecting African Americans during the 1927 Mississippi floods. Sterling Brown (1932) in *Southern Road: Poems* shared six poems focused on "the natural and political disaster" (Johnson, 2017, p116).

> *Little Muddy, Big Muddy, Moreau, and Osage,*
> *Little Mary's, Big Mary's, Cedar Creek,*
> *Flood deir muddy water roundabout a man's roots,*
> *Keep him soaked and stranded and git him weak.*
>   Riverbank Blues[20]; Sterling A. Brown (1932)

In such cultural settings, folk poetry and song can be intimately interlinked, forming an important informal flood archive (see poet Dave Reeve's 'Flood Songs for an Unmade Album'; Figure 5.7).

Songs in the American Blues tradition also captured the differential flood impacts within the Mississippi Delta. As floodwaters rose in 1927, many blues artists were inspired to write songs that described experiences of being in an exceptional flood, and environmental justice issues (Mizelle, 2014; Monge, 2022).

FIGURE 5.5   (A) "Flood in the Highlands" (1860), by Sir Edwin Landseer, oil on canvas (reproduced courtesy of Aberdeen Art Gallery and Museums collections); (B) "Sarpsfossen in Norwegen" (1789) by Eric Paulsen, oil on canvas (National Gallery of Denmark)

FIGURE 5.6 Woodcut of 1607 coastal storm surge on the Severn estuary UK (source: Title page of Lamentable newes out of Monmouthshire in Wales, an English news-book of 1607; copyright Stephen Rippon)

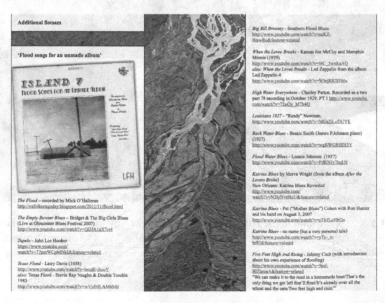

FIGURE 5.7 "Flood Songs for an Unmade Album" (artwork credited to Sage Brice, 2012; playlist credited to David Reeves, 2012; McEwen et al., 2012; background photo credit to Richard Droker (www.flickr.com/photos/83432882@N05/)

The legacy is over 30 Delta blue songs relating to that flood disaster[21]; exemplified by Lonnie Johnson's (1928) 'Broken Levee Blues'

They want me to work on the levee, I have to leave my home. They want me to work on the levee, I have to leave my home. I was so scared the levee might break and I may drown.

or Charlie Patton's (1929) 'High Water Everywhere'

I would go to the hilly country, but they got me barred.

Evans (2006), an ethnomusicologist researching how blues music depicted the 1927 flood disaster, observed, however, how these songs provide varied commentary.

The ones by the few artists that were from the area, who might have actually experienced the flood (like Charlie Patton or Alice Pearson) tend to be the most realistic in their descriptions, the most accurate in their details. Some of the others are inaccurate, based on hearsay, some sentimentalize the flood, some even trivialize it.[22]

This style of song writing has influenced others in their recording of floods in different cultural settings (Figure 5.7).

Floods have also been marked through performance narratives. An example is Vignes (2008) who, as a storyteller in a solo performance script, reflected on "performing ethnography" and "cultural memory" in "Hang it out to Dry" about Hurricane Katrina (2005) in Charrette, Louisiana. As a participant and observer during the floods, she suggested that telling the story of "multiple voices of lived experience can offer insight into the larger community" with implications for its recovery. Her performance depicted the "struggles, hopes, and fears" that many community members were still facing in recovery, several years after the storm (Vignes, 2008, p344).

The significance of this performance speaks to the timeliness of generating new forms of Katrina narratives. Hurricane Katrina consumed our Gulf Coast over two years ago and already the evaporation of stories and interest is in play. This performance traces many events and the stories of those who bore witness. Teller and listener engage the story and enrich the possibility of the stories being transmitted by others. The story is ever-changing and never-ending.

Vignes (2008, p349)

Other art forms can also visually mark floods. For example, sculpture can commemorate disasters within community settings (e.g., Cwmcarn dam burst disaster of 1875, Wales; Figure 5.8).

FIGURE 5.8   Steel bench sculpture as memorial to the Cwmcarn flood victims (1875), Ebbw valley, Wales, UK (photograph copyright author)

5.6 A place-based, flood heritage approach to building community resilience

Some floodplain, deltaic, estuarine, and coastal communities, like those along the Mississippi, USA, the Rhine delta, Netherlands or the Severn estuary, UK, have extensive flood archives for specific floods that draw across multiple sources, reflecting a strong 'watery sense of place' (McEwen et al., 2016). Here, it is important to validate across sources, voices, and experiences. In other settings, the flood archive may be more hidden and limited.

Key questions are what knowledges for resilience are contained in such flood archives, their degree of community connection, and the accessibility of their knowledge for present-day communities in social learning for resilience (Chapters 4 and 9). Where there has not been an extreme flood locally in living memory, establishing and sharing the nature and impact of historic events can have high value in (re)connecting local communities with their flood heritage.

So how can a flood heritage approach to building community resilience at-a-place over time be developed that draws on local flood archives? This involves engaging communities with their (changing) relationships with floods, and local folk memories and common wisdom from learning to live with risk and actual floods – of both short-term coping and longer-term adaptive strategies. There is potential for an engaged arts and humanities approach to work alongside that of the natural and social sciences to deepen our understanding of relations between watery memories and resilience to the wet (McEwen et al., 2012, 2014). A place-based, flood heritage approach integrates the following characteristics (Box 5.1), with strong emphasis on community embedding in research and practice.

BOX 5.1   CHARACTERISTICS OF A PLACE-BASED, FLOOD HERITAGE APPROACH

- Brings together disparate conceptual frameworks around lay knowledge, a watery sense of place, and heritage from below.
- Values local knowledge and develops its links with sustainable flood memory (Chapter 4; see below).
- Draws on interdisciplinary expertise, bringing together environmental science, cultural geography, memory/media, heritage, creative arts, etc.
- Is transdisciplinary and inter-professional, co-working with the local cultural sector.
- Links concern for present, past, and future risk, and longer-term approaches to citizen agency for resilience at-a-place.
- Draws both on, and brings together, diverse flood archives and different archiving practices (creative, cultural, technological) in materialising local knowledge.
- Promotes social, cultural, and community memory of floods, including those extremes beyond human lifespans.
- Values different forms of 'data', embracing oral and visual narratives. Oral history archives take the form of stories, vignettes, and anecdotes of, for example, hardship and agency.
- Needs to be owned or co-owned by local communities, recognising critical roles of citizen, citizen archivist, and collective agency in local FRM and in developing community resilience.
- Has a strong emphasis on participatory engagements around local flood heritage.

5.6.1   Recognising value of flood narratives

The impact of public archiving of disaster narratives can be analysed. For example, Veer et al. (2016) explored the large numbers of stories that were shared online during a disaster (in their case study, earthquakes in Canterbury, New Zealand), and the importance and power of narration, and sharing stories online for emotional well-being and catharsis in recovery. In these storying practices, it is not necessary to know the listener; notions of community were found to extend to virtual community networks online for solidarity.

Storytelling has also been developed as practice, with oral history/cultural heritage specialists in local universities[23] working with flood-impacted communities and adapting digital forms to document flood impacts (e.g., *Hurricane Digital Memory Bank, Collecting and Preserving the Stories of Katrina and Rita*[24]; see Sloan, 2008). Digital storytelling (2–3 minutes of audio with images selected by the storyteller; Meadows, 2003) has been developed and applied to capturing critical reflections in communities after floods (e.g., Klaebe, 2013 in Australia; Holmes and McEwen, 2020 in the UK). For example, in the Sustainable Flood Memories[25] project, creative experiments were undertaken in co-creating an archive of digital stories about residents' experiences of the 1947 and 2007 floods on the River Severn, UK in community flood risk settings. Here, Holmes and McEwen (2020) explored the value of capturing personal 'flood stories' about preparedness for the storyteller, listener, and community. Such stories, conceived as 'gifts', could promote reflection on resilience through different lenses (infrastructural, emotional, etc.) when shared for social learning within and between flood risk communities (Chapter 8).

However, flood heritage narratives can be contested in governance settings. Carmichael et al. (2020, p300) warned about different heritage narratives and their use in lobbying, noting "limited understanding of how heritage narratives mediate local dialogue about climate resiliency". This is a particular concern in heritage narratives of blame, with implications for environmental justice.

5.6.2 Archiving as community practice

Flood memory, associated lay knowledge, its archiving, and resilience have complex relationships (Chapter 4). McEwen et al. (2016, p17) proposed the concept of 'Sustainable Flood Memory' (SFM) conceived as:

> an approach to memory work that is both individual and community-focused, taking account of materialised memories, e.g., in landscape, technology, social media, formal and informal archives. It integrates individual (personal) and collective (community) experiences across different media and materialities. Such memory is 'sustainable' and persistent in creating and supporting conditions for its furtherance, with strong attention to inter- and intra-generational exchanges and social learning. It generates strategies for associated lay knowledges (adaptive; building capital) for dealing with flood risk.

They distinguished between the stages of building, organising, and sharing archives (Table 5.2). In constructing a framework for conceptualised SFM as process-practice, McEwen et al. (2016) asked a series of questions, with associated actions, which aim to build SFM within communities. This starts from questioning "does lay flood memory exist?"

TABLE 5.2 A Conceptualised Framework for SFM as Process-Practice within FRM Decision-Making for Local Resilience (McEwen, et al., 2016, Table II)

Questions	Actions
Building SFM	
1. Does lay flood memory exist?	If so, interrogate its use/usefulness, availability, and accessibility. If no, go to 4.
2. Who are the marginalised groups disconnected from local memory and lay knowledge systems?	Establish which residents have no potential for links between memory and experiential lay knowledge. Establish whether young and older citizens are connected for communicative memory processes.
3. What are the opportunities and barriers for connection of horizontal and vertical axes of memory?	Rethink how intergenerational communities of memory develop locally.
Organising SFM	
4. Is lay flood memory being archived? (NB. transient groups)	Provide active encouragement for lay and organisational archiving of flood experiences and their integration.
5. Where are the archives?	Work to signpost archivists and archives. Connect archives at different scales (local, global).
6. Who are the archivists and gatekeepers to archival memory and lay knowledges?	
7. What are opportunities and barriers for flood stories to be organised horizontally – not only during floods but in recovery and preparedness for future floods?	Revisit how FRM agencies deal with storytelling and anecdote at institutional and inter-agency levels; how knowledge conflicts are collectively reconstructed.
Sharing SFM	
8. Where are the effective entry points for sharing SFM and SFM practices?	Encourage 'flood friends' in individual memory practices as valuable 'entry points' in SFM within individual/ household level planning. Flood action groups have similar potential for collective practice in SFM; mix archives and 'flood friends' in remembering.
9. Where are the connections and disconnections between flood memory and lay knowledge?	Ensure spaces for critical reflection and social learning within communities; and between communities and FRM agencies.
10. How can memory and lay knowledge be integrated within more formal FRM decision-making processes?	Ensure protected and trusted spaces for sharing and reconstituting different knowledges (lay, expert).

FIGURE 5.9 Bringing flood heritage into conversation with other flood knowledges (copyright Lindsey McEwen; ESRC Sustainable Flood Memories project)

Such thinking has important implications for local FRM practice (McEwen and Holmes, 2017) and how different types of place-based flood knowledge (science data, heritage, stories, plans) are brought together (Figure 5.9). This also poses questions about the relationships between lay (local) and statutory agencies archiving of organisational flood experiences – their connection and accessibility (Chapter 4). The theme of community social learning about long-standing place-based ways of coping and adapting to floods from local archival knowledge integrates with flood education (Chapter 9).

A flood heritage approach to learning for community resilience has potential to bring in, and value, other types of evidence and professionals. This poses questions about the role of the cultural sector strategically and in sustainable solutions within local climate resilience (Chapter 10). Flood heritage practice requires implementation at a hierarchy of scales. At an international level, UNESCO's 'Water Museums: Global Network'[26] promotes "connections with water and its natural, cultural, tangible and intangible heritage" and "ancestral techniques, legacies, and traditional [water] knowledge".[27]

However, at national level, differences exist in the extent to which flood heritage is linked to statutory policy and practice in public awareness raising about risk. In

France, importance is given to recording flood histories and promoting local flood memory of past extremes. Indeed, such marking in urban settings is now legally required in France for local FRM. The French environmental regulator also has a national Historic Flood Database (Base de Données Historiques sur les Inonda-tions[28] BDHI) that uses Geographical Information Systems to map historic flood outlines (from Middle Ages to present) and bring together archival evidence on impacts ("physical" and "human, socio-economic, environmental and cultural") from scientific reports to newspaper accounts (Lang et al., 2016). This public-facing website encourages exploration and visualisation of past extremes.

At a local and regional scale, creation and implementation of a flood heritage approach requires co-working between communities and organisations with community-facing responsibilities across sectors. Table 5.3 shares some community-focused, local flood heritage activities that can involve different key actors. For example, naming of individual extreme floods as both government and local practice aids public recall and reminiscence. Such local engagement requires careful negotiation between active remembering and active forgetting within communities (see Figure 4.4 for possible reasons). It also involves navigating economic tensions from being branded a 'flood town', tourism opportunities from flood heritage, and

TABLE 5.3 Community Activities That Promote Learning about Local Flood Heritage

Strategies and activities	Activities	Key actors
Promote ability to read landscape evidence of risk and resilience	Flood heritage trails linked to river	C, HE
Highlight previous flood levels	Physical flood marking in community settings	C, B, LG
Make extreme floods easier to remember	Naming extreme weather events as cognitive markers	F
Advancing campaigns for memorials	Physical reminders in landscape	LG
Commemorating event anniversaries	Practices to aid collective remembering	LG, HE
Co-researching flood heritage	Resource guides developed by archives	A, C, S
Community flood reminiscence sessions	Intergenerational sharing	M
Local citizen researchers sharing archival knowledge	Sharing flood knowledge with local community organisations	C
Linking citizens to evidence in institutional flood archives	Engagement activities within research projects	A; HE

C, Citizens; B, Businesses; F, Statutory FRM agencies; LG, Local government; A, Archives; HE, Higher education institutions; S, Schools; M, Museums.

benefits from promoting wider societal learning from places with long histories of flood experience. As Puzyreva and de Vries (2021, p1) highlighted "different readings of the flood history of an area may coexist and compete within one flood-prone community" impacting relationships within a community and with statutory FRM agencies.

While important potential exists for the cultural sector's role in increasing knowledge about extreme events with local, and sometimes national, resonance, many communities can have limited connections with their formal flood archives. Some government-owned archives can be seen as distant places rather than repositories for community flood memory. Examples of formal archives proactively collating and promoting resources for community/school engagement include local government work to bring together sources that allowed public study of "the greatest civilian disaster of Victorian Britain" – *The Sheffield Flood 1864 UK*[29] (see Sheffield City Council, 2013). This now "forgotten flood" was caused by a collapsed dam embankment (Chapter 2). The study guide cites the Chief Constable's records showing that 240 people were drowned, 100 buildings and 15 bridges were destroyed, and around 4,000 houses flooded. Flood museums also act as important fora for promoting social learning about flood heritage (Table 5.1), although their sustainability may require volunteer effort.

5.7 Conclusions

This chapter has explored the value of flood heritage as both concept and practice linked to a sense of place and community engagement with its local flood histories. Some historic extreme floods like the 1927 Mississippi floods, USA and the 1953 storm surge, the Netherlands have strong resonance and persistence in the national public psyche. These possess established flood archives with flood knowledges that draw extensively across science and stories. Archives for more localised or regional floods can have more variable connections with their at-risk communities. The chapter cautions on the importance of recognising whose voices are captured within archives – past, present, and future – and the potential for selective representation of a community's flood heritage. This requires a focus on garnering voices that reflect the experiences of minority and marginalised groups, with attention to the ethics of the archive in terms of past archival practices. Investigation of flood heritage has potential to both perpetuate and surface inequalities with implications for environmental justice. However, democratisation of the archive, with citizen archivists, brings valuable opportunities for inclusive local knowledge sharing, but less scrutiny about the validity and reliability of what is shared (Chapter 8).

It is a flood myth that resilience at-a-place only increases over time, with potential for changing social vulnerabilities (Chapter 2). Valuable 'flood heritage' capital can also be latent within place-based communities and cultural organisations. It is important to seize opportunities for communities to (re)connect with such lay knowledge in social learning for resilience. Much can potentially be learnt – about coping, adapting, and transforming – from place-based flood archives that capture how past communities lived with risk, whilst being aware that possible futures may be different. This includes low technology and simple behavioural solutions (e.g., seasonal resilience practices to properties; building access walkways) that traditionally kept riparian communities functioning during moderate floods. More can be done to evidence and share how historical examples of resilient communities have been used to inform and shape local flood management practices today, and what 'travels' over time and space.

It is also important to develop organisational strategies, policies, and practices to support community engagement with its flood heritage, attuning types of approach, alignment, and activities to community character. Integration of local flood heritage and its embedded lay knowledge with science communication provides a valuable opportunity in promoting living flood histories and a watery sense of place (Chapter 9), whilst still recognising that a flood archive can be read in different ways.

Notes

1 https://www.neh.gov/humanities/2013/mayjune/feature/the-blue-humanities.
2 https://www.eea.europa.eu/data-and-maps/data/external/european-floods-database.
3 'large flood' in Scottish vernacular.
4 https://lichfields.uk/blog/2021/november/17/what-can-we-learn-from-the-past-about-flood-resilience/.
5 https://scmemory.org/collection/the-2015-south-carolina-flood-oral-history-collection/#page-content.
6 A play re-performed in commemoration of 2005 floods. "Based on actual interviews, collected stories, and found texts, the play provides a voice to the greatest natural disaster in U.S. recorded history". https://www.usm.edu/news/2015/release/special-hurricane-katrina-project-performances-planned-august.php.
7 http://multi-story-shipley.co.uk.
8 https://issuu.com/martinharriscentre/docs/drama_multi-story_water_report.
9 https://rarehistoricalphotos.com/the-johnstown-flood-in-rare-pictures-1889/.
10 See American Experience 'The Fatal Flood'; Programme transcript (combination of oral history and narration); http://www.shoppbs.pbs.org/wgbh/amex/flood/filmmore/index.html.
11 https://www.pbs.org/video/american-experience-flood-film-clips/.
12 http://www.shoppbs.pbs.org/wgbh/amex/flood/filmmore/pt.html.
13 https://vimeo.com/300380363.
14 https://youtu.be/y83qrouRZa8.
15 https://tmhc.ca/heritage_flooding.
16 https://www.abc.net.au/news/2011-07-08/memorial-to-honour-lockyer-valley-flood-victims/2787032.
17 https://monumentaustralia.org.au/display/108498-"after-the-past".
18 https://www.nickydavis.com/pages/mural-art.
19 https://www.livinglevels.org.uk/stories/2018/12/10/the-great-flood-1607.
20 "Riverbank Blues," from *The Collected Poems of Sterling. A. Brown*, selected by Michael S. Harper. Copyright© 1980 by Sterling A. Brown. Reprinted by permission of the John L. Dennis Revocable Trust.

21 *American experience 'FATAL FLOOD: Voices from the Flood. https://www.pbs.org/ wgbh/americanexperience/features/flood-voices-flood/*
22 https://www.pbs.org/wgbh/americanexperience/features/flood-delta-blues/.
23 The Centre for Oral History and Cultural Heritage (University of Southern Mississippi).
24 https://hurricanearchive.org/.
25 https://esrcfloodmemories.wordpress.com/.
26 https://www.watermuseums.net/project/.
27 https://www.watermuseums.net/about/mission/.
28 https://bdhi.developpement-durable.gouv.fr.
29 https://www.bbc.co.uk/news/uk-england-south-yorkshire-26478728.

References

Allan, R., Endfield, G., Damodaran, V. et al. (2016) Toward integrated historical climate research: The example of atmospheric circulation reconstructions over the earth. *WIREs Climate Change*, 2(7), 164–174. https://doi.org/10.1002/wcc.379

Barry, J.M. (1997) *Rising Tide: The Great Mississippi Flood of 1927 and How It Changed America*. Simon & Schuster, New York.

Benito, G., Brázdil, R., Herget, J. and Machado, M.J. (2015) Quantitative historical hydrology in Europe. *Hydrology and Earth System Sciences*, 19, 3517–3539. https://doi.org/ 10.5194/hess-19-3517-2015

Benito, G., Lang, M., Barriendos, M. et al. (2004) Use of systematic, palaeoflood and historical data for the improvement of flood risk estimation. Review of scientific methods. *Natural Hazards*, 31, 623–643.

Bennett, J. (2004) The force of things: Steps toward an ecology of matter. *Political Theory*, 32(3), 347–372.

Bicksler, R. (2019) The role of heritage conservation in disaster mitigation: A conceptual framework for connecting heritage and flood management in Chiang Mai, Thailand. *Urban Geography*, 40(2), 257–265. https://doi.org/10.1080/02723638.2018.1534568

Black, A.R. and Law, F.M. (2004) Development and utilization of a national web-based chronology of hydrological events. *Hydrological Sciences Journal*, 49(2), 246. https:// doi.org/10.1623/hysj.49.2.237.34835

Blöschl, G., Hall, J., Viglione, A., Perdigão, R.A.P. et al. (2019) Changing climate both increases and decreases European river floods. *Nature*, 573(7772), 108–111. https://doi. org/10.1038/s41586-019-1495-6

Bottoms, S. and McEwen, L. (2014) *Multi-Story Water: Sited Performance in Urban River Communities*. University of Manchester. https://issuu.com/martinharriscentre/docs/ drama_multi-story_water_report/1

British Hydrological Society *Chronology of British Hydrological Events*. Available at http:// cbhe.hydrology.org.uk/

Brown, S.A. (1932) *Southern Road: Poems*. Harcourt, San Diego.

Campbell, A. and Paye, M. (2020) Water enclosure and world-literature: New perspectives on hydro-power and world-ecology. *Humanities*, 9, 106. https://doi.org/10.3390/h9030106

Carey, M. (2012) Climate and history: A critical review of historical climatology and climate change historiography. *WIREs Climate Change*, 3, 233–249. https://doi.org/10.1002/wcc.171

Carmichael, C., Danks, C. and Vatovec, C. (2020) Assigning blame: How local narratives shape community responses to extreme flooding events in Detroit, Michigan and Waterbury, Vermont. *Environmental Communication*, 14(3), 300–315. https://doi.org/10.1080/ 17524032.2019.1659840

Crowden, J. and Wright, G. (1996) *In Time of Flood: The Somerset Levels – The River Parrett*. Parrett Trail Partnership, Somerset.

Davies, G.L. (1969) *The Earth in Decay: A History of British Geomorphology 1578–1878*. MacDonald Technical and Scientific, London.

Davis, N. (2017) Houston Strong – A Mural. https://houstonmuralmap.com/murals/houston-strong/

Davis, L., Harde, T.M., Muñoz, S.E., Godaire, J.E. and O'Connor, J.E. (2018) Preface to historic and paleoflood analyses: New perspectives on climate, extreme flood risk, and the geomorphic effects of large floods. *Geomorphology*, 327, 610–612. https://doi.org/10.1016/j.geomorph.2018.10.021

Derrida, J. (1996) *Archive Fever: A Freudian Impression*. University of Chicago Press, Chicago. Translated by Eric Prenowitz. 113pp.

Dicks, B. (2000) *Heritage, Place and Community*. University of Wales Press, Cardiff.

Dournel, S. (2016) La vulnérabilité d'un territoire, un héritage complexe à révéler: démonstration à travers l'étude géohistorique du risque d'inondation en Val d'Orléans (France) *VertigO*, 16(3). https://doi.org/10.4000/vertigo.18040

Engeland, K., Wilson, D., Borsányi, P., Roald, R. and Holmqvist, E. (2018) Use of historical data in flood frequency analysis: A case study for four catchments in Norway. *Hydrology Research*, 49(2), 466–486. https://doi.org/10.2166/nh.2017.069

Eugene, N. (2006) Bridges of Katrina: Three survivors, one interview. *Callaloo*, 29(4), 1507–1512. https://doi.org/10.1353/cal.2007.0017

Evans, D. (2006) "High Water Everywhere": Blues and Gospel Commentary on the 1927 Mississippi River Flood. In R. Springer (ed.)*"Nobody Knows Where the Blues Come From": Lyrics and History*. University Press of Mississippi, Jackson, pp3–75.

Fanta, V., Šálek, M. and Sklenicka, P. (2019) How long do floods throughout the millennium remain in the collective memory? *Nature Communications* 10(1105) https://doi.org/10.1038/s41467-019-09102-3

Ganje, L. and Kenney, l (2004) Come hell and high water; Newspaper photographs, minority communities and the Greater Grand Forks Flood. *Race, Gender, and Class in Media*, 11(2), 78–89.

Gillis, J.R. (2013) The blue humanities in studying the sea, we are returning to our beginnings. *Humanities*, May/June 2013, 34(3). https://www.neh.gov/humanities/2013/mayjune/feature/the-blue-humanities

Hall, J., Arheimer, B., Aronica, G.T. et al. (2015) A European flood database: Facilitating comprehensive flood research beyond administrative boundaries. *Proceedings of IAHS*, 370, 89–95. https://doi.org/10.5194/piahs-370-89-2015

Haslett, S.K. and Wong, B.R. (2021) Recalculation of minimum wave heights from coastal boulder deposits in the Bristol Channel and Severn Estuary, UK: Implications for understanding the high magnitude flood event of AD 1607. *Atlantic Geology*, 57, 193–206.

Hein, C. (ed.) (2020) *Adaptive Strategies for Water Heritage Past, Present and Future*. Springer Open, Switzerland.

Hewison, R. (1998) The *Heritage Industry: Britain in a Climate of Decline*. Methuen, London.

Holm, P., Adamson, J., Huang, H., Kirdan, L., Kitch, S., McCalman, I., Ogude, J., Ronan, M., Scott, D., Thompson, K., Travis, C. and Wehner, K. (2015) Humanities for the environment – A manifesto for research and action. *Humanities*, 4, 977–992. https://doi.org/10.3390/h4040977

Holmes, A. and McEwen, L. (2020) How to exchange stories of local flood resilience from flood rich areas to the flooded areas of the future. *Environmental Communication*, 14(5), 597–613. https://doi.org/10.1080/17524032.2019.1697325

Hulme, M., Dessai, S., Lorenzoni, I. and Nelson, D.R. (2009) Unstable climates: Exploring the statistical and social constructions of 'normal' climate. *Geoforum*, 40(2), 197–206. https://doi.org/10.1016/j.geoforum.2008.09.010

Itsukushima, R., Ohtsuki, K. and Sato, T. (2021) Learning from the past: Common sense, traditional wisdom, and technology for flood risk reduction developed in Japan. *Regional Environmental Change*, 21, 89 (2021). https://doi.org/10.1007/s10113-021-01820-z

Johnson, B. (2017) Modernity, authenticity, and the blues in Sterling Brown's flood poems. *MELUS*, 42(2), 115–135. https://doi.org/10.1093/melus/mlx039

Kjeldsen, T.R., Macdonald, N., Lang, M. et al. (2014) Documentary evidence of past floods in Europe and their utility in flood frequency estimation. *Journal of Hydrology*, 517, 963–973. https://doi.org/10.1016/j.jhydrol.2014.06.038

Klaebe, H.G. (2013) Facilitating local stories in post-disaster regional communities: Evaluation in narrative-driven oral history projects. *Oral History Journal of South Africa*, 1, 125–142.

Lang, M., Coeur, D., Audouard, A., Villanova-Oliver, M. and Pène, J. (2016) BDHI: A French national database on historical floods. *E3S Web of Conferences*, 7, 04010. https://doi.org/10.1051/e3sconf/20160704010

Lauder, T.D. (1830) *An Account of the Great Floods of August, 1829 in the Province of Moray and Adjoining Districts* (2nd edn). Adam Black, Edinburgh.

Johnson, L. (1928) Broken levee blues. *Okeh Electric* (8618) https://archive.org/details/78_broken-levee-blues_lonnie-johnson_gbia0365912b

Lowenthal, D. (1985) *The Past Is a Foreign Country*. Cambridge University Press, Cambridge.

Lowenthal, D. (1988) *The Heritage Crusade and the Spoils of History*. Cambridge University Press, Cambridge.

Macdonald, N. (2007) Epigraphic records: A valuable resource in reassessing flood risk and long-term climate variability. *Environmental History*, 12(1), 136–140. https://doi-org.ezproxy.uwe.ac.uk/10.1093/envhis/12.1.136

Macdonald, N. and Sangster, H. (2017) High-magnitude flooding across Britain since AD 1750. *Hydrology and Earth System Sciences*, 21, 1631–1650. https://doi.org/10.5194/hess-21-1631-2017

McEwen, L.J. (2006) Seasonality and generating conditions of floods on the River Tay: 1200 to present. *Area*, 38(1), 47–64.

McEwen, L.J., Garde-Hansen, J., Holmes, A., Jones, O. and Krause, F. (2016) Sustainable flood memories, lay knowledges and the development of community resilience to future flood risk. *Transactions of the Institute of British Geographers* 42 (1), 14–28.

McEwen, L.J., Garde-Hansen, J., Robertson, I. and Holmes, A. (2018) Exploring the changing nature of flood archives: Community capital for flood resilience. In A. Metzger and J. Linton (eds.) *La Crue, l'inondation: un patrimoine*. L'Harmattan Publishing House, France.

McEwen, L.J. and Holmes, A. (2017) "Sustainable Flood Memories": Developing the concept, process and practice in flood risk. In F. Vinet (ed.) *Floods Vol 2: Risk Management*. Editions ISTE, 141–153.

McEwen, L.J., Jones, O. and Robertson, I. (2014) "A Glorious Time?" Reflections on flooding in the Somerset Levels. *The Geographical Journal*, 180, 326–337.

McEwen, L.J., Reeves, D., Brice, J., Meadley, F.K., Lewis, K. and Macdonald, N. (2012) Archiving flood memories of changing flood risk: Interdisciplinary explorations around knowledge for resilience. *Journal of Arts and Communities*, 4, 46–75.

McEwen, L.J. and Werritty, A. (1988) The hydrology and long-term geomorphic significance of a flash flood in the Cairngorm mountains, Scotland. *Catena*, 15, 361–377.

McEwen, L.J. and Werritty, A. (2007) 'the Muckle Spate of 1829': The physical and human impact of a catastrophic nineteenth century flood on the River Findhorn, Scottish Highlands. *Transactions of the Institute of British Geographers* NS, 32, 66–89.

Meadows, D. (2003) Digital storytelling: Research-based practice in new media. *Visual Communication*, 2(2), 189–193.

Metzger, A. and Linton, J. (eds.) (2017) *La Crue, l'inondation: un patrimoine*. L'Harmattan Publishing House, France.

Mizelle, R.M. (2014) *Backwater Blues: The Mississippi Flood of 1927 in the African American Imagination*. University of Minnesota Press, Minnesota, 224pp.

Monge, L. (2022) *Wasn't That a Mighty Day: African American Blues and Gospel Songs on Disaster*. University Press of Mississippi, Jackson.

Moran, C. (2009) *The Royal Court's Caravan: a flood victim writes*. February 2009. https://www.theguardian.com/stage/theatreblog/2009/feb/19/royal-court-caravan-flood-victim

Morgan, J.M. (2015) Understanding flooding in early modern England. *Journal of Historical Geography*, 50, 37–50.

Muzaini, H. and Minca, C. (2018) *After Heritage: Critical Perspectives on Heritage from Below*. Edward Elgar Publishing, Cheltenham

Naylor, L.A., Spencer, T., Lane, S.N., Darby, S.E., Magilligan, F.J., Macklin, M.G. and Möller, I. (2017) Stormy geomorphology: Geomorphic contributions in an age of climate extremes. *Earth Surface Processes and Landforms*, 42(1), 238–241. https://doi-org.ezproxy.uwe.ac.uk/10.1002/esp.4062

Palm, R.I. (1990) *Natural Hazards: An Integrative Framework for Research and Planning*. John Hopkins University Press, Baltimore

Patton, C. (1929) High Water Everywhere Parts 1 and 2 (with subtitles). Floating Castle Digital Workshop. https://www.youtube.com › watch?v=xL6UgyKItSo

Poole, S. (2014) 'So much loss and misery': taking the long view of the West Country flooding. *The Regional Historian*. https://regionalhistorianuwe.org/2014/

Puzyreva, K. and de Vries, D.H. (2021) 'A low and watery place': A case study of flood history and sustainable community engagement in flood risk management in the County of Berkshire, England. *International Journal of Disaster Risk Reduction*, 52, 101980. https://doi.org/10.1016/j.ijdrr.2020.101980

Rashid, H. (2011) Interpreting flood disasters and flood hazard perceptions from newspaper discourse: Tale of two floods in the Red River valley, Manitoba, Canada. *Applied Geography*, 31(1), 35–45. http://dx.doi.org/10.1016/j.apgeog.2010.03.010

Renes, H. et al. (2020) Water Meadows as European Agricultural Heritage. In C. Hein (ed.) *Adaptive Strategies for Water Heritage*. Springer, Cham. https://doi.org/10.1007/978-3-030-00268-8_6

Restemeyer, B., van den Brink, M. and Woltjer, J. (2017) Between adaptability and the urge to control: Making long-term water policies in the Netherlands. *Journal of Environmental Planning and Management*, 60(5), 920–940. https://doi.org/10.1080/09640568.2016.1189403

Retsö, D. (2015) Documentary evidence of historical floods and extreme rainfall events in Sweden 1400–1800. *Hydrology and Earth System Sciences*, 19(3), 1307–1323. https://doi.org/10.5194/hess-19-1307-2015

Rivera, J.D. and Miller, D.S. (2007) Continually neglected: Situating natural disasters in the African American experience. *Journal of Black Studies*, 37(4), 502–522.

Roald, L.A. (2013) *Flom I Norge. Foraget Tom & Tom*. Vestfossen, Norway.

Robertson, I. (ed.) (2012) *Heritage From Below*. Ashgate, Farnham.

Samuel, R. (1994) *Theatres of Memory*. Volume 1. Verso, London.

Sheffield City Council (2013) Sources for the study of the Sheffield Flood 1864. *Libraries, Archives and Information.* https://www.sheffield.gov.uk/sites/default/files/docs/libraries-and-archives/archives-and-local-studies/research/Flood study guide v1-6.pdf

Sheftel, S. and Zembrzycki, S. (2016) Who's afraid of oral history? Fifty years of debates and anxiety about ethics. *The Oral History Review,* 43(2), 338–366.

Sloan, S. (2008) Oral history and Hurricane Katrina: Reflections on shouts and silences. *The Oral History Review,* 35(2), 176–186. https://doi.org/10.1093/ohr/ohn027

Smith, L. (2006) *Uses of Heritage.* Routledge, London.

Thomas, G. and Wilson, S. (2007) *The Gloucestershire Floods 2007.* Sutton Publishing, Stroud.

Thorndycraft, V.R., Benito, G., Barriendos, M. and Llasat, C. (2002) *Palaeofloods, Historical Data & Climatic Variability: Applications in Flood Risk Assessment* (Proceedings of the International PHEFRA SPHERE Workshop, Barcelona, Spain, October 2002).

Van Loon-Steensma, J.M. and Vellinga, P. (2019) How "wide green dikes" were reintroduced in The Netherlands: A case study of the uptake of an innovative measure in long-term strategic delta planning. *Journal of Environmental Planning and Management,* 62(9), 1525–1544. https://doi.org/10.1080/09640568.2018.1557039

Vansina, J. (1985) *Oral Tradition as History.* James Currey, London.

Veer, E., Ozanne, L.K. and Hall, C.M. (2016) Sharing cathartic stories online: The internet as a means of expression following a crisis event. *Journal of Consumer Behavior,* 15, 314–324. https://doi.org/10.1002/cb.1569

Vignes, D.S. (2008) "Hang It Out to Dry": Performing ethnography, cultural memory, and Hurricane Katrina in Chalmette, Louisiana. *Text and Performance Quarterly,* 28(3), 344–350. https://doi.org/10.1080/10462930802120517

Zavar, E.M. and Schumann, R.L. (2019) Patterns of disaster commemoration in long-term recovery. *Geographical Review,* 109(2), 157–179. https://doi.org/10.1111/gere.12316

6

COMMUNICATING FLOOD SCIENCE FOR COMMUNITY RESILIENCE

6.1 Introduction

Science communication is a burgeoning area of research and practice, with distinctive development in different national settings (e.g., Gascoigne et al., 2020). Flood risk communication (FRC) has specific requirements, for working with individuals and, importantly, with and within communities. In a risk-based approach, understanding elements of flood risk science, like 'probabilities', can act as critical 'steppingstones' towards more accurate risk perception. Lack of risk awareness, or under-estimation or denial of risk, are well-recognised inhibitors to action (Chapter 3). This need for a basic understanding of flood risk science (hereafter 'flood science') also applies to other stakeholders in local flood risk management (FRM) such as formal news media.

A thorough appraisal of capacity-building needs in flood science for different stakeholders is therefore essential to establishing good practice in communication. Such territory sits at a nexus of different agendas: of changing models of communication processes; movement to 'make science public'; and in determining 'what works' in communication of environmental risk and climate change science. In addition, different academic disciplines, including risk sciences, decision sciences, behavioural sciences, social sciences, and creative arts, bring distinct theoretical frames and preferred practices. As Demeritt and Nobert (2014, p313) observed

> Risk communication plays an increasingly central role in flood risk management, but there is a variety of conflicting advice about what does – and should – get transmitted, why, how, and to whom.

'Science capital' can be defined as "science-related forms of cultural and social capital" (Archer et al., 2015, p922), as manifested in "knowledge, experiences,

DOI: 10.4324/9781315666914-6

attitudes, behaviours and practices".[1] This chapter explores issues in effective FRC with citizens and communities, recognising diverse prior science and water capital, and dispositions for engagement. To communicate flood risk science effectively with communities, statutory agencies and academic researchers need scientific knowledge and expertise, but also well-honed skills in engaging with people who may value different knowledges. Such communication has different purposes: to develop awareness, knowledge, and understanding and to promote individual and collective actions for flood preparedness. Importantly, it helps communities shape local risk management and thereby become invested in it.

The chapter asks:

- What are the opportunities and challenges for effective flood science communication?
- What can be learnt from science communication in other risk domains that also require effective community engagement (e.g., in public health)?
- How are diverse community groups – with different science capital, attitudes, interests, values, and concerns – best engaged with local flood science?
- What guidance can be distilled for the development of effective community-focused, FRC strategies?

6.2 Engagement and communication with communities about flood risk

As well as paradigm shifts in flood management, there have been changes in the approaches and language used by scientists and science communication practitioners to engage and communicate with citizens and communities about science. These include science transfer, knowledge exchange, and knowledge co-generation (Table 6.1; Chapter 4).

Traditionally, agencies with statutory FRM responsibilities have tended to use the language and practice of one-way transfer of flood risk science to at-risk

TABLE 6.1 Different Models for Public Engagement with Science

Terminology	Character	Model of engagement
Knowledge transfer	Hierarchical	One-way 'knowledge deficit' model; assumes action and requires good scientific or technical knowledge to be transferred better
Knowledge exchange	Horizontal	Two-way flow of knowledge (specialist scientific or technical with local/lay). Interaction between knowledge producers and users results in mutual learning
Knowledge co-production	Participatory, sustained, iterative	Longitudinal engagement involving participatory activities that integrate science

residents. This is embodied by strategies of mailshotting materials about risk to residents in flood risk zones or posting guidance information (e.g., flood zone maps) online. In media research, this is described as the 'broadcast' or 'knowledge deficit' model of organisational communication – the classic sender-receiver model (Shannon and Weaver, 1949). This approach has limited effectiveness, frequently failing in its objectives to alter public behaviour in clearly defined or predictable ways (e.g., Stewart and Rashid, 2011). Its key flaw is assuming that the problem is with the information receiver and that once the 'information gets through', the receiver will 'make the right decision'. This fails to take account of risk perception (e.g., cognitive dissonance), how information is processed, and constraints upon the receiver's agency (Chapter 3; Wood and Miller, 2020). Recognising these problems, alongside lack of recipient feedback within such approaches, has led to increased emphasis placed on interactive, two-way, knowledge exchanges and longitudinal dialogue between organisational actors and 'the public', as a heterogenous group.

Alongside these changes in general communication models, gradual change in risk communication has occurred over almost 50 years. Leiss (1996, p85) identified three early phases involving progressively increased engagement and dialogue: Phase I (ca. 1975–1984) focused on "quantitative expressions of risk estimate"; Phase II (ca. 1985–1994) emphasised characteristics of "successful communication"; and Phase III (1995–) promoted organisational responsibility in bringing both elements together, conceiving "sound risk communication as a matter of good business practice". This latter phase recognised issues of pervasive lack of public trust in risk issues (Chapter 3).

6.2.1 Modelling risk communication

More recently, conceptual models of risk communication have also progressed. Building on Wardman (2008), Demeritt and Nobert (2014) outlined four models of risk communication based on their underlying rationale (instrumental versus normative) and forms of communication (one-way versus two-way engagement). These distinct models are: "risk message", "risk instrument", "risk dialogue", and "risk government" (Figure 6.1; Table 6.2).

Not only do these four models define the "basic purpose, practice and future prospects of flood risk communication in quite different ways" (p313), but they also have implications for understanding what success looks like (knowledge transfer or exchange), and the role of communities (passive, active) within communication processes and practices. They require scientists to use different skills and approaches in presenting risk, with contrasting understandings of what might constitute 'best practice'. For example, dialogue requires attention to possible interactions with communities after an initial message gains a response. This could involve answering questions, addressing flood myths, advising on actions, and supporting emotional responses.

FIGURE 6.1 Four conceptual models of risk communication (after Wardman, 2008 in Demeritt and Nobert, 2014: reproduced by permission of Taylor and Francis Ltd)

TABLE 6.2 Detail on Four Conceptual Models of Risk Communication in Figure 6.1; (*edited into a table from Demeritt and Norbert, 2014; Wardman, 2008)

Model	Character*	Terms for assessing success*	Issues in practice	Implications for role of community
1. Risk message model of information transfer	Based on "the belief that 'good' risk communication is about faithfully transmitting risk information without distortion, bias or misunderstanding" (p315)	Information transfer to resolve information deficit (informing rather than influencing)	Potential for breakdown between stages	Passive receiver
2. Risk instrument model of behavioural change	Sees risk communication as "conscious instrument for changing attitudes and behaviours of message recipients" (p317)	Information transfer (informing and influencing)	Issues with strategies that aim to increase fear to motivate responsive action	Expectation to act rationally on message receipt
3. Risk dialogue model of participatory deliberation	"Based upon two-way exchanges that blur the sharp distinction between senders and recipients" (p319)	Information exchanges; feedback essential	Terminology as a barrier Requires time, expertise, and disposition for meaningful communication	Community active
4. Risk government model of self-regulation and normalisation	Sees risk communication as "an exercise of political power" (p321)	Aim "academic explanation rather than practical application" (p322)	Reframing risk as acceptable that "institutions could not reasonably be expected to prevent" (p321)	Community active

6.2.2 Communicating science with the public

Having knowledge can be construed as power,[2] determining who is seen as having authority and influence in local politics (Chapter 4). As Wardman (2008, p1621) warned:

Risk communicators draw from a wide body of techniques without necessarily being aware that the techniques they adopt bear the imprints of broader scientific, political, economic, or social theory [and]… are permeated by power/ knowledge relations.

The potential exists for flood risk communicators to learn from wider debates about the efficacy of science communication strategies from different stakeholder perspectives. For example, Burke (2015, in the USA) noted that many scientists hold misperceptions about the public and their understanding of science. She identified eight common myths that can prevail (e.g., disagreements are about facts rather than values), concluding that "when scientists communicate well, their opinions are valued more". Trust alongside expertise is also important to a communicator's credibility (Fiske and Dupree, 2014).

From a community-driven perspective, the desire for citizens to acquire science capital – to support their agency and activism – can be strong. Also significant in flood risk settings is the ability of highly motivated and emotionally driven citizens to engage with, and contest, specialist science (Fischhoff, 1995). Hence, risk communication with at-risk communities requires careful negotiation; trust in the messenger and clarity of message are critical. As the UK Government Office of Science (2011, p25) stated:

Existing government guidance rightly stresses the importance of: providing a trusted source of consistent information; being transparent about what is known and unknown; clearly distinguishing 'worst case scenarios' and what is expected; providing regular updates; and giving clear guidance as to suitable action for people to take.

The same report recommended that the UK government should work more closely with risk communication experts and behavioural scientists to develop internal and external communication strategies.

6.2.3 Making science public

Another key shift in research thinking is the drive to 'make science public' and accountable – opening-up the assumptions and processes of specialist science to public scrutiny. This involves questioning its integrity and authority as the evidence base for local decision-making in policy and practice. This movement sits within an era of contested expertise and growing emphasis on democratic participation.

One research project that critically explored this territory was *Making science public: challenges and opportunities*[3] (Nerlich et al., 2018). It focused on

> the challenges involved in making science public; making public science; making science in public; making science more public; making science private ... How are such activities changing the relationship between science, politics and publics, and what are the normative implications for problems relating to political legitimacy, scientific authority and democratic participation?[4]

While a solution to these problems in principle is for greater openness, transparency, and democracy in science processes underpinning policy-making and political participation, this can have pitfalls and dangers. Nerlich et al. (2018) highlighted the importance of transparency and openness; different understandings of expertise, responsibility, and justice; and awareness of impacts of faith.

This new positioning of science needs to connect with changing public attitudes to science, scientists, and confidence in 'experts'. In the UK, shocks to the integrity of science included 'Climategate',[5] where scientists were falsely accused of manipulating climate data to show warming trends. The status given to expert evidence in political decision-making is increasingly challenged (UK examples include "Who needs experts?", Portes, 2017; BEIS,[6] 2019) and, while it might be argued that the COVID-19 pandemic has helped restore the status of scientists and science high in political realms in the UK, the opposite might be seen to be the case in some other parts of the world.

6.3 Flood science communication: Background contexts

Against this backdrop, FRC can have different purposes and timings, both before a flood as a possible future situation and during an acute event. These situations are distinguished as 'static risk' versus 'live risk' (Table 6.3) but can have blurred and diffuse boundaries (Chapter 2).

To establish what flood science to communicate and how to do this, it is important to be aware of its distinctive character, alongside notions of threshold concepts and flood myths. Flood science does not adhere to the logic of mathematical theories and absolute truth or falsehood. As Baker (2007, p164) reflected

> Our empirical understanding of this world may in large part be true, but because the possibility of falsehood remains for specific cases, these must always remain undecided.

This brings challenges; it is critical to establish what flood risk knowledge is needed to build science capital within community groups with different social vulnerabilities. This will vary with the nature and scale of risk, prior (water) science

TABLE 6.3 Purposes of Flood Risk Communications (adapted from Orr et al., 2015; Table 1.1; Section 2.3; Contains public sector information licensed under the Open Government Licence v3.0.)

Concern	Before a flood (static risk) 'distant'	During a flood (live risk) 'immediate'
Risk factors	Living in flood risk area; whether property has ever been flooded	Physical conditions create new situation of heightened risk
Purpose of communication	Create awareness of flood risk and encourage action to prepare for future flooding	Encourage specific actions to prepare for flood event
Context	Area at flood risk but facing no immediate risk	Area at flood risk, with possibility of flooding of property from current or imminent flood event
Type of communication	Visual prompts (e.g., flood marks, road signage) Flood risk mentioned in routine transactions (information from estate agents, builders, DIY shops, and insurance companies) and relevant information from public services (local authority, other health, and education) Flood risk mentioned in information for vulnerable groups or those who engage with them (e.g., care homes and tourists)	Flood warnings (phone, text, tweet, email) Warnings on TV/radio Online flood maps and information Word of mouth (e.g., neighbours, local organisations)

capital and experience, and values. Key questions include what constitutes basic flood science as a foundation for more advanced understanding and what are the critical concepts? The science needs of local communities will also vary with specific flood risk settings – their causal, intensifying, and compounding factors.

It is also important to determine the capacity-building needs of statutory and non-statutory organisations that work with communities in local FRM. For example, McEwen et al. (2014) provided a mind map of priority specialist flood risk science needs, co-produced with local government FRM practitioners in SW UK (Figure 6.2). These included (p283):

- "Basics of flood science: terminology and language; risk and probability; measurement – as building blocks for more advanced flood science knowledge and understanding.
- Flood mapping: map history and creation; catchment 3D view; floodplain mapping; impacts of development on flood risk.
- Flood risk modelling: climate change impacts and predictions."

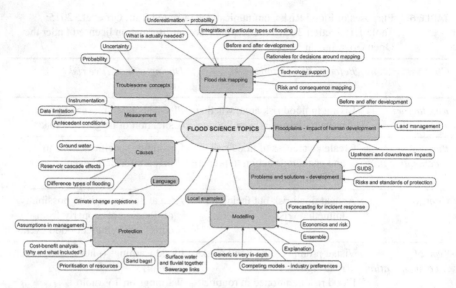

FIGURE 6.2 Mind map of flood risk science needs in local government co-generated with project partners, SW UK (McEwen et al., 2014; reproduced by permission of Taylor and Francis Ltd)

These local government participants considered that a good understanding of the concept of uncertainty, in context of specialist flood risk modelling and climate change predictions, was particularly important in increasing their skills in strategic use of flood science, and their confidence in communicating it with the public.

6.3.1 Troublesome knowledge and threshold concepts

Any theorising of FRC needs to be aware of the notions of 'troublesome knowledge' and 'threshold concepts' in science learning involving different groups of citizens and FRM agencies (Chapter 9). These include flood 'probabilities' and 'return periods' – critical in understanding risk. Here, the focus is on the receiver of knowledge and their ability to engage effectively. Knowledges or ideas can be 'troublesome' to this receiver because they are complex, counterintuitive, from a different culture or discourse, and/or the receiver is unwilling to alter the way they normally see or interpret data or a situation (Land et al., 2005). The notion of a threshold concept (Perkins, 1999; Meyer and Land, 2003) proposes that within some academic disciplines, key concepts or ideas exist that can be considered ways into higher levels of thinking about that subject. Such a 'portal' opens up

> a new and previously inaccessible way of thinking about something. It represents a transformed way of understanding, or interpreting, or viewing something without which the learner cannot progress.
>
> *Meyer and Land (2003, p1)*

This is not just about "mastery": grasping threshold concepts may involve acquiring knowledge that is 'troublesome' to the process (e.g., relating to emotional issues and anxieties; Cousin, 2006, p4). Meyer and Land (2005, p373) identified three main characteristics of threshold concepts as being: transformative (causing "a significant shift in the perception of a subject"); irreversible ("unlikely to be forgotten or unlearned only through particular effort"); and integrative ("exposing the previously hidden interrelatedness of something"). Threshold concepts in flood science include understanding probabilities and uncertainty (Section 6.4).

6.3.2 Communication and dispelling myths

Pervasive myths present another challenge to both transfer and dialogic models of FRC. McEwen (2011) identified flood risk myths (Box 6.1) that surface regularly in community engagement activities and public fora (e.g., on social media blogs and newspaper media narratives). Flood myths, circulating in communities, are frequently focused on the operation and effectiveness of upstream flood mitigation assets during floods (e.g., embankments, sluices), effects of land use and management changes, and climate change impacts on flood patterns. Such myths concur with Pielke's (1999, p413) listing of flood fallacies in the USA that

> are not universal, with many flood experts, decision-makers, and sectors of the public escaping their seductive logic. But enough people do fall prey to these fallacies of floods so as to create obstacles to improved utilization of the lessons of experience.
>
> *Pielke (1999, p413)*

Such fallacies include "understanding the science" – that "we know the wrong things about the nature of the problem" (Box 6.2).

BOX 6.1 EXAMPLES OF RECURRENT FLOOD SCIENCE MYTHS AMONG THE PUBLIC (ADAPTED FROM MCEWEN, 2011)

- Extreme floods are unnatural
- If a '1 in 50-year' flood occurs, there will be no flood of similar size for 50 years
- Extreme floods occur because rivers have not been dredged
- Rivers flood because upstream sluice gates/lock gates have been opened
- Embankments (levees) will protect us from floods
- Extreme floods are caused by land use changes
- One extreme flood provides evidence of climate change

BOX 6.2 PIELKE'S (1999) FLOOD FALLACIES BASED ON SCIENTIFIC UNDERSTANDING (WITH PERMISSION: ROGER PIELKE)

Fallacy 1: Flood frequencies are well understood.

Fallacy 2: Damaging flooding in recent years is unprecedented because of 'global warming'.

Fallacy 3: Levees 'prevent' damages.

Fallacy 4: Flood forecasts are universally available.

Such myths or fallacies can act as apparently impenetrable barriers to individual and community learning that are difficult to surmount or dispel. Lanza and Negrete (2007, p61), however, noted that myths can act as a hook for public engagement and learning (Chapter 9).

Wider deliberations exist on social constructions of science, and about when an individual's understanding of flooding is scientifically in error (e.g., around changing floods with climate change). Does this vernacular science knowledge still have value to that person or community as capital (Chapter 4; Table 4.2)? For example, there can be at least awareness of the issue. Wagner (2007, p7) argued that such science knowledge is "acceptable as long as [scientific facts] are able to serve as acceptable and legitimate belief systems in discourses with other lay people". This chapter now explores the nature, level, and complexity of flood science to be communicated and exchanged with communities, and its challenges.

6.4 Flood science for communication

Developing FRC strategies for community engagement requires an understanding of the importance and conceptual challenge of basic definitions. These include river catchments or basins, water systems, rivers as transporters and floodplains, along with underpinning threshold concepts like uncertainty.

6.4.1 Communicating rivers as complex systems: Definitions

The term river 'catchment' or 'basin' has a physical meaning in terms of the pathway of an individual raindrop and the flows it ends up in. While rivers are visible water transfers within a river basin, they can be conceived as complex physical systems involving many inter-relationships and feedbacks between variables that control how rainfall relates to water flows over space and time. A key concept is the 'hydrological cycle' as a system of water flows and storage; important relationships exist between the character and scale of flows in the upper, middle, and lower stretches of rivers within a catchment. For large catchments, it is often not intense local rainfall that causes large river floods, sometimes countering local risk perception (Chapter 3). Additionally, many infographics of the hydrological cycle omit impacts of human activities in enhancement or mitigation of flood risk. Complexity in water flows to be communicated also increases when human systems are overlaid.

Another important concept for FRC is 'rivers as transporters', not just of water but also of sediment and potential pollutants (Chapter 2). Rivers carry a load, whether dissolved, suspended, or coarser bedload – organic and/or inorganic – that naturally deposits on floodplains during floods. It can be invisible effluent load, a hidden risk that exacerbates local flood impacts in community settings, leading to the need to engage with health risk warnings in 'clear up' during recovery.

A further concept is that of a 'floodplain' and its definition locally (Section 4.3 of Chapter 4). A 'floodplain' and its mapped zones of risk are not static entities; definitions can change over time with lived experience of the spatial extent of extreme floods and re-estimation of their probabilities.

6.4.2   Communicating flood risk: Threshold concepts

A key skill in FRC is identification of possible threshold concepts (Section 6.3) that may act as barriers to understanding for target groups within communities with differing science capital. These might include 'probabilities' and 'return periods', 'uncertainty', 'confidence', system 'complexity', and the 'flood-drought continuum' (Table 6.4; McEwen, 2011).

Hydrological models are always abstractions of natural systems, so there is a need to make uncertainties explicit when modelling is used in decision-support. Martini and Loat (EXCIMAP, 2007), in outlining good practices for flood risk mapping, observed that uncertainty enters through: uncertainty related to natural phenomena, hydrology, and climate; lack of observations (data gaps; length of record); measurement inaccuracies in data; and limitations of modelling in capturing the 'real world' including scale effects (Moss and Schneider, 2000). This multi-faceted concept is a communication challenge: "communicating deeper

TABLE 6.4 Threshold Concepts in Flood Risk Communication

Concept	Definition	References
Probability and return periods	Frequency of flood of a given magnitude	Bell and Tobin (2007)
Uncertainty	Variability in scientific measurement	Hall (2010)
Scientific confidence	Reproducibility and integrity of science	AAAS[a]
Complexity	How systems behave rather than how they should behave	Bucknall and Hitch (2018)
Flood and drought continuum	Recognition of potential flooding overlaid with underlying drought	Weitkamp et al. (2020)

[a] https://www.scienceintheclassroom.org/collections/scientific-confidence.

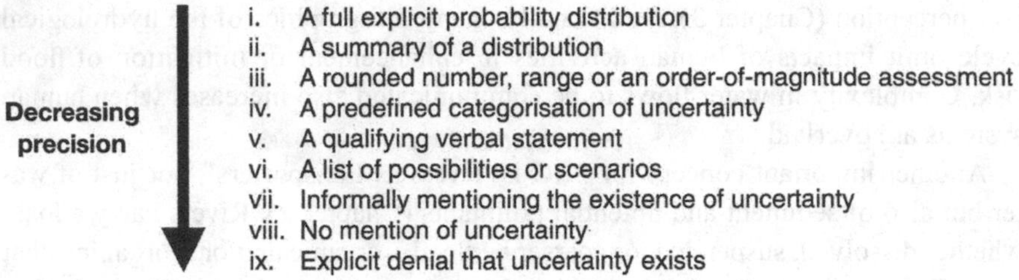

Decreasing precision

i. A full explicit probability distribution
ii. A summary of a distribution
iii. A rounded number, range or an order-of-magnitude assessment
iv. A predefined categorisation of uncertainty
v. A qualifying verbal statement
vi. A list of possibilities or scenarios
vii. Informally mentioning the existence of uncertainty
viii. No mention of uncertainty
ix. Explicit denial that uncertainty exists

FIGURE 6.3 Alternative expressions for communicating direct uncertainty about a fact, number or scientific hypothesis (van der Bles et al., 2019; © 2019 The Authors. Attribution 4.0 International (CC BY 4.0))

uncertainties resulting from incomplete or disputed knowledge – or from essential indeterminacy about the future" (Spiegelhalter et al., 2011, p1393).

This links to notions of, and concern for, 'ramped uncertainty' in predictive flood risk modelling 'cascades' that are undertaken sequentially. This might scale down from global scale circulation models, through their regional manifestations, their impacts on regional rainfall, then catchment runoff and human use systems to local flood losses (cf. Felder et al., 2018). In final stages of modelling, actual uncertainty may significantly exceed reported uncertainty (Faulkner et al., 2008). Openly conveying uncertainty in models to diverse citizens can be challenging as this can sometimes be equated with data that are 'no use', leading to concerns about the resulting audience response (van der Bles et al., 2019).

Debate is live within specialist flood science about 'communicating uncertainty', with pleas for scientists to be more upfront and rigorous about uncertainty in their scientific predictions (Beven, 2016). It has been argued that there is no room for "sloppiness" in expressing the confidence levels that scientists are able to have in their findings (Moss and Schneider, 2000; Giles, 2002), with a need to firm up the "language around 'low confidence'" (Giles, 2002, p476). van der Bles et al. (2019, p9) distinguished between alternative expressions for communicating uncertainty with decreasing precision. These ranged from a "full explicit probability distribution" to "no mention of uncertainty" and "explicit denial that uncertainty exists" (Figure 6.3).

This spectrum of use has strong implications for how a risk message may be received and accessed by citizens and communities. This highlights importance of dialogue, and opportunities for participatory science, where discussion about uncertainty is integral to community engagement processes. Related challenges include how to visualise uncertainty for communities (Spiegelhalter et al., 2011, e.g., on flood risk maps).

6.4.3 Communicating risk: Floods as statistical events

Communications about floods are frequently defined by qualitative descriptors – 'large', 'severe', or 'extreme'. However, perceptions of scale are conditioned by place and experience. People in a specific flood risk setting need to be able to differentiate between 'routine' or seasonal flooding at-a-place and extreme floods. Flood clusters, flood rich, or flood poor (or 'acquiescent') periods also need distinguishing from trends. Moreover, using the term 'flood poor' can be problematic in community engagement (Chapter 2). Identification of these periods, as an expected part of a flood series, is particularly critical as they can lull at-risk communities into false senses of security – with fading collective flood memory (Chapters 4 and 5).

Also critical is public engagement with the notion of probabilities in dealing with risk. The language used to communicate the likelihood of a particular flood size and severity occurring has been a continued challenge for statutory FRM agencies. While traditionally relative size of events has been described in terms of 're-turn period' or 'recurrence interval' (i.e., long-term average time in years between successive floods of the same magnitude), this has led to public misunderstanding and risk underestimation. For example, the '100-year flood' can be confusing for those with limited statistical awareness, leading to misinterpretation that extreme floods come in regular sequences so perpetuating optimism biases (Chapter 3). This brings issues for both public-facing flood forecasts (in real-time) and flood predictions (not in real-time) that involve probabilistic warnings of event severity. The latter applies to communities understanding the design limits of local flood mitigation infrastructure (e.g., a 100-year flood standard) as shared by engineering companies implementing new measures. For a resident who benefits from such local flood mitigating infrastructure, what does that mean for their residual risk (Chapter 1), and for their insurance cover? These long-standing problems with language and its interpretation are well articulated by Baker (2000, p371):

> The public perception issue of the 100-year flood is simply stated. The concept has nothing to do with real years, and it has very little to do with real floods [...]. The 100-year flood is an idealization. Human perception is grounded, not in idealizations, but in experience of real years and real floods.

At present, better practice in FRC by statutory agencies articulates flood frequency in terms of probabilities (e.g., an event with 0.01 probability of occurrence), alongside a qualitative descriptor like 'extreme'. However, in the UK, reference to return periods still often pervades news media reporting during floods.

Flood science communication also has the new challenge of 'unforeseen' events with extreme impacts – those assumed that cannot happen until they do. Flage and Aven (2015) distinguished between concepts of "emerging risk" (with "known unknowns") and "black swan" events (Taleb, 2010; with "unknown knowns", "unknown unknowns", and a subset of "known knowns" with very low probabilities). Aven (2013) favoured defining a black swan event as "as an extreme, surprising event relative to the present knowledge/beliefs" (Flage and Aven, 2015, p63). Unfortunately, black swan events in hydrology have high societal impact (Blöschl et al., 2015). For example, Hurricane Ida (in 2021),[7] causing a catastrophic flash flood emergency in New York City, might be described as a 'black swan' event. Previously unexperienced rainfall intensities vastly exceeded the capacities of infrastructure, and floodwater surged rapidly into the subway and flooded basement apartments. Over 40 people died. Both types of events – emerging and black swan – are important in changing contexts (climate, urban development), have implications for risk assessment and management, and provide challenges in FRC.

6.5 Communicating uncertainty in practice

This section considers the challenges of communicating uncertainty in community-focused practice within different phases of the FRM spiral: risk mapping, flood early warning, and floods in context of climate change.

6.5.1 Visualisations of flood risk: Flood risk mapping and its contestation

One way of communicating flood risk is through the visualisation of flood zone mapping that captures a floodplain's risk dynamics, with the accompanying challenge of how to communicate its uncertainty. However, official flood maps are often contested knowledge by community flood groups. In part, this reflects the extent of communication, and public understanding, of underpinning information about map resolution, data sources and their limitations; their mode of construction (whether real or modelled data); differential risk within a mapped grid square; and uncertainties in map production (e.g., date of last map update; Martini and Loat, 2007). Connected issues relate to the validity of what is shared (e.g., flood maps based on out-of-date data[8]; Lehmann, 2020 in the USA), linked to occurrence of recent extremes and climate change impacts, and the extent of available mapping for communities to use in local FRM (US Association of State Floodplain Managers, 2020).

Hence issues can exist with public understanding and interpretation of available flood maps posted online by statutory FRM agencies (e.g., UK Environment Agency's 'Check your long-term flood risk'[9]; US FEMA's Flood Map Service Center[10]). This represents a major problem when maps act critical evidence bases for individual and community action (Figure 6.4).

Maps can act as valuable 'boundary objects' or entities that help translate or negotiate knowledge across culturally defined boundaries – here between scientists and communities. The practice of exchanging different flood risk maps of the same locale, if undertaken sensitively, can be valuable for knowledge exchange, as those created by FRM organisations (environmental regulators, insurance companies) and civil society (e.g., community flood groups) frequently differ. These can be underpinned by different knowledges – specialist based on hydrological modelling versus lay experiential or intergenerational drawing on oral histories and vernacular flood science often visualised on maps by community flood groups (Chapters 4 and 8). Without negotiation, their discrepancies can lead to public

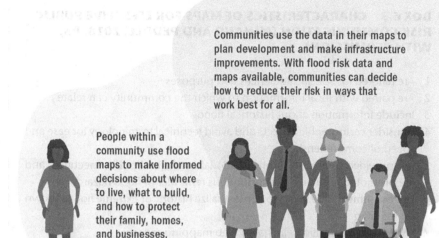

People within a community use flood maps to make informed decisions about where to live, what to build, and how to protect their family, homes, and businesses.

Communities use the data in their maps to plan development and make infrastructure improvements. With flood risk data and maps available, communities can decide how to reduce their risk in ways that work best for all.

FIGURE 6.4   Using flood maps (source: https://www.fema.gov/flood-maps)

contestation, particularly when linked to planned urban development on flood-plains. An example is Severn and Avon Valley Combined Flood Group's (UK, 2008) report *Building on the floodplain is misguided (a proof of evidence)*, with its community-drawn maps of known local flood risk.

The location of mapped boundary lines around flood zones is inevitably sensitive territory for communities, with implications for the flood risk status of individual properties (their economic value; owner's ability to sell property; insurance premiums; associated anxiety and stigma). This also poses questions about the effective visualisation of uncertainty (e.g., the efficacy of solid lines demarking zones of different scales of risk). Haughton and White (2017, in New Zealand) provided a useful synthesis of issues in "creating, contesting and communicating lines on environmental hazard maps". Notably, tensions and controversies can exist in the production, presentation, and revision of hazard maps, when they are used to change government policy.

Using maps to communicate risk to communities with high levels of illiteracy can be problematic (Donovan, 2010). Hence, Hagemeier-Klose and Wagner (2009) emphasised the importance of achieving balance between simplicity and complexity, and the readability and usability of public-facing risk maps. They observed (p563)

Well designed and associative maps (e.g., using blue colours for water depths) which can be compared with past local flood events and which can create empathy in viewers, can help to raise awareness, to heighten the activity and knowledge level or can lead to further information seeking.

BOX 6.3 CHARACTERISTICS OF MAPS FOR EFFECTIVE PUBLIC RISK COMMUNICATION (MINANO AND PEDDLE, 2018, P3, WITH PERMISSION)

1 Are tailored for specific audiences and purposes
2 Are paired with local information to which the community can relate
3 Include information about historical floods
4 Consider cartographic aspects and avoid technical terminology for ease and speed of comprehension
5 Are provided online, through traditional media and public meetings, and are promoted regularly as a continuous reminder of flood hazards
6 Use real-time gauge levels to contextualize historic or extreme floods shown on the map
7 Use property-specific, searchable Web mapping services
8 Are complemented with information about the consequences of flooding and tangible protective actions

---

Maps are effective for community-focused risk communication (Minano and Peddle, 2018) if they have characteristics outlined in Box 6.3.

Map producers in government also need to understand how different publics read maps (e.g., EXCIMAP, 2007) in order to identify how they can be improved. As Kate Marks (2016; UK Environment Agency's Deputy Director of Flood Risk Mapping, Modelling and Data[11]) reflected

> For at least the 13 years I've been with the [environmental regulator], maps have been a good way of communicating flood risk to the public. But we know they don't work for everyone. Take many of the people I teach, when they start they'd be hard pushed to find their own property or local river on a map, let alone use the map to judge their flood risk.

Hence, statutory FRM agencies need to undergo continual processes of development in their visualisation and communication of flood risk mapping, in response to user feedback. Co-production of maps (Chapter 4) provides opportunities and co-benefits. For example, Luke et al. (2018) described co-production processes involving agencies and communities used for diversifying the use and utility of US flood maps, from a primary objective of defining insurance premiums. Through this process, they identified important end-user preferences that aided understanding, e.g.

(1) legends that frame flood intensity both qualitatively and quantitatively, and (2) flood scenario descriptions that report flood magnitude in terms of rainfall, streamflow, and its relation to an historic event.

p1097

The UK Environment Agency now differentiates longer-term and immediate flood risk on its maps, combining a simple view of flood risk areas and a more detailed view for information (e.g., likely depth and velocity of flood water). This involves linking mapping zones explicitly to citizen and community preparedness actions and "listening, learning and evolving the service" (Marks, 2016[11]; UK HM Government's, 2016 *National Flood Resilience Review*). While researchers are exploring how other digital media could be used in delivering real-time interactive flood visualisation for different FRM stakeholders (e.g., Kuser Olsen et al., 2016; Haynes et al., 2018), there is also strong potential to co-create and trial immersive communications with at-risk communities.

6.5.2 Communicating uncertainty in flood early warning

Communication of uncertainty in the likelihood of a flood event and the range of potential severity levels is also an issue in real-time probabilistic flood forecasts (Kuller et al., 2021). This can have significant negative impacts on the message recipients' motivation to prepare within warnings designed to promote action. They found disagreement within research about how best to represent uncertainties within warnings for understanding and response.

> Although transparency concerning the inherent uncertainty associated with a flood risk (e.g., likelihood of occurring, severity and timing) were [sic] often associated with increased understanding, some studies highlighted the complexity associated with interpreting such information (Mileti et al., 2004; Spiegelhalter and Riesch, 2011; Shanahan et al., 2019), inhibiting improvement of response measures taken by recipients.
>
> *p4*

There are alternative ways forward – in changing the message's emphasis or improving the science communication as a stimulus for dialogue and increased understanding. Some propose that uncertainty is best explained in conjunction with recommendations for action (Wood et al., 2012). Hogan Carr et al. (2016, p1660) experimented with visualisation of confidence in timelines of forecasted river levels; their community participants found figures with this addition "easier to understand, sparking discussion of uncertainty in weather forecast".

6.5.3 Communicating uncertainty: Floods in climate change science

Communication of changing extreme weather risks in climate science is an additional challenge. Corner et al. (2015) provided guidance on how best to communicate uncertainty under climate change, including the importance of not conflating different types of uncertainty, and use of positive messaging, human stories, and case studies to drive messages home.

The more that the risks of climate change can be brought to life through vivid 'mental models', the better. This means using clear practical examples of the risk of a village flooding, or a farmer's crops being destroyed, or a coastal building slipping into the ocean.

Corner et al. (2015, p9)

They also give instances of 'Dos' and 'Don'ts' in uncertainty communication around climate change (Box 6.4), using floods as an example. This has implications for how floods are communicated in news media.

BOX 6.4 COMMUNICATING UNCERTAINTY ABOUT 'CLIMATE IMPACTS' (CORNER ET AL., 2015, P9, WITH AUTHOR PERMISSION)

DO say: "As the Earth warms there is more moisture in the air, which increases the chances of intense rainfall. So this flood is consistent with what scientists have long been predicting."

DON'T say: "No single extreme weather event can be attributed to climate change."

6.6 Issues in flood risk communication with communities

Risk communication gaps can lead to community inaction or poorly attuned action, leading to vulnerability to future flooding (Stewart and Rashid, 2011). Several guidance documents, produced for FRC in different institutional and national settings, identify principles and success factors. Orr et al.'s (2015) review of types of communications that improve public understanding of flood risk and encourage people to act concluded by highlighting six key issues (Table 6.5).

Recent guidance documents on FRC tend to offer principles or 'top tips', frequently garnered through literature reviews. However, the proactive role and contribution of flood-active citizens with event experience and awareness in co-developing good FRC practices are also recognised (Fisher, 2015). These focused on the meaning of flood risk messaging, including links between understanding risk and acting; and innovative methods and techniques to help individuals and communities understand their flood risk. Table 6.6 combines principles and advice to inform FRC from Fisher (2015) and the UK National Flood Resilience Review (HM Government, 2016).

This need to communicate risk effectively poses questions about possible roles – both realised and potential – for other academic disciplines (e.g., arts and humanities) and professions (e.g., cultural sector) as risk communicators using dialogue and knowledge exchange (Section 6.2).

TABLE 6.5 Six Key Issues for Practitioners to Address in Flood Risk Communication (Adapted from Orr et al., 2015, Chapter 6, permission Environment Agency)

Domain	Issue
Coverage	• Some marginalised groups are less likely to receive information or warnings as they do not use any channels used for communication (older traditional or new social media) • Online storage of useful information in multiple places leading to lack of ease in finding relevant information (cf. a single, joined up portal)
Communicating risk	• Failure to implement good practice (e.g., in relation to language used or flood alerts) may be associated with lack of training for operational staff
Understanding risk	• Many practitioners perceive issues with people's understanding of risk. Suggestion that some social groups will never engage with flood risk
Moving from awareness of flood risk to response	• Use of social media suggests change in relationships between key providers and members of the public, who now have greater involvement in sharing and creating information. Implications for changes in people's response are uncertain
Improving warnings	• Research indicates appetite for probabilistic warnings among people who have experienced flooding. New probabilistic flood warning materials need developmental testing with different groups (with and without flood experience) to appraise implications for response
Role of dialogue and participation in increasing understanding of and response to flood risks	• Public dialogue helps in encouraging engagement with flood risk and provides insights into risk communication issues. There is a need to model dialogue-based risk communication approaches • Involvement of local, non-technical individuals to support the development of FRC materials may increase their success

TABLE 6.6 Principles and Advice to Inform Flood Risk Communication (Adapted from Fisher, 2015, piv; HM Government, 2016, Annex 6; pp 85–89)

Concern	Advice
Audience	Think carefully about needs of different audiences for any communication; do not address 'the public' as an undifferentiated aggregate of individuals
Avoid broadcast	Avoid implying that target audiences are ignorant and simply require 'education'
Emotion	Don't assume a little bit of information will scare people – telling the truth about risk and impacts is more likely to lead to action
Open data	Make data public and collect as well as disseminate information
Logic	Provide an early explanation of the logic and structure of the central tenets and argument of any communication

(Continued)

TABLE 6.6 (Continued)

Concern	Advice
Claiming	Don't over claim
Risk language	Express estimations of the likelihood of events in intuitive, consistent, and unambiguous ways. Stop talking about probability and risk in mathematical language as it means very little to a lot of people
Uncertainty	Make uncertainties and levels of confidence in the estimations transparent
Language	Take particular care with terminologies that have a more vernacular use
Link to action	Be very clear with people on what is happening before, during and after a flood, and what actions they should take
Actions of others	If you are asking people to take individual actions, tell them in the same communication what local/national organisations are doing too
Link to place	Focus on making information local, with historical context
Emphasis	Don't just focus on the negative impacts of flooding – focus on what people can do about it

6.7 Other approaches to community risk communication

Different forms of writing and visualisation involving storying of science can assist, and potentially transform, models of science communication as two-way interaction (Tables 6.1 and 6.2). As Gotham et al. (2018, p354) reflected

Since experiential processing is affective and based on subjective interpretations,[...] outreach and communications personnel should consider emotions, metaphors, stories, and images when discussing flood risks.[...] Highly technical, tedious, and dull presentations of scientific and analytical facts are less likely to appeal to people who are processing information experientially.

The communicators' challenge is to connect with people's experiential processors while retaining "credibility and trustworthiness" (Carlton and Jacobson 2013, p37).

Shanahan et al. (2019, p19) explored the construction of science messages and how risk communication may be more effective when scientific information is embedded in narratives – what they described as "the science of stories in risk communication". This involves construction of narrative science messages with "characters in action" (p8) that through "narrative transportation" ("feeling the story experience" p3) stimulate emotion and affect as a bridging language with communities. Opportunities exist to bring skills in storying science or science storytelling (Joubert et al., 2019) into "messages to shape risk perceptions and improve hazard preparedness" (Shanahan et al., 2019, p19).

Strong potential exists for embedding local culture in risk communications and exploration of connectivity at-a-place over time (e.g., through immersive engagements with past, present, and future flood risk; Chapter 5). This includes the role of disruptive communications that challenge people's value systems, and more oblique engagements through the arts that share risk science in socially and culturally attuned and sensory ways. For example, the arts are increasingly exploring, materialising, and re-visioning challenging concepts, such as understandings of complexity in climate change (e.g., Blue Action/Creative Carbon Scotland's "Understanding climate complexity: science through art"[12]). The role of the creative arts in changing risk communication is potentially impactful, drawing on varied senses, with different messengers, and taking diverse participatory forms. Government can sponsor flood-themed community arts exhibits to target community awareness, with increasing recognition of the communicative value of community arts practice. For example, Arts Work's 'juried show', with its visual flood stories shared in Virginia's Flood Awareness Week 2021 (USA),[13] aimed to promote awareness of need for flood preparedness within the community.

Socially engaged arts and cultural practitioners are increasingly involved in co-creative approaches to awareness raising in local flood risk and climate change communications. Such practices can have high degrees of community embedding – involving collaborative working *within* a community, or co-creating and curating *with* communities, sometimes co-working with academic researchers and statutory FRM practitioners. Table 6.7 provides examples of how arts practitioners, in different roles (e.g., brokering or supporting visioning), have engaged communities effectively about past and future flood risk. Such practices can generate new and inclusive conversations within communities about living with risk, explore different values, and promote creative thinking about future possibilities without constraints of personal worldview and prior science capital (Chapter 4). These bring valuable opportunities for participatory engagements with science, providing oblique ways into, and fresh safe spaces for, dialogue.

This collaborative working occurs at differing scales. Specialist scientists are seizing opportunities to engage with the communicative potential of co-working with the arts and humanities in creative, place-based approaches (e.g., integration of film and performance within community flood risk engagement in international flood conference programmes like ICFM7[14]; Table 6.7). For example, Scott-Bottoms in his one-man play "Too Much of Water" drew from stories garnered in his interviews with residents who had property flooded during the extreme Boxing Day floods (2015) in the town of Shipley, UK. His storytelling explored the flood's impacts on homes and well-being (Figure 6.5), with the performance – "hard-hitting" around official warning messaging and with "black humour" – shared back to those river-side communities and to statutory agencies.

Several initiatives have brought science communication and creative practice together in co-production within at-risk communities. For example, facilitator Creative Carbon Scotland, an NGO "connecting arts and sustainability",[14] co-created

TABLE 6.7 Examples of Arts Practice in Communication for Flood Risk Awareness

Role of art	Example
As broker to local oral flood histories	Steve Scott-Bottoms' (academic dramatist) performance of his one-man play *Too Much of Water*[a] (delivered in flood risk community settings and ICFM7[b])
As broker to visual flood histories	*Back to the Future?* – A photographic overlay of past historic postcards and photographs of flooding with contemporary flood risk communities by Joanna Brown (shared at ICFM7)
As sharer of local flood archives	Community exhibition, *Flood Response* with "photographs, stories and artistic responses" curated by people of Leeds within its flood-affected Industrial Museum
As negotiator between stakeholders	Simon Read[c] (artist within local estuary partnership, UK) worked with communities 'to foster an understanding of coastal and estuarine change' (e.g., 'A map of the river Deben to explore its systems and susceptibility for them to fail 1999–2009'[d])
As stimulus in visioning future risk	HighWaterLine[e] 'visualizing climate change' conceived by US artist Eve Mosher
As provocation to think about changing values and risk	Socially and environmentally engaged artist, Sage Brice's *Future Museum* project[f] involved sharing a travelling "cabinet of curiosities" including an "Artist's Reconstruction of a Coastal Settlement in the Early Post-Catalystic Period".

[a] https://vimeo.com/463039624.
[b] The 7th International Conference on Flood Management.
[c] simonread.info.
[d] https://www.simonread.info/portfolio-items/a-map-of-the-river-deben/.
[e] https://highwaterline.org/.
[f] https://sagebrice.com/archives-2/environmental-art/futuremuseum/.

FIGURE 6.5   Stephen Scott-Bottoms performing 'Too Much of Water' to a community audience on the floodplain of the River Aire, UK (Photo courtesy of Stephen Scott-Bottoms)

two songs with a primary school about flooding on their local river: "The Flood Kit" and "The Burnie Journey",[15] the latter with the refrain: "The Den Burn is flooding, don't panic". These promoted community flood risk awareness in collaboration with the Scottish environmental regulator (see 'Creative Approaches to Engaging Flood Risk Communities'[16]).

Examples of large-scale, public/community arts projects include HighWaterLine[17] that aims to promote local conversations about changing flood risk and climate change (Table 6.7). This project, conceived by US artist Eve Mosher as a replicable social practice public art project, involves community participation in flood zone mapping around the city using a football pitch chalk marker, as a stimulus for local dialogue as the marking process played out. Participants mark the continuous "flood zone" boundary to identify areas that could be submerged under future storm surge flooding. Mosher and Quante's (2018) Action Guide[18] supports four stages: 'UNDERSTAND climate change; MAP the HighWaterLine, TAKE the HighWaterLine to the Streets and ADVOCATE for Climate Change'. The original performance took place in Manhattan Brooklyn and then was cascaded to other US cities (Miami; Philadelphia) and Bristol, UK (Figure 6.6; Tarr and Haydock-Wilson, 2015).

The humanities also have a valuable role to play in linking flood heritage to FRC (Chapter 5). For example, an archive-based public understanding of science project[19] developed by the author involved co-working with communities along the lower River Severn, England to research their local river flood histories in government archives. This project was developed as a new space for dialogue around long-term local risk during a flood acquiescent period (McEwen, 2011).

FIGURE 6.6   Conversations about tidal flooding while chalking during HighWaterLine in Bristol, UK (2014) (credit: Richard Clutterbuck)

6.8 Communicating flood risk in the context of climate change

Communicating complex changing risks is challenging but essential in future FRM. This includes communication of 'emerging risk' and 'black swan' events (Section 6.4). Armstrong et al. (2018) drew on environmental psychology and climate change communication to develop guidance "on framing, metaphor and messengers". Here, as in FRC, it is important to understand risk perception in how people interact with climate change information (Chapter 3). Indeed, Carlton and Jacobson (2013) proposed that focusing on specific risks may increase message salience in climate change communications. In the UK, for example, Messling et al. (2015) noted growing evidence that flooding and climate change are linked in the public psyche, with opportunities for connecting messages.

However, specific issues exist in FRC in context of climate change, for example, in conceiving maps with flood severity zones as static entities (Section 6.5). As Lehmann (2020, p4) pointed out

> ... updated maps only reflect past flood experiences. They do not reflect the degree to which climate change and sea-level rise are expected to heighten the risk of flooding and expand the areas that will be subject to flooding in the future.

In relation to new messengers and messages connecting flood risk and climate change communications, guerrilla communication campaigns about climate impacts can appropriate official warning symbols. For example, creative agency Phantom, in its 'Flood the Streets' campaign,[20] reimagined the official English flood risk symbol (a generic house in rising water) with some of the UK's iconic monuments submerged in water (e.g., Big Ben, London; Figure 6.7), distributing through stickers and social media campaigns and during climate strikes.

This campaign was specifically targeted at increasing young people's awareness of potential climate change impacts through emotional connections with places and shock. Urban graffiti can also be a powerful communication tool, as in "Flood Level 2030?" street art[21] in Plymouth, UK – with the appearance of yellow lettering thought to depict those areas inundated by a 1-m sea-level rise.

6.9 Developing a framework for community flood risk communication

As Spiekermann et al. (2015, p97) observed in the context of disaster risk reduction (DRR):

> Many aspects of the complex interface between information sharing, knowledge-making and decision-making are still unexplored and better appraisal is needed to effectively integrate information, knowledge, and expertise into the efforts directed at DRR, in particular with regard to mechanisms for positive exchange between science, policy, practise [sic], and the public.

FIGURE 6.7 Appropriated official icons from Phantom's 'Flood the Streets' campaign (www.phantom.land, with permission)

Any FRC framework needs to draw on good practice for wider communication and engagement, including inclusive participatory communication models involving co-production of knowledge *with* diverse publics (Alexander et al., 2014; Chapter 4). Science communicators are increasingly diverse – whether FRM professionals with statutory responsibilities, NGOs, communities themselves, artists or academics in research and knowledge exchange involving science engagement. Figure 6.8 adapts the Construction Industry Research and Information Association's (CIRIA) (Daly et al., 2015a) reflective cycle for good communication and engagement for collaborative working with and within communities. This is essential for ownership of scientific knowledge, and its integration with local knowledge to help empower communities to plan and take action.

In summary, many key considerations exist in developing community-focused FRC strategies (e.g., Daly et al., 2015a, 2015b). An essential starting point is awareness of the national history of FRC and its linkage to flood management paradigms. This involves understanding what has worked – or otherwise – in the past within a particular community with a specific flood risk and socio-cultural context. For example, Ping et al. (2016) provided an overview of progression in FRC strategies in the UK (Figure 6.9), against improvements to detection and forecasting, and mitigation, preparedness and warning.

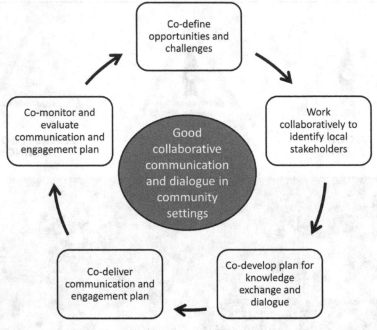

FIGURE 6.8 The reflective cycle of good collaborative communication and dialogue (adapted from CIRIA's Framework for communication and engagement, Daly et al., 2015a, p22; with permission)

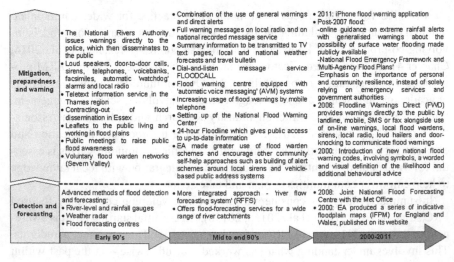

FIGURE 6.9 Overview of the progression of flood risk communication strategies over the decades in the UK (source: Ping et al., 2016, Figure 2; compiled from Parker et al., 1995, Haggett, 1998, Handmer, 2001, McCarthy, 2007, and Parker and Priest, 2012) (Reprinted from Journal of Water & Climate Change volume 7, issue number 4, pages 651-664, with permission from the copyright holders, IWA Publishing.)

TABLE 6.8 Opportunities and Inhibitors in Flood Risk Communication Strategies (adapted from Ping et al., 2016, Table 2)

Stakeholders	Influences on flood risk communication strategies	
	Positive	*Negative*
FRM organisations	• Technological advancement in flood forecasting • Ability to target multiple groups through multiple channels	• Lack of sustainable budget allocation to support activities at local level
Community level	• Some groups developing closer relations with local government • Different messengers about risk and action (e.g., local flood wardens)	• Interaction limited to a sub-group • Lack of attention to other groups (e.g., in terms of language, technology access, demographics, and marginalisation)

This highlights developments in the use of information and communication technologies (ICT), the increasingly tailored nature of messaging to community needs, and importantly, the use of multiple-channel, risk communication strategies that support two-way communication (Chapter 10).

The information disseminated through some of the more advanced communication strategies appear to be increasingly real-time and interactive in nature. There has been a rapid development of ICTs, which enables communities to provide valuable feedback about their local situation, instead of being only at the receiving end of the communication chain.

Ping et al. (2016, p655)

Hence, FRC strategies and community resilience are influenced by positive and negative factors (Table 6.8) including the need to attend to diversity and equity.

Finally, in developing a collaborative framework for FRC, 11 important imperatives are now proposed drawing across sources (Table 6.9).

6.10 Conclusions

This chapter has explored the complex territory of FRC needs when working with local communities to increase their resilience. While there is already guidance about risk communication in general, and flood risk in particular, effective practices that bring scientific knowledge into meaningful two-way exchanges with local knowledge and values to support community empowerment for preparation still require further exploration. Importantly, actors are increasingly diverse, encompassing specialist flood scientists, policymakers, FRM practitioners, news media and artists, as are their interactions with citizens and embedding in communities.

TABLE 6.9 Imperatives for Effective Flood Science Communication (Drawing from Orr et al., 2015; Ping et al., 2016 and Other Referenced Sources)

Imperative	Actions	Additional information
1 Understand your community(ies) from within or outside	• Ask who are the local stakeholders, communicators, and influencers? • Identify and work with key groups and gatekeepers (e.g., resident, faith, community groups)	Understanding risk perception (Chapter 3) among different groups is critical underpinning in developing effective FRC strategies
2 *Be aware of key groups as well as individuals*	• Attend to how the message cascades within social networks and promotes local dialogue	FRC for, and within, communities can be different to that targeted at individual citizens
3 *Be aware of communications continuum and likely resistance at citizen and community levels*	• Engage with different communication stages: moving from awareness through understanding then acceptance to behaviour change • Identify varying degrees of resistance to messaging and action within target community (Orr et al., 2015)	In any setting, likely to be continuum from "early adopters" to "laggards" and "sceptics" (O'Neill, 2004; cited in Höppner et al., 2010, p29)
4 *Tailor communication to needs of both communities and risk setting:*	• If using probabilities, link to 'the known' (e.g., lottery cards; Daly et al., 2015a) • Focus messaging on local impact rather than likelihood. Risk can be abstract • Be aware role of affect, emotion, and empathy in risk communication narratives • Tailor different messaging for different subgroups in communities	"Tailored, people-centred, flood risk communication" (Haer et al., 2016) more likely to be effective Communications around impact can be stronger stimulus for community action than likelihood of occurrence (Daly et al., 2015a), particularly for groups with less science capital Emotional connection can be important trigger for action to protect self, family, and others
5 *Make strong links between risk messaging and whole disaster risk management spiral*	• Pay attention to timing of communications • Bring risk messaging out from acute events into preparedness and adaptation (Ping et al., 2016)	See "hydro-illogical cycle" and relative timing of flood and drought communications (Wilhite, 2012)

(Continued)

TABLE 6.9 (Continued)

Imperative	Actions	Additional information
6 Take time to build relationships and trust	• Invest in relationship building within communications strategy	Knowledge exchange between at-risk communities, scientists, and those who communicate science in FRM organisations takes time
7 (Re)think the messenger(s)	• Adopt flatter, less hierarchical structures of communication and greater emphasis on participation and social learning within communities (Chapters 4, 8, and 9)	One model is citizen science activities (e.g., local people monitoring their own rainfall and water courses, with potential to share science locally; Chapter 10)
8 (Re)think settings for communication	• Provide opportunities for different groups to bump into knowledge about flood risk and action	This involves promoting flood risk at local community events, set up for participants with other interests and purposes
9 Develop use of different channels, routes, and techniques for risk communication	• Recognise that some communication methods and techniques (e.g., social media) will date with technological advances and changing societal contexts • Different target groups may need different channels (Feldman et al., 2016)	Tabulations of different options in risk communication exist (e.g., Daly et al., 2015b) See CIRIA's (Daly et al., 2015a) mapping of communication strategies against spectrum of stakeholder participation and impact: Inform, Consult, Involve, Collaborate, and Empower
10 Use accessible language and be aware of threshold concepts	• Recognise that terms used by the public and FRM agencies to describe risk are often different, so need careful consideration to achieve mutual understanding	"Recognise that the public 'users' of the risk assessments and risk owners do not usually think in the same way as expert risk assessors" (UK Government Office of Science, 2011, p24) This does not necessarily mean ignorance or lack of understanding, but rather differences of language used for expression
11 Co-evaluate with communities	• Undertake formative and summative evaluation of any community-based, FRC strategy	This element of implementation is frequently omitted (e.g., due to cost, time)

Opportunities exist for innovation in FRC: to rethink the message, messenger, and facilitator beyond the statutory regulator; to experiment with new media and technologies; and to co-work with the arts and humanities in communicating, and promoting dialogue about, local flood risk. Importantly, this also involves moving communication out from the acute event phase, alongside tailoring for increasingly diverse flood types and communities (Chapter 2). Challenges include how to communicate threshold concepts like uncertainty and changing risks (compound, emerging) and how to work with more marginalised communities with different science capital who have fewer resources and less time to engage.

Within the paradigm of FRM, civic agency in scientific knowledge exchange has become increasingly important. The best messengers are not always specialist scientists; longitudinal working with locally trusted citizens such as embedded citizen scientists, flood wardens, or community flood groups can provide fresh ways forward in some settings. Other risk domains also bring valuable insights to FRC, particularly science communication in public health sciences with its expertise in engaging with diverse, marginalised, and vulnerable groups. Importantly, risk communication is not 'one size fits all'; multi-channel modes of communication are essential.

Navigating this territory requires integration of different expertise, particularly co-working between science communication specialists and social behavioural scientists (e.g., in how risk is perceived; in how to persuade) to communicate different aspects of risk, in real-time and outside acute events, in ways that engage and empower. Paramount is co-working between FRC professionals and at-risk communities in trials to develop models of scientific and lay knowledge exchange that are tailored and inclusive to the needs of different socio-economic and cultural community settings.

Notes

1 https://www.stem.org.uk/sites/default/files/pages/downloads/Science-Capital-Made-Clear.pdf.
2 https://conversational-leadership.net/knowledge-and-power/.
3 Funded by Leverhulme Trust (2012–2018).
4 https://www.nottingham.ac.uk/sociology/research/projects/making-science-public/index.aspx.
5 https://www.factcheck.org/2009/12/climategate/.
6 UK Department for Business, Energy, and Industrial Strategy.
7 https://www.theguardian.com/us-news/2021/sep/02/new-york-flooding-state-of-emergency-ny-city-flash-flood-nyc-hurricane-ida-remnants.
8 https://www.scientificamerican.com/article/studies-sound-alarm-on-badly-out-of-date-fema-flood maps/.
9 https://check-long-term-flood-risk.service.gov.uk/risk.
10 https://www.fema.gov/flood-maps.
11 https://environmentagency.blog.gov.uk/author/kate-marks/.
12 https://www.creativecarbonscotland.com/green-tease-reflections-understanding-climate-complexity/ (includes event recording).

13 Flood Awareness Week Exhibit at Artworks Gallery; https://youtu.be/5mr0QZvskGI.
14 https://www.creativecarbonscotland.com/.
15 'The Burnie Journey'; https://www.youtube.com/watch?v=c_OIdbz7j_w.
16 https://www.creativecarbonscotland.com/the-den-burn-is-flooding-dont-panic/.
17 https://highwaterline.org/.
18 https://highwaterline.org/join/action-guide/.
19 'Community engagement with its flood history – understanding risk' – funded through Royal Society's Connecting People to Science. https://www.gov.uk/flood-and-coastal-erosion-risk-management-research-reports/community-engagement-with-its-flood-history-understanding-risk.
20 https://www.thedrum.com/creative-works/project/phantom-flood-the-streets.
21 https://www.plymouthherald.co.uk/news/plymouth-news/plymouth-graffiti-artists-hit-area-4516551.

References

Alexander, M., Viavattene, C., Faulkner, H. and Priest, S. (2014) Translating the complexities of flood risk science using KEEPER – a knowledge exchange exploratory tool for professionals in emergency response. *Journal of Flood Risk Management*, 7(3), 205–216. https://doi.org/10.1111/jfr3.12042

Archer, L., Dawson, E., DeWitt, J., Seakins, A. and Wong, B. (2015) "Science capital": A conceptual, methodological, and empirical argument for extending bourdieusian notions of capital beyond the arts. *Journal of Research in Science Teaching*, 52, 922–948. https://doi.org/10.1002/tea.21227

Armstrong, A.K., Krasny, M.E. and Schuldt, J.P. (2018) *Communicating Climate Change: A Guide for Educators*. Cornell University Press, New York.

Association of State Floodplain Managers (2020) *Flood Mapping for the Nation: A Cost Analysis for Completing and Maintaining the Nation's NFIP Flood Mapping Inventory*. https://asfpm-library.s3-us-west-2.amazonaws.com/FSC/MapNation/ASFPM_MaptheNation_Report_2020.pdf

Aven, T. (2013) On the meaning of a black swan in a risk context. *Safety Science*, 57, 44–51.

Baker, V.R. (2000) Paleoflood hydrology and the estimation of extreme floods. In E. Wohl (ed.) *Inland Flood Hazards: Human, Riparian, and Aquatic Communities*. Cambridge University Press, New York.

Baker, V.R. (2007) Flood hazard science, policy, and values: A pragmatist stance. *Technology in Society*, 29, 161–168.

Bell, H.M. and Tobin, G.A. (2007) Efficient and effective? The 100-year flood in the communication and perception of flood risk. *Environmental Hazards*, 7(4), 302–311.

Beven, K. (2016) Facets of uncertainty: Epistemic uncertainty, non-stationarity, likelihood, hypothesis testing, and communication. *Hydrological Sciences Journal*, 61(9), 1652–1665. https://doi.org/10.1080/02626667.2015.1031761

Blöschl, G., Gaál, L., Hall, J. et al. (2015) Increasing river floods: Fiction or reality? *WIREs Water*, 2, 329–344. https://doi.org/10.1002/wat2.1079

Bucknall, T. and Hitch, D. (2018) Connections, communication and collaboration in healthcare's complex adaptive systems: Comment on "Using complexity and network concepts to inform healthcare knowledge translation". *International Journal of Health Policy and Management*, 7(6), 556–559. https://doi.org/10.15171/ijhpm.2017.138

Burke, K.L. (2015) *8 myths about public understanding of science*. https://www.american-scientist.org/blog/from-the-staff/8-myths-about-public-understanding-of-science

Carlton, S.J. and Jacobson, S.K. (2013) Climate change and coastal environmental risk perceptions in Florida. *Journal of Environmental Management*, 130, 32–39.

Corner, A., Lewandowsky, S., Phillips, M. and Roberts, O. (2015) *The Uncertainty Handbook*. University of Bristol, Bristol.

Cousin, G. (2006) An introduction to threshold concepts. *Planet*, 17(1), 4–5.

Daly, D., Jodieri, R., McCarthy, S., Pygott, K. and Wright, M. (2015a) *Communication and Engagement Techniques in Local Flood Risk Management*. Report C751. CIRIA, London.

Daly, D., Jodieri, R., McCarthy, S., Pygott, K. and Wright, M. (2015b) *Communication and Engagement Techniques in Local Flood Risk Management: Companion Guide*, Report C752. CIRIA, London.

Demeritt, D. and Nobert, S. (2014) Models of best practice in flood risk communication and management. *Environmental Hazards – Human and Policy Dimensions*, 13(4), 313–328. https://doi.org/10.1080/17477891.2014.924897

Department for Business, Energy and Industrial Strategy BEIS (2019) *Public attitudes to science 2019* Main report. BEIS Research Paper Number 2020/012

Donovan, K. (2010) *Cultural responses to volcanic hazards on Mt Merapi, Indonesia*. Unpublished PhD thesis, University of Plymouth.

Faulkner, H., Parker, D., Green, C. and Beven, K. (2008) Developing a translational discourse to communicate uncertainty in flood risk between science and the practitioner. *Ambio*, 36, 692–703. https://doi.org/DO-10.1579/0044-7447(2007)36[692:DATDTC] 2.0.CO;2

Felder, G., Gómez-Navarro, J.J., Zischg, A.P., Raible, C.C., Röthlisberger, V., Bozhinova, D., Martius, O. and Weingartner, R. (2018) From global circulation to local flood loss: Coupling models across the scales. *Science of the Total Environment*, 635, 1225–1239. https://doi.org/10.1016/j.scitotenv.2018.04.170

Feldman, D., Contreras, S., Karlin, B. et al. (2016) Communicating flood risk: Looking back and forward at traditional and social media outlets. *International Journal of Disaster Risk Reduction*, 15, 43–51 https://doi.org/10.1016/j.ijdrr.2015.12.004

Fischhoff, B. (1995) Risk perception and communication unplugged: Twenty years of process. *Risk Analysis*, 15(2), 137–145.

Fisher, H. (2015) *Public dialogues on flood risk communication*. Joint Defra/Environment Agency Flood and Coastal Erosion Risk Management R&D Programme, Final Report – SC120010/R1. Environment Agency, Bristol.

Fiske, S.T. and Dupree, C. (2014) Gaining audiences' trust and respect about science. *Proceedings of the National Academy of Sciences*, 111(Supplement 4), 13593–13597. https://doi.org/10.1073/pnas.1317505111

Flage, R. and Aven, T. (2015) Emerging risk – Conceptual definition and a relation to black swan type of events. *Reliability Engineering and System Safety*, 144, 61–67.

Gascoigne, T., Schiele, B., Leach, J., Riedlinger, M., Lewenstein, B.V., Massarani, L. and Broks, P. (eds.) (2020) *Communicating Science: A Global Perspective*. ANU Press, Canberra. http://doi.org/10.22459/CS.2020

Giles, J. (2002) Scientific uncertainty: When doubt is a sure thing. *Nature*, 418, 476–478.

Gotham, K.F., Campanella, R., Lauve-Moon, K. and Powers, B. (2018) Hazard experience, geophysical vulnerability, and flood risk perceptions in a postdisaster city, the case of New Orleans. *Risk Analysis*, 38, 345–356. https://doi.org/10.1111/risa.12830

Haer, T., Botzen, W.J.W. and Aerts, J.C.J.H. (2016) The effectiveness of flood risk communication strategies and the influence of social networks—Insights from an agent-based model. *Environmental Science & Policy*, 60, 44–52.

Hagemeier-Klose, M. and Wagner, K. (2009) Evaluation of flood hazard maps in print and web mapping services as information tools in flood risk communication. *Natural Hazards and Earth System Sciences*, 9, 563–574. www.nat-hazards-earth-syst-sci.net/9/563/2009/

Haggett, C. (1998) An integrated approach to flood forecasting and warning in England and Wales. *Water and Environment Journal* 12, 425–432.

Hall, B. (2010) *Teaching Uncertainty: The Case of Climate Change*. Unpublished PhD, University of Gloucestershire.

Handmer, J. (2001) Improving flood warnings in Europe: A research and policy agenda. *Global Environmental Change Part B: Environmental Hazards* 3, 19–28.

McCarthy, S. S. (2007) Contextual influences on national level flood risk communication. *Environmental Hazards* 7, 128–140.

Haughton, G. and White, I. (2017) Risky spaces: Creating, contesting and communicating lines on environmental hazard maps. *Transactions of the Institute of British Geographers*, 43, 1475–5661. https://doi.org/10.1111/tran.12227

Haynes, P., Hehl-Lange, S. and Lange, E. (2018) Mobile augmented reality for flood visualisation. *Environmental Modelling and Software*, 109, 380–389.

HM Government (2016) *UK National Flood Resilience Review*. https://www.gov.uk/government/publications/national-flood-resilience-review

Hogan Carr, R., Montz, B., Maxfield, K., Hoekstra, S., Semmens, K. and Goldman, E. (2016) Effectively communicating risk and uncertainty to the public: Assessing the National Weather Service's flood forecast and warning tools. *Bulletin American Meteorological Society*, 1649–1666. https://doi.org/10.1175/BAMS-D-14-00248.1

Höppner, C., Buchecker, M. and Bründl, M. (2010) *Risk communication and natural hazards*. CapHaz-Net WP5 Report. Birmensdorf, Switzerland: Swiss Federal Research Institute.

Joubert, M., Davis, L. and Metcalfe, J. (2019) Storytelling: The soul of science communication. *JCOM*, 18 (05), E. https://doi.org/10.22323/2.18050501

Kuller, M., Schoenholzer, K. and Lienert, J. (2021) Creating effective flood warnings: A framework from a critical review. *Journal of Hydrology*, 602, 126708.

Kuser Olsen, V.B., Momen, B., Langsdale, S.M., Galloway, G.E., Link, E., Brubaker, K.L., Ruth, M. and Hill, R.L. (2016) An approach for improving flood risk communication using realistic interactive visualisation. *Journal of Flood Risk Management*, 11, S783–S793. https://doi.org/10.1111/jfr3.12257

Land, R., Cousin, G., Meyer, J.H.F. and Davies, P. (2005) Threshold concepts and troublesome knowledge (3): Implications for course design and evaluation. In C. Rust (ed.) *Improving Student Learning – Diversity and Inclusivity*. Oxford Centre for Staff and Learning Development (OCSLD), Oxford, pp53–64 https://www.ee.ucl.ac.uk/~mflanaga/ISL04-pp53-64-Land-et-al.pdf

Lanza, T. and Negrete, A. (2007) From myth to earth education and science communication. *Geological Society, London, Special Publications*, 273, 61–66.

Lehmann, R.J. (2020) Do no harm: Managing retreat by ending new subsidies. *R Street Policy Study No. 195*. https://www.rstreet.org/wp-content/uploads/2020/02/195.pdf

Leiss, W. (1996) Three phases in the evolution of risk communication practice. *The Annals of the American Academy of Political and Social Science*, 545, 85–94.

Luke, A., Sanders, B.F., Goodrich, K.A. et al. (2018) Going beyond the flood insurance rate map: Insights from flood hazard map co-production. *Natural Hazards and Earth System Sciences*, 18, 1097–1120, https://doi.org/10.5194/nhess-18-1097-2018

Marks, K. (2016) Communicating flood risk. https://environmentagency.blog.gov.uk/2016/11/08/communicating-flood-risk/

Martini, F. and Loat, R. (2007) *Handbook on good practices for flood mapping in Europe*. European exchange circle on flood mapping EXCIMAP. European Commission. https://www.preventionweb.net/publication/handbook-good-practices-flood-mapping-europe

McEwen, L.J. (2011) Approaches to community flood science engagement: The Lower River Severn catchment, UK as case-study. *International Journal of Science in Society*, 2(4), 159–179.

McEwen, L.J., Stokes, A., Crowley, K. and Roberts, C.R. (2014) Using role-play for expert science communication with professional stakeholders in flood risk management. *Journal of Geography in Higher Education*, 38, 277–300. https://doi.org/10.1080/03098265.2014.911827

Messling, L., Corner, A., Clarke, J., Pidgeon, N.F., Demski, C. and Capstick, S. (2015) *Communicating Flood Risks in a Changing Climate*. Climate Outreach, Oxford. https://climateoutreach.org/reports/communicating-flood-risks-in-a-changing-climate/

Meyer, J.H.F. and Land, R. (2003) Threshold concepts and troublesome knowledge: Linkages to ways of thinking and practising within the disciplines. In C. Rust (ed.) *ISL10 Improving Student Learning: Theory and Practice Ten Years On*. OCSLD, Oxford. pp.412–424.

Meyer, J.H.F. and Land, R. (2005) Threshold concepts and troublesome knowledge (2): Epistemological considerations and a conceptual framework for teaching and learning. *Higher Education*, 49, 273–288.

Mileti, D., Nathe, S., Gori, P., Greene, M. and Lemersal, E. (2004) *Public Hazards Communication and Education: the State of the Art*. Natural Hazards Center University of Colorado..

Minano, A. and Peddle, S. (2018) *Using Flood Maps for Community Flood Risk Communication*. Report prepared for Natural Resources Canada (NRCan-30006050733). Waterloo, Ontario: Partners for Action.

Mosher, E. and Quante, H. (2018) *Highwaterline Action Guide*. https://www.evemosher.com/highwaterline-guide

Moss, R.H. and Schneider, S.H. (2000) Uncertainties in the IPCC TAR: Recommendations to lead authors for more consistent assessment and reporting. In R. Pachauri, T. Taniguchi, and K. Tanaka (eds.) *Cross Cutting Issues of the Third Assessment Report of the IPCC*, World Meteorol. Org., Geneva, pp33–51.

Nerlich, B., Hartley, S., Raman, S. and Smith, A.T.T. (2018) Introduction in *Science and the politics of openness*. https://doi.org/10.7765/9781526106476.00005

O'Neill, P. (2004) *Developing a risk communication model to encourage community safety from natural hazards*. Presented at the 4th NSW Safe Communities Symposium, Sydney, NSW, 2004. Wollongong, New South Wales, Australia: NSW Government, State Emergency Service.

Orr, P., Forrest, S., Brooks, K. and Twigger-Ross, C. (2015) *Public dialogues on flood risk communication, Literature review*. SC120010/R3 Technical Report.

Parker, D. J. and Priest, S. J. (2012) The fallibility of flood warning chains: can Europe's flood warnings be effective? *Water Resources Management* 26, 2927–2950.

Parker, D., Fordham, M., Tunstall, S. and Ketteridge, A.-M. (1995) Flood warning systems under stress in the United Kingdom. *Disaster Prevention and Management* 4, 32–42.

Perkins, D. (1999) The many faces of constructivism. *Educational Leadership*, 57(3), 6–11. http://www.scribd.com/doc/32920521/Perkins-The-Many-Faces-of-Constructivism

Pielke, R.A. (1999) Nine fallacies of floods. *Climatic Change*, 42, 413–438.

Ping, N.S., Wehn, U., Zevenbergen, C. and van der Zaag, P. (2016) Towards two-way flood risk communication in a community in the UK. *Journal of Water and Climate Change*, 07(4), 651–664.

Portes, R. (2017) "I think the people of this country have had enough of experts". Think at London Business School. https://www.london.edu/think/who-needs-experts

Severn and Avon Valley Combined Flood Group (2008) *Building on the floodplain is misguided (a proof of evidence).* https://flooding.london/jrp/Report%202008.pdf

Shanahan, E.A., Reinhold, A.M., Raile, E.D. et al. (2019) Characters matter: How narratives shape affective responses to risk communication. *PloS One*, 14(12) e0225968.

Shannon, C.E. and Weaver, W. (1949) *The Mathematical Theory of Communication.* University of Illinois Press, Champaign.

Spiegelhalter, D., Pearson, M. and Short, I. (2011) Visualizing uncertainty about the future. *Science*, 09, 1393–1400.

Spiegelhalter, D.J. and Riesch, H. (2011) Don't know, can't know: Embracing deeper uncertainties when analysing risks. *Philosophical Transactions Royal Society A: Mathematical, Physical and Engineering Sciences*, 369(1956), 4730–4750.

Spiekermann, R., Kienberger, S., Norton, J., Briones, F. and Weichenselgarter, J. (2015) Disaster-knowledge matrix: Reframing and evaluating the knowledge challenges in disaster risk reduction. *International Journal of Disaster Risk Reduction*, 13, 96–108. https://doi.org/10.1016/j.ijdrr.2015.05.002

Stewart, R.M. and Rashid, H. (2011) Community strategies to improve flood risk communication in the Red River Basin, Manitoba, Canada. *Disasters*, 35(3), 554–576.

Taleb, N.N. (2010) *The Black Swan: The Impact of the Highly Improbable* (2nd edn). Penguin, London.

Tarr, I. and Haydock-Wilson, A. (2015) *Report for HighWaterLine Bristol.* https://annahaydockwilson.com/highwaterline/

UK Government Office for Science (2011) *Blackett review of high impact low probability risks.* https://assets.publishing.service.gov.uk/government/uploads/system/uploads/attachment_data/file/278526/12-519-blackett-review-high-impact-low-probability-risks.pdf

US Association of State Floodplain Managers (2020) *Flood Mapping for the Nation: A Cost Analysis for Completing and Maintaining the Nation's NFIP Flood Map Inventory.* Association of State Floodplain Managers, Inc, Madison.

van der Bles, A.M., van der Linden, S., Freeman, A.L.J., Mitchell, J., Galvao, A.B., Zaval, L. and Spiegelhalter, D.J. (2019) Communicating uncertainty about facts, numbers and science. *Royal Society. Open Science*, 6, 181870. http://dx.doi.org/10.1098/rsos.181870

Wagner, W. (2007) Vernacular science knowledge: Its role in everyday life communication. *Public Understanding of Science*, 16, 7–22.

Wardman, J.K. (2008) The constitution of risk communication in advanced liberal societies. *Risk Analysis*, 28, 1619–1637. https://doi.org/10.1111/j.1539-6924.2008.01108.x

Weitkamp, E., McEwen, L.J. and Ramirez, P. (2020) Communicating the hidden: Towards a framework for drought risk communication in maritime climates. *Climatic Change*, 163, 831–850. https://doi.org/10.1007/s10584-020-02906-z

Wilhite, D.A. (2012) Breaking the Hydro-Illogical Cycle: Changing the Paradigm for Drought Management. Drought Mitigation Center Faculty Publications, 53. http://digitalcommons.unl.edu/droughtfacpub/53

Wood, M.M., Mileti, D.S., Kano, M., Kelley, M.M., Regan, R. and Bourque, L.B. (2012) Communicating actionable risk for terrorism and other hazards. *Risk Analysis*, 32(4), 601–15. https://doi.org/10.1111/j.1539-6924.2011.01645.x.

Wood, E. and Miller, S.K. (2020) Cognitive dissonance and disaster risk communication. *Journal of Emergency Management and Disaster Communications*, 2(1), 39–56. https://doi.org/10.1142/S2689980920500062

7

FLOOD MANAGEMENT STRATEGIES AND THEIR RELATION TO COMMUNITY AWARENESS AND ACTION

7.1 Introduction

Within a 'risk society' (Beck, 1992), an adaptation approach is needed where impacts of changing risk are prevented where practical, mitigated where possible, and where attention is paid to preparing for quicker recovery. Over the past 75 years, several researchers have evaluated different approaches to flood management from engineering and/or behavioural and social vulnerability perspectives (e.g., White, 1945; Wisner et al., 2004; Pender and Faulkner, 2011). This chapter aims to consider different measures within integrated and sustainable flood risk management (FRM) through an explicit community lens for More Economically Developed Countries (MEDCs). It focuses on the extent that different approaches rely on, embrace or have the potential to involve, individual or collective actions by empowered citizens. It asks:

- What is the changing nature of the inputs needed from local communities with shifts between different flood management paradigms?
- What can be (re)learned from historical approaches to local flood management used by communities and other stakeholders for local resilience building?
- What opportunities do new ways of preparing for, mitigating, and adapting to floods bring to building co-benefits for community resilience?

Academic disciplines that contribute to this socio-technical territory include geography, psychology, sociology, water engineering, and technology/design. Potential exists for valuable knowledge exchange about effective community agency in specific risk management measures between different cultural and socio-economic settings, including Less Economically Developed Countries (LEDCs) and MEDCs.

DOI: 10.4324/9781315666914-7

7.2 Different paradigms for managing floods and the role of communities

Chapter 2 outlined different paradigms or shifts in flood management and their environmental, economic, and societal drivers; each paradigm has implications for the roles of citizens and diverse communities. There is accompanying growth in concern for the role of, and in securing, citizen and community agency in FRM and in dealing with residual risk once any measures are implemented (Chapters 1 and 2). But as floods play out, there can be a strong sense of 'paradigm lock' (Chapter 4), with the dominant discourse in both community campaigning and media coverage still articulated solely in terms of 'defence'. This can mean neglect of integrated flood risk solutions that actively promote awareness of the design limits of protective structures; the nature and scale of residual risk (which may also be increasing due to climate change); and the status of personal and collective responsibility in FRM.

To be effective, many flood preparedness interventions for dealing with residual risk require strong positive behavioural elements involving the agency of at-risk citizens and communities in their implementation. Examples include emergency management planning, early flood warning systems, flood insurance, and property-level protection (PLP). A range of management options operating at different scales, and used individually or in combination, chip away at levels of risk, after any structural measures have been put in place (Figure 2.8). Hence, flood science communication is not just about local flood prediction and forecasting (Chapter 6); rather, communities need to be empowered through improved scientific and technical knowledge, and understanding of how interventions function to reduce risk, with their strengths and limitations. The role of active community participation in the design, implementation, monitoring, and evaluation of innovations is critical (Chapter 8).

There is a strong drive within the UNDRR's *Sendai Framework for Disaster Risk Reduction* 2015–2030 (hereafter 'Sendai Framework'; Priority 4) that "back to normal is not enough" in disaster recovery (Section 8.2 of Chapter 8). Countries should be "investing in disaster reduction for resilience", "enhancing disaster preparedness for effective response, and to 'Build Back Better' in recovery, rehabilitation and reconstruction" to deal proactively with societal inequalities. Community-focused disaster risk reduction needs to be an integral part of that strategy (Chapters 1, 2, and 8).

In safeguarding wider community functioning, it is not just risk to residential properties and businesses that needs considering. Also important during floods is protection of local critical infrastructure. To quote the Sendai Framework (Priority 4, 33C), it is necessary

> To promote the resilience of new and existing critical infrastructure, including water, transportation and telecommunications infrastructure, educational facilities, hospitals and other health facilities, to ensure that they remain safe,

effective and operational during and after disasters in order to provide life-saving and essential services.

Depending on scale and the role of the state, local-level community inputs to planning can help protect these critical resources and plan local contingencies for when infrastructure systems fail (Chapter 8). The chapter now considers the diverse options for managing floods and their interfaces with communities.

7.3 Interactions of infrastructural measures and communities

Several 'mitigation myths' pervade public thinking about structural flood management as a panacea (Chapter 2). These include that flood defence measures protect against the maximum probable flood within a river catchment or coastal setting and that the same measures or interventions can protect from all scales of flooding. As Beven (2014, np) states: "total flood defence is a myth". While levees, dikes, or embankments can protect up to their design limits, their failure or breaching can exacerbate floods with devastating impacts on communities (e.g., in China, Plate 2002). In addition, channel straightening can speed water flows, protecting an upstream community but exporting the flood problem to other downstream communities (e.g., along the Mississippi, Emerson, 1971). Thus, structural measures can offer a level of protection to one community at the expense of others.

Many urbanised rivers and deltas have long traditions of flood protection. Historically, flood defence would have been implemented by experimenting with what works, with individuals and communities 'learning by doing' (Van Loon-Steensma and Vellinga, 2019). Structural measures – such as building of levees – would have been implemented by individual riparian landowners or communities for local flood protection (e.g., Bagus, 2007). Such measures have historically been an adaptive response of community leaders to extreme floods (e.g., 'Let's build the Ike Dike' – a seawall in response to the 1900 Great Storm on Galveston Island, USA[1]). In Germany, the first dykes were built over 1,000 years ago without any government assistance, starting a long and complex history of subsequent building that eventually moved from private to public dyking. In the 20th century, the last three private dikes were built (Bagus, 2006).

Over time, large-scale embankment building in many countries has become a collective rather than an individual endeavour, designed and built by trained government engineers. This is seen on a large scale in the Rhine Delta (Netherlands) or along the Yangtze River (China). There are also complicated issues in the 'economics of dikes' in relation to the roles of the public and the state (Bagus, 2006). These include non-rivalrous consumption (what is built for one person, benefits others) and non-excludability of public goods (everyone gains whether they paid or not). Different national systems of responsibility for protection still exist, for example, in Denmark where property protection from coastal flooding is a private responsibility (Baron, 2020). Baron found that conflicts in creating and maintaining dikes were frequently strongly linked to their relationships with other "controversies" within communities that linked to place-specific values and experiences. These included liveability and population decline of an area. Such embedded community-focused issues tend not to be considered in science-based, flood risk assessments.

In the late 20th and early 21st centuries, not all communities want hard engineering structures in their townscapes. With rapidly advancing, new flood resistance technologies[2] and associated business entrepreneurship, more attention has been given to less visually disruptive, protective structures that combine design and engineering in urban riverside settings. This takes account of the everyday aesthetics of riverfront locations, recognising that businesses and communities rely on such amenity resources for tourism. Varied options exist, with new types of less visually intrusive, permanent structures. These include glass barriers set above a flood wall to maintain sensitively the visual appeal of historic waterfronts (e.g., Figure 7.1A at Upton-upon-Severn, UK).

FIGURE 7.1   (A) Glass flood defence wall along riverside in action (https://www.geograph.org.uk/photo/3848842 Copyright Bob Embleton and licensed for reuse under Creative Commons Licence Attribution-ShareAlike 2.0 Generic (CC BY-SA 2.0)); (B) example of an earlier dismountable barrier (copyright Environment Agency, 2009) – both Upton-upon-Severn, UK (© Environment Agency, 2009)

Alternatively, a temporary barrier can be installed only when high flood levels are forecast; dismountable or removable flood barriers normally require built-in parts and hence may only normally be used at one location. A range of smaller-scale, creatively designed inventions (e.g., Scandinavian) can be rapidly installed in preparation for forecasted floods, with some using force of water to maintain their physical integrity. These structures can be completely removed between floods (e.g., Figure 7.1B; same UK town) so returning the riverside and its community relations to their prior state.

Not all structural measures need to be permanent and aesthetically displeasing, as seen in many historically canalised sections of urban rivers. Strategies for integrated FRM need regular revisiting to consider opportunities offered by new or advancing technical knowledge, as well as changing values of lay and specialist stakeholders (Chapter 4). In developing integrated flood management that is 'sustainable' (socially, economically, and environmentally), three elements are critical: understanding how communities interact with management options where the state has a role; understanding how communities might engage with residual risk to minimise individual or collective losses; and communities understanding the extent of personal versus collective (community) responsibility for action.

Different approaches are needed for managing routine moderate floods, as opposed to extreme floods and outliers. These highlight the importance of lateral, systemic (non-linear), and creative thinking in adaptive FRM. The efficacy of different types and scales of flood management invention requires different socio-technical and socio-environmental relationships with communities – their awareness and agency. This chapter now explores a suite of adaptive options – not normally used in isolation – that involve contrasting scales of intervention and different requirements for community buy-in and cooperative working of multiple professions and organisations (AIDR, 2017; Tyler et al., 2019). These include nature-based solutions (NbS) among farmers; sustainable drainage systems (SuDs) for integrated (urban) land and water management among urban dwellers and planners; and property-level flood resilience measures among at-risk citizens and communities.

7.4 Catchment-based solutions and their communities

Multiple benefits for both communities and biodiversity can be potentially secured from interventions at wide-ranging scales, from site-specific (floodplain) measures to those implemented as whole catchment solutions.

7.4.1 Floodplain-based approaches

Historically, building on any locally higher ground has been the traditional way that key community actors, such as medieval religious leaders, managed flood risk on floodplains. In the 21st century, designated flood risk zones frequently have different permitted building codes, and hence accurate floodplain zoning is needed. For example, in the USA, such zoning is critical to the functioning of the federal insurance programme, along with community awareness of its implications. However, as Pralle (2019, p227) noted such line drawing on maps can have political motivations, expressing concerns that:

> Because mapping takes place within the context of the National Flood Insurance Program, the conversation at a local level often centers on the costs of revising the flood hazard zones rather than the risks associated with flooding.
>
> *author's emphasis*

Recognising these challenges, more recently, approaches to floodplain management have developed, with alternative systemic thinking about floodplain-human interactions like 'making space for water' and river 'daylighting'.

7.4.1.1 Making space for water

A different narrative about local FRM is of 'making room for the river', as opposed to continuing to 'battle against water' (de Groot and de Groot, 2009). This approach recognises that structural protection brings challenges and that allowing space for rivers can have co-benefits. For example, European rivers have undergone long histories of transformation, with alteration from medieval times and industrialisation of channels through hard engineering measures. Alternative holistic thinking is about river restoration or rehabilitation that removes the confines of artificially bounded channels, releasing rivers from barriers and allowing them to 'rewild'. This approach recognises the importance of riverine ecosystem services for local communities (Brown et al., 2018). Ecosystem services are the benefits people obtain from ecosystems. However, such strategies require local community 'buy in' in financial and other terms, whilst promoting wild rivers for their biodiversity, conservation, and cultural (aesthetic, spiritual, educational, and recreational) benefits. The Millennium Ecosystem Assessment's (2005) conceptual model highlighted the complex web of relationships between ecosystem services and different aspects of individual and collective human well-being (Figure 7.2).

Such approaches do, however, pose questions as to whether the values and river relations of riparian communities also change. For example, de Groot and de Groot (2009, p1), surveying communities along the River Waal floodplain, Germany, set the scene:

> Dutch river management is moving from traditional dike reinforcements toward 'room for river" measures to assure flood protection and to serve other societal goals. With that, the engineering paradigm shifts from mastership over nature to a more partnership-like attitude toward nature. Are these values shared by Dutch riverside residents? Are the measures proposed under the new paradigm accepted?

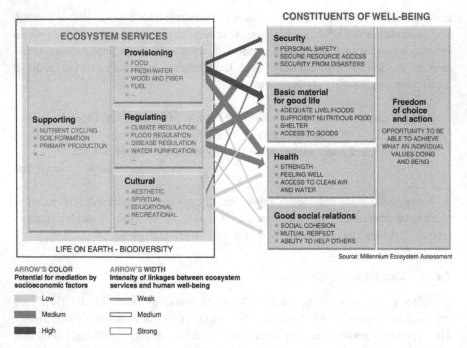

FIGURE 7.2   Conceptual model of linkages between ecosystem services and human well-being (Millennium Ecosystem Assessment, 2005) (credit: Millennium Ecosystem Assessment; Figure 1.1)

They found that despite eco-centric visions of nature, partnership with nature was not a strongly held view among riparian dwellers. This poses questions about how values of communities and farmers might need to change with the implementation of such measures.

7.4.1.2 'Daylighting' rivers

Linked to the developments just described, there is an international movement to de-culvert or 'daylight' urban rivers that were originally constrained for flood control and to hide polluted urban streams from urban residents. This radical drive initially focused on the environmental outcomes and ecosystem benefits of river restoration (Broadhead and Lerner, 2013; Kaufman, 2013). Wild et al. (2011), in their international review of daylighting, cited environmental, economic, and social objectives against outcomes, including flood risk reduction. More recently, however, a stronger emphasis has been placed on its multiple societal co-benefits in regenerating blighted urban settings. Now, daylighting agendas are frequently framed in emotive language around imperatives for "rethinking urban rivers" (cf. Kibel, 2007) and "breathing life into urban streams and communities" (American Rivers, undated). High-profile examples of successful daylighting in urban regeneration in different cultural settings include the Sawmill River (New York, USA) and Cheonggyecheon River (South Korea).

However, this "opening up subterranean, culverted waterways often forgotten by communities above ground" (Usher et al., 2020, p1) frequently only involves communities later in its implementation processes. They (p1) highlighted the multiple benefits of engaging communities early:

> Daylighting can unleash the social 'stickiness' of water, its proclivity to draw and bind together, to revitalize the park, enhancing connection to wildness, attachment to place and sense of community.

This requires attention to meaningful community participation (Arnstein, 1969; Chapter 8) in river visioning processes. Lost Rivers (2012, np) noted shifting practice:

> "Participation" and citizen engagement are more frequently included in [daylighting] its strategies, processes, and practices, in a move from "informing" to "citizen control." Local citizens are increasingly involved in coworking, and politically as lead proponents/activists in campaigning for their local watercourse. This recognises "daylighting" as "having a new vision about rivers in the city".

River 'daylighting' offers potential to engage all citizens in its conception. For example, McEwen et al. (2020) explored the possible role of the socially engaged arts in deliberative participatory daylighting processes of river visioning involving children (Chapter 8). They found primary children suggested "playful, innovative, but realistic solutions to real problems, juxtaposing different ideas (e.g., 'flood park')" (p790). A growing number of researched examples exist of daylighting where urban rivers have been successfully reconnected with their riparian communities, with significant health and well-being benefits. This includes the River Quaggie scheme, London, UK[3] (Wild et al., 2011); as CIWEM (2007, p6) observed, benefits obtained from associated increased physical activity were estimated "as potentially being capable of offsetting the cost of the whole scheme in three years". Transformative possibilities include use of these new blue-green spaces in social prescribing for better local health and well-being of place-based communities (Chapter 10).

7.4.2 Communities engaging with buffers and managed retreat

Other local flood management strategies include buffers and retreat from flood risk zones; both require effective community engagement around their rationale and functioning. A buffer is a "defined area by a river or coast that is protected from development to preserve natural benefits of its ecosystems and reduce flood or erosion risks".[4] As Costanza (2021, np) proclaimed "Forget massive seawalls, coastal wetlands offer the best storm protection money can buy". Wetlands and marshes absorb storm energy acting as natural buffers that protect coastal property and their communities from storm surges and sea-level rise. Protecting and preserving such natural habitats can make important contributions to enhancing coastal resilience. Rezaie et al. (2020), working in New Jersey, USA, noted that such liminal zones are not static, with greater prevalence of floods due to climate change. Their modelling results indicated that salt marshes can be effective during less severe storm events, reducing flood depth and property damage by up to 14%. They argued that this demonstrates the efficacy of natural flood protection as a strategy for more moderate storms, an important shift from structural defence thinking to valuing natural habitats for their resilience function.

As Alexander et al. (2012, p409) reflected from work in Australia: "The threat of accelerated sea-level rise (SLR) presents coastal communities with a host of social, legal, economic and environmental challenges". Another adaptive option for riverine and coastal communities in response to repeated flooding, increasing flood risk, and sea-level rise associated with climate change, is managed retreat (Hino et al., 2017; Doberstein et al., 2019; Siders et al., 2019). Other related terms include "managed realignment" or "deliberate process of altering flood defences to allow flooding of a presently defended area" (UN Climate Technology Centre and Network[5]). These approaches involve "the purposeful, coordinated movement of people and assets out of harm's way" (Siders, 2019a, p216) and provide an alternative long-term, potentially transformative and sustainable strategy to resistance or accommodation measures. They can "buy time for the thoughtful planning and social reform needed to engage in large-scale managed retreat" (Siders, 2019a, p218). Such thinking is gaining significant research and policy interest, particularly in its interactions with at-risk communities. Ultimately managed retreat may be "one of the few policy options available for coastal communities facing long-term risks from accelerated sea-level rise" (Alexander et al., 2012, p405; Bevan (BBC News, 2020)). However, where coastal towns after repeat flooding are asked to abandon waterside properties, communities may prefer to support expensive beach replenishment schemes aimed at protecting property values (e.g., Anderson, 2019).

Despite its recent traction, this is not a new strategy. Pinter (2021a) reconstructed histories of managed retreat, and wholesale or partial community relocation in the USA, giving ca. 50 examples from over 140 years. The earliest cited case occurred after ice dam flooding in Niobrara, Nebraska (in 1881) when the whole town elected to move (Pinter, 2021b). This local community action occurred before the US federal government gained statutory responsibility for disaster management. Pinter (2021a) proffered his database of historic examples of retreat as a learning resource for "disaster management professionals, and local stakeholders contemplating retreat" (p1). Completely successful examples of relocation are rare but one is Valmeyer, IL – now New Valmeyer – which moved to higher ground after the extreme 1993 Mississippi floods, USA, although even here, some community members relocated out of the area altogether.

Valmeyer's relocation has gained recent media attention; for instance, CNN news in 2019[6] asked, "could it be a model for other towns hit by the climate crisis?". This approach involving "not repairing structures or building higher ones" (offering increased standards of protection) is gaining some momentum as a way of working in exposed river and coastal environments. However, as Carey (2020, p3) noted, "moving an entire community out of harm's way, while keeping its culture and cohesiveness intact is extraordinarily difficult". He (2020, p1) reflected:

> … managed retreat presents numerous complex challenges – legal, logistical, ethical, political, financial, and architectural. Communities, and community cultures, aren't easily transported and retained. But all indications are that researchers, policymakers, and the general public will need to confront these challenges with increasing frequency in the coming decades.

Barriers to retreat and relocation can include "place attachment, potential loss of livelihoods, and lack of funding, suitable land, community consensus, and governance procedures" (Dannenberg et al., 2019, p1). Siders (2019a, p219) differentiated psychological, institutional, and practical barriers, emphasising their interactions may "reinforce one another, requiring transformation of social, legal and economic systems to address" (Table 7.1). All are increased by persistence of a "lack of co-ordinating vision". Siders also proposed possible multiple elements of an engaged participatory process with flood-affected communities to overcome these barriers for transformation (Figure 7.3). Notably, this includes roles for creativity (arts) and social learning.

This is complex conceptual and operational territory, with social justice implications in how property acquisition programmes operate and whether buyouts "promote or reduce existing social inequities" (Siders, 2019b, p239). Similar issues occur in other hazard settings, e.g., Nguyen's (2020) appraisal of homeowners' choices for earthquake hazards in New Zealand. Here, factors affecting choice of options for buyout schemes included property value, damage levels, and household demographics. Nguyen also found that peer influence in communities also had a role in property owners' sub-optimal decision-making. Dannenberg et al. (2019, p1), reviewing the literature on the health implications of managed retreat on small island or coastal communities, also cited "disruptive health, socio-cultural and economic impacts" for communities that relocate.

TABLE 7.1 Barriers to Managed Retreat (Adapted from Siders, 2019a, Figure 3, p221, permission from Elsevier and author)

Psychological	Fear, optimism bias, place attachment, equating retreat with defeat
Institutional	Subsidised risk, disincentives, authority mismatch, lack of fit
Practical	Lack of learning, lack of evaluation, inequity and logistics

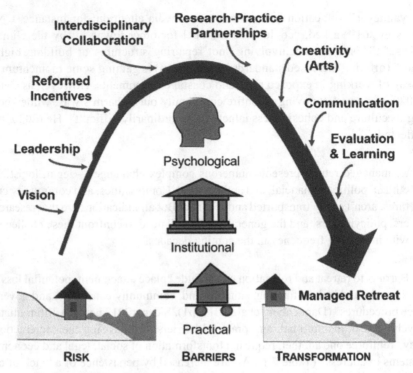

FIGURE 7.3 Overcoming barriers to managed retreat requires diverse actors coordinated by leadership and vision (source: Siders, 2019a, Figure 4, p221; reproduced by permission of Elsevier and author)

Hino et al. (2017) provided a conceptual model for 'managed retreat' with axes reflecting: residents' initial desire to move; and the scale of who benefits, working out from those residents. This appraisal of key influencing factors included the residents' feelings about risk – whether "tolerable" or "intolerable", and whether they or others initiate the move. Effective local community engagement about these core aspects is critical to the likely success of any managed retreat process. Despite its challenges, Siders et al. (2019, p761) observed that by

reconceptualizing retreat as a set of tools used to achieve societal goals, communities and nations gain additional adaptation options and a better chance of choosing the actions most likely to help their communities thrive.

7.4.3 Catchment-based solutions: Managing land and water together

Scaling up further to integrated catchment-based approaches, since the early 1990s, there has been increased recognition of the importance of managing land and water together in holistic ways to mitigate systemic impacts (e.g., Newson, 1992). This involves changing human relationships with the river catchment as a critical spatial unit. These have been articulated in various ways, including 'thinking like a watershed', interweaving cultural relationships to the land and resilience (cf. Loeffler and Loeffler, 2012); 'upstream thinking' (e.g., in catchment-based approach – CaBA[7]); and 'catchment systems thinking' (CaST) that integrates flood and water management alongside other risks and co-benefits.

Since the 2000s, multi-stakeholder interest has grown in the potential of nature-based solutions (NbS) in wider climate change adaptation in urban and rural settings (e.g., Kabisch et al., 2016; Strout et al., 2021). NbS are based on a wider philosophy of co-benefits and a set of approaches to meet complex challenges (e.g., extreme weather events). The European Commission defined NbS[8] as

Solutions that are inspired and supported by nature, which are cost-effective, simultaneously provide environmental, social and economic benefits and help build resilience. Such solutions bring more, and more diverse, nature and natural features and processes into cities, landscapes and seascapes, through locally adapted, resource-efficient and systemic interventions.

In NbS, diverse interests and motivations of civil society sit alongside those of other stakeholders, including commercial, media, political, and academic (European Environment Agency, 2021; Strout et al., 2021). Multi-stakeholder awareness of the parameters of effectiveness of NbS is essential; using the term 'solutions' can be problematic in promoting community awareness of limits of land use-based measures and remaining residual risk after their implementation. Also, NbS cannot always meet expectations of being resilient to change, when applied to evolving systems and higher stresses (Krauze and Wagner, 2019).

Where flooding is identified as a specific function, the term Natural Flood Management (NFM) is adopted. This is defined as: "when natural processes are used to reduce the risk of flooding and coastal erosion".[9] NFM involves a large suite of potential flexible, customisable measures that can be used in rural, suburban, and urban settings at varying scales (Figure 7.4). This emphasis on working with nature sits alongside increased concern for community engagement around attitudes to NFM, and possible roles for communities and farmers (e.g., co-designer, guardian) in their implementation and maintenance within catchment partnerships. This includes how action can be incentivised (e.g., state/EU subsidies that support rewilding of rivers and floodplains).

Effective NFM can therefore be conceived as a multi- and inter-disciplinary challenge requiring diverse expertise for success. Alongside technical skills, NFM measures also require

interdisciplinary insights from land-use planning, economics, property rights, sociology, landscape planning, ecology, hydrology, agriculture and other disciplines to address the challenges of implementing them.

Hartmann et al. (2019, np)

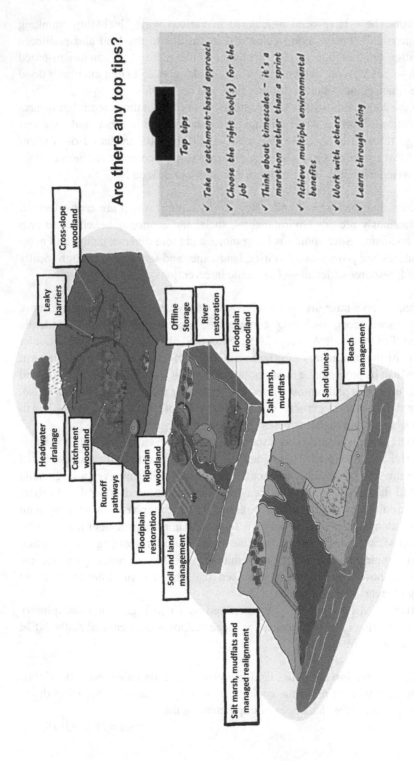

FIGURE 7.4 Working with natural processes to reduce flood risk (Burgess-Gamble et al., 2017). https://assets.publishing.service.gov.uk/media/6036c730d3bf7f0aac939a47/Working_with_natural_processes_one_page_summaries.pdf (permission: Lydia Burgess Gamble)

NFM also allows exploration of roles for the non-human, for example, beavers and their leaky dam-building, in catchment-wide FRM. This is being trialled at different scales (e.g., on River Otter, a small rapid response catchment, SW UK; in Las Vegas, Nevada, USA and north London, UK). However, such (re)introductions can be locally controversial, needing negotiation and broking with different stakeholder (e.g., farmer) perceptions of the species involved.

Awareness of the strengths and limitations of NFM is essential. Strategies for 'slowing the flow' through altering land use within catchments can have potential benefits for managing flood and drought risk together. However, land use manipulation tends to work better in attenuation of more moderate flows rather than extreme floods (Chapter 2). There are scientific, policy, and public debates about the magnitude and frequency of flows affected by NFM, due to a limited scientific evidence base that has monitored the effectiveness of interventions. This can lead to a mismatch with community expectations of a quick and total FRM solution.

7.4.3.1 Sustainable drainage systems (SuDS)

At a smaller urban scale, SuDS can be considered a toolbox of distributed, green-engineered features that manage surface water locally to mitigate surface water flooding. SuDS aim to mimic 'natural' drainage, collecting, storing, and cleaning water as close as possible to where rain falls, using surface interventions like reed beds and swales. The design of SuDS pays strong attention to integrating concern for water quantity and quality, conservation (increasing wildlife, biodiversity) and amenity (enhancing green spaces), with multiple benefits for both communities and nature.

(Re)engaging communities in local surface water management planning is a critical imperative, with engineers increasingly needing softer skills in community engagement. See, for example, the public-facing Construction Industry Research and Information Association (CIRIA) animation "Ever wondered where the rain goes?".[10] Community involvement in design of local SuDS can increase a sense of ownership, critical for the monitoring and maintenance that ensures long-term performance after building contractors have gone.

Engaging young citizens in SuDS developments is a strategy that can reap return in their learning to value water. This can involve children designing local SuDS in primary school activities (e.g., Ward et al., 2022 in SW UK) or in young learners in London schools engaging in the design, implementation, and care of local SuDS in climate change adaptation to increased surface runoff (CIRIA et al., ud). Pilot projects include: 'SuDS for schools'[11] where the UK Wetlands and Wildfowl Trust, with the environmental regulator and a water supply company, installed SuDS in ten schools in the same catchment in London, UK. Valuable co-benefits of such green-blue spaces included quick adoption for young people's outdoor learning and recreation, and cascading of adaptive learning through local student, parent, and teacher networks.

7.4.3.2 Thinking 'Blue-Green Cities'

At the broader city scale, new design thinking focuses on benefits of nature-based solutions in 'Blue-Green' Cities[12] for urban water management, designed to store water during floods. As O'Donnell and Thorne (2020, p2) articulated

> The concept of a Blue–Green City moves beyond its constituent infrastructure towards a philosophy that embraces the creation of multiple co-benefits for the environment, society and economy through natural, resilient and adaptive management of water in all its forms and in ways that are socially equitable and valued by residents, communities and decision-makers.
>
> *Lawson et al. (2014)*

This approach is not new. Since ancient times, China has had 'water towns' defined by their "distinct architecture and way of life, centred on a dense network of waterbodies and canals" (Qiu, 2013 in Tang et al., 2018, p3), with a strong sense of living with water in their cultural practices. More recently, China's piloting of 'sponge city' approaches, for sustainable urban stormwater management, makes space for water storage, using China's long-established "Blue-Green" principles. Stormwater parks deliver multiple ecosystem services (e.g., in Harbin City, northern China; World Future Council, 2016). However, Jiang et al. (2017, p525) noted the implementation challenges, including need for "the engagement in knowledge co-creation of stakeholders committed to common objectives". This concern, specifically related to community involvement, is reinforced by Tang et al. (2018, p12) who observed that planned networks of Blue-Green Infrastructure (BGI) have been identified to have higher benefits

> when and where public participation is strong and local communities are fully engaged in decision-making from the planning stages through to construction and maintenance. This is the case because, for the co-benefits of BGI to be realised in practice, the beneficiaries need to appreciate the multiple benefits that it can provide.

This poses questions about the role of citizens – their knowledge, values, and agency – alongside other stakeholders in BGI processes within different cultural settings. For example, the Australian focus on 'water-sensitive urban design' acknowledges the need to engage with local communities. Sharma et al. (2016) investigated barriers to the uptake of Water Sensitive Urban Drainage Systems (WSUD systems) in Greater Adelaide, South Australia. Their project explored community perceptions of these systems and their potential "to meet local objectives for water harvesting, stormwater quantity management and flood mitigation" (p3). While the latter roles of WSUDs were well recognised by the community, water quality improvement aspects were much less so. They also highlighted that "the landscape amenity benefits … can enable a more active, attractive and connected community" (Sharma et al., 2016, p13). These authors emphasised the need for community engagement and education on sustainable water use, alongside opportunities for local people to feedback on effectiveness of local systems so informing design improvements.

This poses questions about society's relations with NbS in promoting blue-green urban areas. From a community perspective, knowledge gaps in NbS focus on how to involve stakeholders (local residents as well as planners), availability and inclusivity of benefits to *all* citizens, issues of displacement (with associated environmental justice issues and 'eco-gentrification'), and how to communicate about the strengths and limitations of NbS. Importantly, indicators of effectiveness need to consider citizen involvement and health and well-being, alongside performance and monitoring (Kabisch et al., 2016; Figure 7.5).

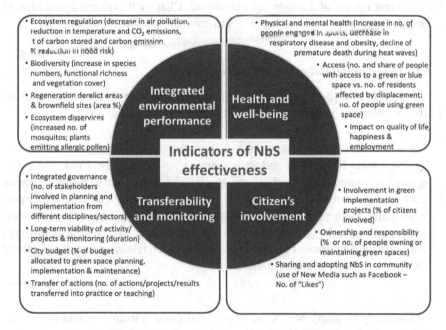

FIGURE 7.5 Potential indicators of the effectiveness of NbS for climate change adaptation and mitigation and associated co-benefits (Kabisch et al., 2016, Figure 1)

They highlighted strong potential for such proactive adaptive measures to help mitigate effects of climate change and decrease vulnerability of communities, so increasing urban resilience. However, these are *long-term* adaptive solutions requiring *long-term* community engagement as a common feature "in order to co-design and operate BGI that delivers multiple, locally valued co-benefits" (Lamond et al., 2020, p99).

Opportunities also exist to use new approaches to capture differing community experiences of urban water management as complex, multi-faceted territory with many gains (e.g., for community health and well-being). For example, various socially engaged artists have explored connectedness and relationships between water, urban settings and citizens, linking local and global concerns (e.g., Bouyer and Lydon, ud, "Artists in Conversation with Water in Cities[13]").

7.5 Chipping away at residual flood risk: Other measures involving communities

Another suite of approaches within local FRM combines infrastructural, institutional, or governance concerns with management of human behaviours. These include flood warnings, flood insurance, and PLP. The key challenge when using these measures, and in their development, is the design of socio-technical interfaces so that uptake and action are maximised. Chapter 9 explores integration of these strategies with learning as a non-structural or behavioural mechanism.

7.5.1 Improving flood warnings and communities

The Sendai Framework, in achieving Priority 4 *Enhancing disaster Preparedness* (33b), emphasises the importance of early warning, stating the need:

> To invest in, develop, maintain and strengthen <u>people-centred</u> multi-hazard, multi-sectoral forecasting and early warning systems, disaster risk and emergency communications mechanisms, <u>social technologies</u> and hazard-monitoring telecommunications systems; develop such systems through a participatory process; <u>tailor them to the needs of users, including social and cultural requirements, in particular, gender</u>; promote the application of <u>simple and low-cost</u> early warning equipment and facilities; and broaden release channels for natural disaster early warning information.
>
> *author's emphasis*

Historically in many MEDC settings, local flood warnings would have been community-led and tailored to local needs, with local people warning each other. This would be based on their observations and local (environmental) knowledge of their river – its changing levels and indicators of impending floods – from experience of past events. Even in modern settings, effective flood warning systems are challenging to set up when attending to both technical and socio-cultural elements and where responsibilities may shift between state and communities. This includes combating early warning myths (International Federation of Red Cross and Red Crescent Societies – IFRC, 2012), for example, that 'information is enough'.

Most research and practice literature is focused on effective warnings for river or coastal flooding. Effective community-focused flood warning depends not only on whether communities can trust the validity and accuracy of hydrological modelling, but also how warnings are communicated. As Key acknowledged in 1997 (p1):

> In most countries, flood warning systems are incomplete and fragmented and, tend to function sub-optimally. The full complexity of the ideal system is not widely recognised, and severe flood events, in particular, illustrate flaws in system operation which are serious and potentially dangerous when viewed from the perspective of the flood liable community.

There are many aspects to consider in planning a formal flood warning system to secure desired community action or 'protective behaviour' on receipt. Should household sign-up for warnings be compulsory or voluntary? How should the warning message be designed (message construction, language, colour coding, icons used, etc.)? How should it be delivered (e.g., by mobile phone alerts, etc.)? And are there socially vulnerable groups who need to be reached via different formats or means (e.g., local flood wardens)?

The concept of the Total Flood Warning System (TFWS) emerged from the Australian experience but has wider applicability (Dufty, 2021). This term recognised that effective flood warning needs to integrate many elements beyond the technical. Importantly, "the need to help those in the path of a flood to understand the warnings they received and take effective action was recognised as central" (Cawood et al., 2018, p47). The Australian Institute for Disaster Resilience (2009) identified a critical sequenced process for effective warning (Figure 7.6).

Keys and Cawood (2009) emphasised that flood managers need to recognise high-quality warning as a critical input to a community's capacity to manage flooding, and that this is predicated on organisational culture and modes of working, rather than financial resources. TFWS recognise the importance of message construction that details likely impact ('impact forecasting') alongside the predicted risk levels ("risk-based warnings") in a "people-centred approach" (Anderson-Berry et al., 2018, p7).

Community input is essential in designing, planning, and reviewing performance of TFWS to inform improved practice. Issues still exist in maximising the positive impacts of warning systems in reducing community flood losses. As Cawood et al. (2018, p47) observed, loss reduction is the damage avoidance to communities that could have occurred

> ... had people heeded the warnings that were provided or had the warnings been better tailored to suit the risk in terms of focus, content and messaging. At least part of the reason is that the agencies involved in forecasting and warning of coming floods have not fully engaged with the communities that their activities are intended to help. The consequence is that the forecasts and warning messages provided are not sufficiently relevant and community members have not understood what has been provided to them.

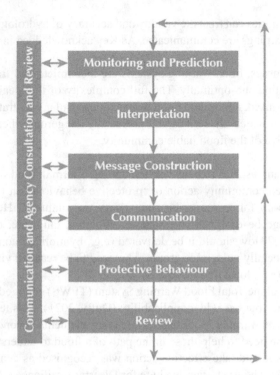

Improvement requires co-working and message testing with the communities that will receive warnings.

7.5.2 Communities as early flood warning systems

The emphasis on flood warning as a key strategy also poses questions about what happens in community settings where the state cannot warn for varied reasons (e.g., because of remoteness or short lead times for floods in rapid response catchments). Two important and contrasting ameliorators in such settings are opportunities from technological advances and the role of indigenous or local flood knowledge. New technologies are being trialled to establish how communities – without formal flood warnings – can monitor their own smaller water courses, alerting local people to impending risk, thus empowering communities and building resilience. For example in Scotland, SEPA,[14] the environmental regulator, working with technology developers RiverTrack®,[15] has explored how remote communities can use relatively cheap technology to monitor their local rivers in real time and be alerted to flooding. These community-maintained, flood warning systems allow communities to set their own tailored river 'stage' thresholds for warning, given the nature of their local activities and infrastructure at risk. Such schemes are being increasingly rolled out to good effect.

While Australia has led development in people-centred flood warning within MEDCs, community-based, early warning systems (CBEWS) have emerged in LEDCs, where historically there has had to be less reliance on the state or where limited lead times exist for early warning. Examples include CBEWS – driven and managed by community actors – for flood risk mitigation in Nepal (Smith et al., 2016). Early examples involved, for example, communities mapping historic floods to establish relationships between "observed" upstream flood height and "expected" downstream flood inundation (p426). Macherera and Chimbari (2016, p1) reviewed CBEWS that

> involve community driven collection and analysis of information that enable warning messages to help a community to react to a hazard and reduce the resulting loss or harm. .... Local communities' capacity to predict weather conditions using indigenous knowledge has been demonstrated in studies focusing on climate change and agriculture in some African countries.

They compared the main characteristics of national versus community early warning systems (EWS) highlighting strengths and differences. For example, the design of the system contrasted "deliberate structure, based on legal mandate by government and other agencies" against "flexible design [based] on need and adapted by trial and error" (IFRC, 2012, p15). In "human resourcing", the national level has "technicians, specialists" and at a community-level ranges from "ad hoc volunteers" to "individuals appointed by community leaders". In the latter, communities are critical active agents. The aim is for seamless integration of national and community EWS, as both are important. In Nepal, CBEWS have become gradually more integrated with EWS delivered by national government as modelling and technology advances, while retaining positive elements of CBEWS (Smith et al., 2016).

IFRC (2012) also distinguished between community-*based* and community-*driven* (or -led) EWS in terms of their key elements. The ability to tailor appropriate technology to different EWS components, scaling up from the local, is important. Table 7.2 summarises IFRC's ideas for local or community-level EWS, noting that the best model is where communities possess a strong understanding of the EWS. Appropriate technology is an important enabler of community-level EWS. Table 7.3 draws on IRFC (2012; column A) while adding learning for MEDC contexts (column B).

Effective community EWS exemplifies the high value in integrating local knowledge across different flood warning components, whatever the socio-cultural setting, and in building long-term relationships between at-risk communities and FRM agencies. This includes participatory citizen science inputs to early warning (Marchezini et al., 2018; Chapters 8 and 10).

TABLE 7.2 Community Involvement in Early Warning Systems (EWS) (IFRC, 2012, Table 1)

Key elements	Community	
	Based EWS	*Driven EWS*
Orientation	With the people	By the people
Character	Democratic	Empowering
Goals	Evocative/ consultative	Based on needs, participatory
Outlook	Community as partners	Community as managers
Views	Community is organised	Community is empowered
Values	Development of people's abilities	Trust in people's capacities
Result/impact	Initiates social reform	Restructures social fabric
Key players	Social entrepreneurs; community workers; and leaders	Everyone in the community
Methodology	Coordinated with technical support	Self-managed
Active early learning components (out of the four)	At least one is active (e.g., response capability)	All are active, especially the monitoring of indicators

* Four components (proposed by UNISDR) are: risk knowledge, monitoring and warning, dissemination/ communication and response capability.

TABLE 7.3 Ideas for Appropriate Technology per EWS Component at Local/Community Level (Column A Adapted from IFRC, 2012, Table 8; Column B by Author)

EWS components	(A) Local/community hazard-scape	(B) Learnings for MEDC settings
Risk knowledge	Maps of hazard-scapes drawn by community members (i.e., through vulnerability and capacity assessment (VCA) process, also known as community risk assessment)	Community input to local risk maps (VCA)
Monitoring	Manual river and rainfall gauges; billboards to announce river levels	Local stage boards or low technology, monitoring devices Connection to continuous monitoring shared online by environmental regulator
Response capability	Evacuation routes signalled by locally made signs and evacuation shelters designed locally	Street level community plans developed and shared for action on receipt of warning
Warning communication	Local devices for communication: word-of mouth, runners, criers, drums, flags, bells, telephone, radio, television, megaphone, mosque speakers	Local social media networks involving flood wardens Role of local news media (radio)

7.5.3   Insurance and communities

The Sendai Framework, in achieving Priority 3 *Investing in disaster risk reduction for resilience* (30b), highlights the importance of effective insurance mechanisms, with the imperative

> To promote mechanisms for disaster risk transfer and insurance, risk-sharing and retention and financial protection, as appropriate, for both public and private investment to reduce the financial impact of disasters on Governments and societies, in urban and rural areas.

However, flood insurance can be prohibitively expensive and a contentious mitigation option for communities. Perennial issues exist in matching premiums to (changing) risk for properties – residential or business – and in establishing how losses are shared. Public awareness of residual flood risk also has an important influence on insurance uptake. Anecdotal evidence indicates limited understanding can exist among residents in flood risk settings about why premiums and excesses can stay high despite local infrastructural measures. This relates to the 'levee effect' (Chapter 3); at-risk households may not take out personal insurance, thinking that flood defence structures give their property complete protection.

Access to flood insurance is supported through various schemes and agreements between national governments and the insurance industry (Oakley, 2018, p7). For example, insurance cover policies variably provide incentives (e.g., reduced premiums) for citizen or community action to make property more resilient. The UK FloodRE scheme[16] has involved (re)negotiation between national government and the insurance industry to ensure future affordable flood insurance cover for at-risk domestic properties that encourages uptake (see 'How FloodRE works'[17]). A recent positive development is change in FloodRE to support residential property owners in 'Building Back Better', motivating the uptake of PLP measures (see Section 7.5) to make properties more flood resilient in preparation for future floods.

In contrast, in the USA, flood insurance is compulsory for homes and businesses in areas at high flood-risk (with mortgages from government-backed lenders) through the National Flood Insurance Program (NFIP[18]). Implemented in 1968, NFIP status determines statutory disaster support for a community and its citizens. The US Federal Emergency Management Agency (FEMA, 2011) identified community floodplain management practices that communities can adopt to participate in NFIP. This involves both setting minimum standards but also that "adopting higher standards will lead to safer, stronger, more resilient communities". However, while this scheme has a long history of government subsidy to reduce socio-economic flood impacts, subsidies are being phased out with likely barriers for lower-income residents and further social injustice issues (Frazier et al., 2020). Less research exists on the interfaces between insurance strategies and communities in terms of education, agency, and equity in insurance policies (e.g., Oakley, 2018).

There are implications for mental health morbidity for those without and with insurance (Mulchandani et al., 2019; Chapter 10). In national settings where voluntary uptake of insurance exists, this inevitably means that neighbouring residents can have very different experiences in flood recovery (e.g., McEwen et al., 2002, concerning impacts of variable previous uptake of insurance by flooded caravan owners). However, insurance company practices can also impact in complex ways; with insurance, flood damaged goods can be compulsorily replaced. People without flood insurance can sometimes fare better emotionally after floods because their sentimental valuables can be salvaged. Stories also abound in flood-risk communities in the UK of large insurance excesses incurred by households after being flooded (e.g., by UK Summer 2007 floods). Mulchandani et al. (2019, in the UK) called for targeted services (local government, NGOs, voluntary) to support communities during insurance processes to reduce mental health impacts. In improving the efficacy of insurance as a community resilience strategy, opportunities exist for strong links to promoting increased PLP.

7.5.4 Property-level protection and communities

The major drivers for progress in developing effective individual PLP measures are impact of failure or overtopping of structural measures during extreme floods (Rose et al., 2016) or the need to provide individual property protection where no protective structures exist. PLP options combine 'resistance measures' (or 'water exclusion strategies') that prevent water from coming into a property and 'resilience measures' (Property Flood Resilience, PFR) that recognise water cannot always be kept out, and so focus on minimising damage once water enters a building. Active measures are implemented by citizens on receipt of a flood warning like proper use of sandbags; passive measures represent longer-term preparation for future floods.

In the UK, traditional seasonal resistance measures included "stanks"[19] installed in property entrances in preparation for when river flood water rises (see digital story 'Twelve things to prepare'[20]). Long-standing traditional approaches to minimise property flood damage and reduce time spent out of buildings include simple infrastructural resilience measures (Chapter 5.4). These are commonly seen in medieval riverside properties and include slab floors, which are easily washed out to reduce the recovery period, and breathable fast-drying lime plaster on walls. Both reduce structural flood damage; in contrast, gypsum plaster, in common modern use, absorbs and retains water. Some long-established 'low technology' measures are mirrored in 21st-century adjustments suggested to flood-affected communities. For example, a UK environmental regulator advises communities on 'Building Back Better' in property repair after floods (Box 7.1).

Business entrepreneurship is evident in new designs for PLP. With technology, many simple-to-implement active measures have emerged that can be put in place by residents on receipt of a flood warning, for example, external flood doors, toilet

---

BOX 7.1   BUILDING BACK BETTER (UK ENVIRONMENT AGENCY ADVICE, UNDATED, P20)

Here are some improvements you can make

Discuss them with your loss adjuster and builder.

- Lay ceramic tiles on your ground floor and use rugs instead of fitted carpets.
- Raise the height of electrical sockets to at least 1.5 metres above ground floor level.
- Use lime plaster instead of gypsum on walls.
- Fit stainless steel or plastic kitchens instead of chipboard ones or have freestanding kitchen units you can move.
- Position any main parts of a heating or ventilation system, like a boiler, up stairs or raised well above the ground floor.
- Fit non-return valves to all drains and water inlet pipes.
- Replace wooden window frames and doors with synthetic ones. They are easier to clean.

---

bungs, and air brick blockers, alongside more permanent passive measures like raising electricity sockets, removable doors on kitchen units, and cement-based plaster. Figure 7.7 provides a public-facing infographic of 'The Flood Repairable House' that combines possible measures. Rose et al. (2016, p610), however, noted that "although the uptake of water exclusion strategies [...] is gradually improving in the UK, the longer-term resilience options that permit water entry are less popular" in terms of public engagement. Arguably, this requires a major transformation in community perception (i.e., acceptance) that floodwater could be permitted to enter a property.

In improving uptake of PLP, attention needs to be given to the nature of flood risk (Chapter 2) in appropriate selection of active and passive measures for a specific property. Issues impacting choice include

the likely speed of onset, as surface water or 'flash' floods typically occur very rapidly and often lack any prior warning system; in such circumstances, passive measures represent the most appropriate solutions, as the building's occupants may be unable to put more planned measures into effect.

Rose et al. (2016 p610)

Implementing flood resistance and resilience policies is not just about citizen intention, as Grothmann and Reusswig (2006, p106) observe. Many actual barriers exist to investment and acting within diverse communities, "such as a lack of

Combined resistance and resilience measures
- keeping water out for as long as possible buys valuable time
  to raise / move your belongings

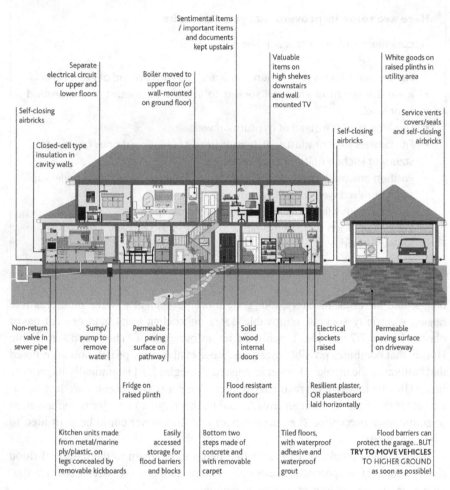

FIGURE 7.7 'The Flood Repairable House' (permission: Long-Dhonau)

resources like time, money, knowledge or social support, not expected at the time of intention forming". Understanding uptake requires behavioural insights into the links between motivation and action (Chapter 3). In Germany, Osberghaus (2015), reporting on a nationwide quantitative survey of the uptake of private mitigation measures, cited that past flood experience as well as expectations of future damage had positive impacts on uptake of PLP measures. He argued that this future thinking and investment can be interpreted as a 'climate adaptation signal' in flood mitigation behaviour. He did not observe (p36) any evidence in his data for moral hazard – or lack of incentive to guard against flood risk – when protected from its consequences.

> Households expecting insurance coverage do not reduce their mitigation efforts. Likewise, the expectation of government relief payments hinders mitigation only for some groups of households.

How to raise awareness of PFR measures in communities is a critical concern. The UK DEFRA *Low cost flood resilience project* (Lamond et al., 2017) undertook a demonstration trial of five different innovations for awareness-raising about low-cost flood resilience options within a case-study flood risk town. These strategies included several possible, locally embedded pathways for sharing knowledge that intersect different stages in the property lifecycle with distinct phases of the FRM spiral (e.g., recovery). Options included: working with local hardware stores to promote flood-resilient products when people are renovating or restoring a property, attention to the property acquisition timeline when people are buying property or refitting kitchens, and involving emergency services as trusted advisers when already visiting properties for fire and security advice. This is complicated territory; advantages included diversification of local messengers for social learning, while challenges included engaging communities, given a lack of recent experience of extreme floods prior to the pilot project. Face-to-face engagements about PLP measures gave better outcomes (Chapter 6).

A critical area of interdisciplinary research links the uptake of possible measures to reduce residual risk with evidence on flood preparedness behaviours (Chapters 3 and 8). Park et al. (2021) identified six steps in "a customer's journey" for resilience actions (Box 7.2).

BOX 7.2 STEPS IN FLOOD PREPAREDNESS BEHAVIOURS (PARK ET AL., 2021, PP2–3, 12; PERMISSION: ENVIRONMENT AGENCY)

Mindsets/attitudes/beliefs/appraisal

"Step one: I know flooding might impact me"
"Step two: I feel able to take action, and feel responsible for taking an action"
"Step three: I am able to assess and access available options"

Target behaviours

"Step four: I adopt resilience measure(s)"
"Step five: I regularly check whether I am sufficiently protected"
"Step six: I take action in critical moments"

They argued that the behavioural social sciences have an important role in improving the uptake of PFR by individuals and communities. One priority is identifying key areas for trialling behaviourally informed interventions to identify possible 'quick wins', alongside longer-term research needs.

Together with these more experimental studies, the arts and humanities also have a role in understanding socio-technical relationships between citizens and communities with flood resilience measures. For example, applied storytelling has strong potential in garnering differing perceptions about PFR of flood-impacted communities and those yet to be flooded. Such stories from flooded property owners can provide valuable insights into the decision-making processes of communities and businesses in whether to adopt measures and can help shift local narratives from 'risk' to 'resilience'. They can also help identify key local mediators or ambassadors, with personal resilience stories, to support uptake. Examples include the digital testimonies by 'Judy'[21] and 'Roger'[22] on their implementation of property-level resilience measures (UK DEFRA *Flood Repairable* Project).

7.6 Conclusions

Each new flood management paradigm changes the nature of stakeholder involvement – from more passive to active – in developing community flood resilience. This shifting territory from coping to adapting, and in some cases transforming, involves an increasingly diverse range of measures and actors. While positive, this is also potentially confusing for communities to navigate. It also contrasts with the strong predominance of engineering expertise within the flood defence paradigm. Integrated FRM involves a progressively wider and scaled range of approaches to dealing with different flood types, and their interactions with diverse communities. Each intervention – whether trialled and established in use or new, innovative, and evolving – benefits from being observed through an explicit 'community lens'. Distinct challenges exist in engaging rural and urban communities around the suite of possible interventions, their role, scale and limits to effectiveness, and in how social and technical elements might interact with communities in individual catchments. These interactions are influenced by wide-ranging factors including characteristics of previous flood events experienced, perceptions of both risk and measures, cultural and governance histories, time and ability to pay, and personal motivations.

The above review highlights the importance of distilling principles for community participation and agency at the socio-technical interface within, and across, different types of intervention. The specific role of communities varies with each measure, with a need for awareness about 'what works' in different cultural and socio-economic settings. There is clear value in international knowledge exchange about the efficacy of different risk management options, with some countries having invested heavily in research on certain measures (e.g., flood warnings in Australia; insurance in the USA; property-level flood resilience measures in the UK). There is also strong potential to learn from countries where the state exerts weaker roles in risk management (e.g., around NFM and community-led flood warning schemes). In MEDCs, evidence also exists of the value in using a historical approach to exploring the strengths and challenges of traditional measures and low technology ways of coping and adapting to 'living with floods', alongside high technology-based appraisals. This recognises that increased technology and funding are not always the answer when changing human behaviour and connection to evidence of risk are involved (Chapter 3).

Finally, there is a need to profile each element of an integrated risk management strategy against community-based criteria, embedding attention to local inequalities, supporting more vulnerable groups, and with strengthened links to learning for resilience (Chapter 9). Again, high value exists in integrating different knowledges (specialist; lay; community-based) to maximise the efficacy of outcomes. This requires actively involving citizens and communities longitudinally in observing and feeding back on 'what works' in planning and implementation of adaptive measures that scale up from the hyperlocal to the whole catchment.

Notes

1 https://www.tamug.edu/ikedike/images_and_documents/Lets_Build_The_Ike_Dike.
2 https://floodcontrolinternational.com/flood-barriers/.
3 https://restorerivers.eu/wiki/index.php?title=Case_study%3AQuaggy_Flood_Alleviation_Scheme.
4 https://planningforhazards.com/stream-buffers-and-setbacks.
5 https://www.ctc-n.org/technologies/managed-realignment.
6 https://edition.cnn.com/2019/07/17/us/valmeyer-flooding-climate-crisis-midwest/index.html.
7 https://catchmentbasedapproach.org/about/.
8 https://research-and-innovation.ec.europa.eu/research-area/environment/nature-based-solutions_en.
9 https://www.gov.uk/government/news/natural-flood-management-part-of-the-nations-flood-resilience.
10 https://www.ciria.org/CIRIA/Resources/Videos.aspx
11 https://www.wwt.org.uk/our-work/projects/suds-for-schools/.
12 http://www.bluegreencities.ac.uk/.
13 https://www.thenatureofcities.com/2018/07/14/artists-conversation-water/.
14 Scottish Environment Protection Agency.
15 http://www.rivertrack.org.
16 https://www.floodre.co.uk/.
17 https://www.floodre.co.uk/how-flood-re-works/.
18 https://www.fema.gov/flood-insurance.
19 A stank was a traditional Gloucestershire (UK) term for a wooden flood barrier on external property doors, slotted in as a flood resistance measure before river floods.
20 https://www.youtube.com/watch?v=QpXS-N_D1nA.
21 https://floodrepairable.wordpress.com/kitchens-and-bathrooms/.
22 https://floodrepairable.wordpress.com/case-study-videos/.

References

Alexander, K.S., Ryan, A. and Measham, T.G. (2012) Managed retreat of coastal communities: Understanding responses to projected sea level rise. *Journal of Environmental Planning and Management*, 55(4), 409–433. https://doi.org/10.1080/09640568.2011. 604193

American Rivers (undated) Daylighting streams: Breathing life into urban streams and communities. Retrieved from https://www.americanrivers.org/wp-content/uploads/2016/05/ AmericanRivers_daylighting-streams-report.pdf

Anderson, E. (2019) Del Mar, California Coastal Commission Clash Over Climate Change Plan, KPBS, October 7. https://www.kpbs.org/news/2019/oct/07/del-mar-and-coastal-commission-clash-climate-chang/. Accessed 26 February 2020.

Anderson-Berry, L., Achilles, T., Panchuk, S., Mackie, B., Canterford, S., Leck, A. and Bird, D.K. (2018) Sending a message: How significant events have influenced the warnings landscape in Australia. *International Journal of Disaster Risk Reduction*, 30A, 5–17. https://doi.org/10.1016/j.ijdrr.2018.03.005

Arnstein, S.R. (1969) A ladder of citizen participation. *Journal of the American Institute of Planners*, 35(4), 216–224. https://doi.org/10.1080/01944366908977225

Australian Institute for Disaster Resilience (2009) *Australian Disaster Resilience Manual 21: Flood Warning*. CC BY-NC. https://knowledge.aidr.org.au/media/1964/manual-21-flood-warning.pdf

Australian Institute for Disaster Resilience (2017) *Managing the floodplain: a guide to best practice in flood risk management in Australia*. Australian Disaster Resilience Handbook Collection 7 (3rd edn).

Bagus, P. (2006) Wresting land from the sea: An argument against public goods theory. *Journal of Libertarian Studies*, 20(4), 21–40.

Bagus, P. (2007) Can dikes be private? An argument against public goods theory. *Journal of Libertarian Studies*. https://mises.org/library/can-dikes-be-private-argument-against-public-goods-theory

Baron, N. (2020) Flood protection beyond protection against floods: How to make sense of controversies related to the building and maintenance of dikes in Denmark. *Natural Hazards*, 103, 967–984. https://doi-org.ezproxy.uwe.ac.uk/10.1007/s11069-020-04021-9

BBC News (2020) Environment Agency chief: Avoid building new homes on flood plains. 25th February 2020. https://www.bbc.co.uk/news/uk-51620992

Beck, U. (1992) *Risk Society: Towards a New Modernity*. Sage Publications, London.

Beven, K. (2014) Total flood defence is a myth. https://www.lancaster.ac.uk/lec/news-and-events/blog/keith-beven/total-flood-defence-is-a-myth/

Broadhead, A.T. and Lerner, D.N. (2013) Case study website supporting research into daylighting urban rivers. *Hydrological Processes*, 27, 1840–1842. www.daylighting.org.uk. https://doi.org/10.1002/hyp.9781

Brown, A.G., Lespez, L., Sear, D.A., Macaire, J., Peter Houben, P., Klimek, K., Brazier, R.E., Van Oost, K. and Pears, B. (2018) Natural vs anthropogenic streams in europe: History, ecology and implications for restoration, River-rewilding and Riverine ecosystem services. *Earth-Science Reviews*, 180, 185–205.

Burgess-Gamble, L. et al. (2017) *Working with natural processes to reduce flood risk: the evidence behind Natural Flood Management*. Environment Agency https://assets. publishing.service.gov.uk/media/6036c730d3bf7f0aac939a47/Working_with_natural_ processes_one_page_summaries.pdf

Carey, J. (2020) Managed retreat increasingly seen as necessary in response to climate change's fury. Core Concepts. *PNAS*, 117(24), 13182–13185. https://doi.org/10.1073/pnas.2008198117.

Cawood, M., Keys, C. and Wright, C. (2018) The total flood warning system: What have we learnt since 1990 and where are the gaps. *Australian Journal of Emergency Management*, 33(2), 47–52. https://knowledge.aidr.org.au/resources/ajem-apr-2018-the-total-flood-warning-system-what-have-we-learnt-since-1990-and-where-are-the-gaps/

CIRIA, Robert Bray Associates and Business in the Community (ud) *Reimagining rainwater in schools*. Greater London Authority, London. https://www.london.gov.uk/sites/default/files/reimagining_rainwater_in_schools_v1_.pdf

CIWEM (2007) Policy Position Statement: De-culverting of watercourses. https://www.ciwem.org/assets/pdf/Policy/Policy Position Statement/Deculverting-of-water-courses.pdf.

Costanza, R. (2021) Forget massive seawalls, coastal wetlands offer the best storm protection money can buy. *The Conversation*. https://theconversation.com/forget-massive-seawalls-coastal-wetlands-offer-the-best-storm-protection-money-can-buy-165872

Dannenberg, A.L., Frumkin, H., Hess, J.J. et al. (2019) Managed retreat as a strategy for climate change adaptation in small communities. Public health implications. *Climatic Change*, 153, 1–14. https://doi-org.ezproxy.uwe.ac.uk/10.1007/s10584-019-02382-0

de Groot, M. and de Groot, W.T. (2009) "Room for river" measures and public visions in the Netherlands: A survey on river perceptions among riverside residents. *Water Resources Research*, 45, W07403. https://doi.org/10.1029/2008WR007339

Doberstein, B., Fitzgibbons, J. and Mitchell, C. (2019) Protect, accommodate, retreat or avoid (PARA): Canadian Community options for flood disaster risk reduction and flood resilience. *Natural Hazards*, 98, 31–50. https://doi-org.ezproxy.uwe.ac.uk/10.1007/s11069-018-3529-z

Dufty, N. (2021) *The Total Flood Warning System: a review of the concept*. https://knowledge.aidr.org.au/media/8447/ajem_18_2021-01.pdf

Emerson, J.W. (1971) Channelization: A case-study. *Science*, 173, 325–326.

Environment Agency (ud) What to do before, during and after a flood. Practical advice on what to do to protect yourself and your property. https://assets.publishing.service.gov.uk/government/uploads/system/uploads/attachment_data/file/403213/LIT_5216.pdf

Environment Agency (2009) *River Severn Catchment Flood Management Plan Summary Report December 2009*. Environment Agency, Solihull

European Environment Agency (2021) *Nature-Based Solutions in Europe: Policy, Knowledge and Practice for Climate Change Adaptation and Disaster Risk Reduction*. Publications Office of the European Union, Luxembourg. https://data.europa.eu/doi/10.2800/919315

Federal Emergency Management Agency FEMA (2011) *History of the National Flood Insurance Program and Reform Efforts*, Roundtable on Flood Risk Management. GFDRR/The World Bank, Washington.

Frazier, T., Boyden, E.E. and Wood, E. (2020) Socioeconomic implications of national flood insurance policy reform and flood insurance rate map revisions. *Natural Hazards*, 103, 329–346. https://doi.org/10.1007/s11069-020-03990-1

Grothmann, T. and Reusswig, F. (2006) People at risk of flooding: Why some residents take precautionary action while others do not. *Natural Hazards*, 38, 101–120. https://doi.org/10.1007/s11069-005-8604-6

Hartmann, T., Slavikova, L. and McCarthy, S. (eds.) (2019) *Nature-Based Flood Risk Management on Private Land. Disciplinary Perspectives on a Multidisciplinary Challenge*. Springer International Publishing, Cham.

Hino, M., Field, C. and Mach, K. (2017) Managed retreat as a response to natural hazard risk. *Nature Climate Change*, 7, 364–370. https://doi-org.ezproxy.uwe.ac.uk/10.1038/nclimate3252

International Federation of Red Cross and Red Crescent Societies (2012) *Community early warning systems: Guiding principles*. Community Preparedness and Risk Reduction Department, Geneva. https://www.ifrc.org/sites/default/files/CEWS-Guiding-Principles-EN.pdf

Jiang, Y., Zevenbergen, C. and Fu, D. (2017) Understanding the challenges for the governance of China's "sponge cities" initiative to sustainably manage urban stormwater and flooding. *Natural Hazards*, 89, 521–529. https://doi-org.ezproxy.uwe.ac.uk/10.1007/s11069-017-2977-1

Kabisch, N., Frantzeskaki, N., Pauleit, S., Naumann, S., Davis, M., Artmann, M., Haase, D., Knapp, S., Korn, H., Stadler, J., Zaunberger, K. and Bonn, A. (2016) Nature-based solutions to climate change mitigation and adaptation in urban areas: Perspectives on indicators, knowledge gaps, barriers, and opportunities for action. *Ecology and Society*, 21(2), 39. http://dx.doi.org/10.5751/ES-08373-210239

Kaufman, R. (2013) Daylighting takes off as cities expose long-buried rivers. https://news.nationalgeographic.com/news/2013/07/130730

Keys, C. and Cawood, M. (2009) Identifying and reducing inadequacies in flood warning processes: An Australian perspective. *Journal of Flood Risk Management*, 2(3), 190–197.

Kibel, P.S. (ed.) (2007) *Rivertown: Rethinking Urban Rivers*. MIT Press, Cambridge.

Krauze, K. and Wagner, I. (2019) From classical water-ecosystem theories to nature-based solutions — Contextualizing nature-based solutions for sustainable city. *Science of the Total Environment*, 655, 697–706. https://doi.org/10.1016/j.scitotenv.2018.11.187

Lamond, J., Everett, G. and England, K. (2020) Understanding citizen and community behaviours and preferences. In C. Thorne (ed.) *Blue-Green Cities: Integrating Urban Flood Risk Management with Green Infrastructure*. ICE, London, pp99–114.

Lamond, J., McEwen, L., Rose, C., Wragg, A., Joseph, R., Twigger-Ross, C., Papadopoulou, L., White, O., Dhonau, M. and Proverbs, D. (2017) *Supporting the Uptake of Low Cost Resilience for Properties at Risk of Flooding: Final Report (FD2682)*. Defra, London.

Lawson, E., Thorne, C. Ahilan, S. et al. (2014) Delivering And evaluating the multiple flood risk benefits in Blue–Green cities: An interdisciplinary approach. In D. Proverbs and C.A. Brebbia (eds.) *Flood Recovery, Innovation and Response IV*. WIT Press, Southampton, pp113–124.

Loeffler, J. and Loeffler, C. (eds.) (2012) *Thinking Like a Watershed: Voices from the West*. University of New Mexico Press, Albuquerque.

Lost Rivers (2012) [Film] Directed by Caroline Bacle. Icarus Pictures, USA.

Macherera, M. and Chimbari, M.J. (2016) A review of studies on community based early warning systems. *Jàmbá: Journal of Disaster Risk Studies*, 8(1), a206. https://dx.doi.org/10.4102/jamba.v8i1.206

Marchezini, V., Horita, F.E.A., Matsuo, P.M., Trajber, R., Trejo-Rangel, M.A. and Olivato, D. (2018) A review of studies on participatory early warning systems (P-EWS): Pathways to support citizen science initiatives. *Frontiers in Earth Science*, 6, https://doi.org/10.3389/feart.2018.00184

McEwen, L.J., Gorell-Barnes, L., Phillips, K. and Biggs, I. (2020) Reweaving urban water-community relations: Creative, participatory river 'daylighting' and local hydrocitizenship. *Transactions of the Institute of British Geographers*, 45(4), 779–801. https://doi.org/10.1111/tran.12375

McEwen, L.J., Hall, T., Hunt, J., Harrison, M. and Dempsey, M. (2002) Flood warning, warning response and planning control issues associated with caravan parks: The April 1998 floods on the lower Avon flood plain, Midlands Region, UK. *Applied Geography*, 22, 271–305.

Millennium Ecosystem Assessment (2005) Volume 1: Current State & Trends. https://www.millenniumassessment.org/documents/document.765.aspx.pdf

Mulchandani, R., Smith, M., Armstrong, B., English National Study of Flooding and Health Study Group, Beck, C.R. and Oliver, I. (2019) Effect of insurance-related factors on the association between flooding and mental health outcomes. *International Journal of Environmental Research and Public Health*, 16(7), 1174. https://doi.org/10.3390/ijerph16071174

Newson, M. (1992) *Land, Water and Development: Sustainable Management of River Basin Systems* (1st edn). Routledge, London.

Nguyen, C.N. (2020) Homeowners' choice when the government proposes a managed retreat. *International Journal of Disaster Risk Reduction*, 47, 101543. https://doi.org/10.1016/j.ijdrr.2020.101543

O'Donnell, E. and Thorne, C. (2020) Chapter 1. Urban flood risk management: The blue–green advantage. In *Blue–Green Cities*. ICE, Thomas Telford Ltd, London, pp1–13. https://doi.org/10.1680/bgc.64195.001

Oakley, M. (2018) *Incentivising Household Action on Flooding: Options for Using Incentives to Increase the Take Up of Flood Resilience and Resistance Measures*. The Social Market Foundation, London.

Osberghaus, D. (2015) The determinants of private flood mitigation measures in Germany — Evidence from a nationwide survey. *Ecological Economics*, 110, 36–50. https://dx.doi.org/10.1016/j.ecolecon.2014.12.010

Park, T., Oakley, M. and Luptakova, V. (2021) *Applying behavioural insights to property flood resilience*. Environment Agency Report FRS17191. Environment Agency, Bristol.

Pender, G. and Faulkner, H. (eds.) (2011) *Flood Risk Science and Management*. Blackwell, Oxford.

Pinter, N. (2021a) The lost history of managed retreat and community relocation in the United States. *Elementa: Science of the Anthropocene*, 9(1), 00036. https://doi.org/10.1525/elementa.2021.00036

Pinter, N. (2021b) True stories of managed retreat from rising waters. Issues in Science and Technology XXXVII, NO.4, SUMMER 2021. https://issues.org/true-stories-managed-retreat-rising-waters-pinter/

Plate, E.J. (2002) Flood risk and flood management. *Journal of Hydrology*, 267(1–2), 2–11. https://doi.org/10.1016/S0022-1694(02)00135-X

Pralle, S. (2019) Drawing lines: FEMA and the politics of mapping flood zones. *Climatic Change*, 152, 227–237. https://doi-org.ezproxy.uwe.ac.uk/10.1007/s10584-018-2287-y

Qiu, Q.F. (2013) Thoughts on the strategies of developing water cultural resources in Ningbo. *Sanjiang Forum*, 7, 40–43 (in Chinese).

Rezaie, A.M., Loerzel, J. and Ferreira, C.M. (2020) Valuing natural habitats for enhancing coastal resilience: Wetlands reduce property damage from storm surge and sea level rise. *PLoS ONE*, 15(1), e0226275. https://doi.org/ 10.1371/journal.pone.0226275.

Rose, C., Lamond, J., Dhonau, M., Joseph, R. and Proverb, D. (2016) Improving the uptake of flood resilience at the individual property level. *International Journal of Safety and Security Engineering*, 6(3), 607–615. https://doi.org/10.2495/SAFE-V6-N3-607-615

Sharma, A.K., Pezzaniti, D., Myers, B., Cook, S., Tjandraatmadja, G., Chacko, P., Chavoshi, S., Kemp, D., Leonard, R., Koth, B. and Walton, A. (2016) Water sensitive urban design:

An investigation of current systems, implementation drivers, community perceptions and potential to supplement urban water services. *Water*, 8(7), 272. https://doi.org/10.3390/w8070272

Siders, A.R. (2019a) Managed retreat in the United States. *One Earth Perspective*, 1(2), 216–225. https://doi.org/10.1016/j.oneear.2019.09.008

Siders, A.R. (2019b) Social justice implications of US managed retreat buyout programs. *Climatic Change*, 152, 239–257. https://doi.org/10.1007/s10584-018-2272-5

Siders, A.R., Hino, M. and Mach, K.J. (2019) The case for strategic and managed climate retreat. *Science*, 365, 761–763.

Smith, P.J., Brown, S. and Dugar, S. (2016) Community-based early warning systems for flood risk mitigation in Nepal. *Natural Hazards and Earth System Sciences*, 17, 423–437. https://doi.org/10.5194/nhess-17-423-2017

Strout, J.M., Oen, A.M.P., Kalsnes, B.G., Solheim, A., Lupp, G., Pugliese, F. and Bernardie, S. (2021) Innovation in NBS co-design and implementation. *Sustainability*, 13(2), 986. https://dx.doi.org/10.3390/su13020986

Tang, Y., Chan, F.K.S., O'Donnell, E.C. et al. (2018) Aligning ancient and modern approaches to sustainable urban water management in China: Ningbo as a "Blue-Green City" in the "Sponge City" campaign. *Journal of Flood Risk Management*, 11, e12451. https://doi.org/10.1111/jfr3.12451

Tyler, J., Sadiq, A.A. and Noonan, D.S. (2019) A review of the community flood risk management literature in the USA: Lessons for improving community resilience to floods. *Natural Hazards*, 96, 1223–1248. https://doi.org/10.1007/s11069-019-03606-3

UNDRR (2015) *Sendai Framework for Disaster Risk Reduction 2015–2030*) https://www.undrr.org/publication/sendai-framework-disaster-risk-reduction-2015-2030

Usher, M., Huck, J., Clay, G., Shuttleworth, E. and Astbury, J. (2020) Broaching the brook: Daylighting, community and the 'stickiness' of water. *Environment and Planning E: Nature and Space*. https://doi.org/10.1177/2514848620959589

Van Loon-Steensma, J.M. and Vellinga, P. (2019) How "wide green dikes" were reintroduced in The Netherlands: A case study of the uptake of an innovative measure in long-term strategic delta planning. *Journal of Environmental Planning and Management*, 62(9), 1525–1544. https://doi.org/10.1080/09640568.2018.1557039

Ward, S., Paling, N. and Rogers, A. (2022) Mobilising sustainable, water-resilient communities in the UK: Evidence and engagement across scales. *Proceedings of the Institution of Civil Engineers – Engineering Sustainability* 1–9. https://doi.org/10.1680/jensu.21.00095

White, G.F. (1945) *Human Adjustment to Floods. Research Paper 29*. Department of Geography, University of Chicago, pp 225.

Wild, T.C., Bernet, J.F., Westling, E.L. and Lerner, D.N. (2011) Deculverting: Reviewing the evidence on the 'daylighting' and restoration of culverted rivers. *Water and Environment Journal*, 25, 412–421. https://doi.org/10.1111/j.1747-6593.2010.00236.x

Wisner, B., Blaikie, P., Cannon, T. and Davis, I. (2004) *At Risk: Natural Hazards, People's Vulnerability and Disasters* (2nd edn). Routledge, New York.

World Future Council (2016) *Sponge Cities: What Is It All About?* January 20, 2016. https://www.worldfuturecouncil.org/sponge-cities-what-is-it-all-about/

8

COMMUNITY PARTICIPATION AND AGENCY IN LOCAL FLOOD RISK MANAGEMENT

8.1 Introduction

Awareness of the roles and interactions between the state and its citizens, in navigating changing responsibilities in local flood risk management (FRM), is critical for developing community resilience. Here distinction can be drawn between 'public/citizen participation' and 'community participation'. The former centres around the relationship between the public and the state, linked to local democratic processes (e.g., Albert and Passmore, 2008). The latter brings on board wider stakeholders, including civil society, linked to participatory approaches to managing risk, natural resources, and health care (e.g., Bath and Wakerman, 2015). This chapter investigates how processes and practices of citizen and community participation might contribute to local FRM. It asks:

- What conceptual framing might contribute to understanding what communities bring to local FRM?
- What does meaningful participation look like, and what could its role be in FRM?
- How can the role of positive local community participation be developed within different phases of the FRM spiral?

Academic disciplines contributing to this interdisciplinary territory include sociology, anthropology, development studies, geography, policy studies, and politics. The chapter considers background contexts to community participation in FRM, then key concepts, models, methods, and processes within participation, before focusing on citizen and community participation within FRM and wider disaster risk management (DRM).

DOI: 10.4324/9781315666914-8

8.2 Background contexts

Chapters 1 and 2 provided background contexts to community-focused FRM. International frameworks for disaster risk reduction (DRR) and shifts in national governance structures are now briefly outlined below.

8.2.1 International context to community involvement in disaster management

Effective environmental participation of citizens is increasingly recognised as good governance (Landström, 2020). A key question is how meaningful community participation is best integrated within local flood risk decision-making and governance, particularly when there are shifts from a structural management paradigm to more integrated approaches (e.g., Vinke-de Kruijf et al., 2015; Chapter 2). This imperative for increased citizen participation in planning, decision-making, and action within local DRR (Section 2.5 of Chapter 2) is evidenced in key international directives of the 21st century that promote increased involvement of civil society and its organisations: the UNDRR's *Hyogo Framework for Action (2005–2015)* followed by the UNDRR's *Sendai Framework for Disaster Risk Reduction (2015–2030)* – hereafter 'Sendai Framework'). Box 8.1 highlights the Framework's important reference to the roles of civil society and communities in volunteering, local knowledge sharing, and co-working, with strong attention to engaging more marginalised groups.

BOX 8.1 REFERENCE TO CIVIL SOCIETY AND LOCAL-LEVEL DRR WITHIN SENDAI FRAMEWORK (UNDRR; AUTHOR'S EMBOLDENING)

Priority 1: Understanding disaster risk at national and local levels

24 (g) To build the **knowledge** of government officials at all levels, **civil society, communities and volunteers**, as well as the private sector, through **sharing experiences, lessons learned, good practices and training and education** on DRR, including the use of existing training and education mechanisms and peer learning

Priority 2: Strengthening disaster risk governance to manage disaster risk at national and local levels

27 (h) To empower local authorities, as appropriate, through regulatory and financial means to work and coordinate with **civil society, communities and indigenous peoples and migrants** in DRM at the local level

27 (j) To promote the development of quality standards, such as certifications and awards for DRM, with the participation of the private sector, **civil society**, professional associations, scientific organizations, and the United Nations

Roles of stakeholders (p23)

36 When determining specific roles and responsibilities for stakeholders, and at the same time building on existing relevant international instruments, States should encourage the following actions on the part of all public and private stakeholders:

(a) **Civil society, volunteers, organized voluntary work organizations and community-based organizations to participate**, in collaboration with public institutions, to, inter alia, provide specific knowledge and pragmatic guidance in the context of the development and implementation of normative frameworks, standards and plans for DRR; engage in the implementation of local, national, regional, and global plans and strategies; contribute to and support public awareness, a culture of prevention and education on disaster risk; and advocate for resilient communities and an inclusive and all-of-society DRM that strengthen synergies across groups, as appropriate.

Focus on: **(i) Women; (ii) children and youth; (iii) persons with disabilities; (iv) older persons; (v) indigenous peoples; and (vii) migrants**

There are continued calls for citizens to be active participants in DRM (e.g., McEwen et al., 2023). For example, implementation of the European Flood Directive 2007/60/EC requires mechanisms that allow effective participatory and collaborative governance to reduce flood hazard costs (Challies et al., 2016). Its rationale is underpinned by recognition that DRR, and programmes to improve public participation and response, can fail to address the specific needs of local communities, including, importantly, socially vulnerable groups. They can also ignore potential use of local resources (framed as 'capitals' or 'assets'; see Section 8.3), and community capacities to reduce flood losses. Such lacunae in FRM may increase local people's vulnerability.

8.2.2 Governance structures and community roles in FRM

Governance cultures in some More Economically Developed Countries (MEDCs) have also changed significantly over past decades, from a reliance on the state to increased devolvement to 'the local' (Chapter 1). Such retreat of the state is a response to broader neoliberalisation agendas (Horst et al., 2020) and provides a backdrop to rising concern for effective citizenship in research, policy, and practice. Increasingly government expectation exists (e.g., in the USA and Europe) that civil society will play more prominent roles in response to, and in planning for, increased weather extremes. In the UK, this means local government has 'power of competence', and communities are actively encouraged to lead, or contribute to, local decision-making. This is evidenced in the principle of subsidiarity[1] for planned flood risk, as found within much EU legislation, where centralised government only undertakes selected actions that cannot be delivered at a different scale (Priest et al., 2016). However, power and resources do not necessarily follow devolved responsibility to the local (Begg, 2018). Cook et al. (2013) argued that traditional models of encouraging citizens to participate[2] in water governance in MEDCs are well recognised as weak and problematic, and hence need adjusting or re-visioning. Many barriers to participation exist, arising from the possibilities and circumstances in which participation is possible, and how engagement is framed (Norris and McLean, 2011; Facer and Enright, 2016). Indeed, the 'desire not to participate' can also be considered a right (Kotus and Sowada, 2017).

Traditionally, attention to citizens' roles in FRM has focused on the acute event phase with floods as local emergencies, where communities inevitably need to have a key role. As the UK Cabinet Office (2008, pxxxiv) stated in its review after the UK summer 2007 floods:

> Community action was one of the most striking impacts of the summer floods. It has considerable potential for the future. In a wide area emergency, the authorities are overwhelmed, and people have little choice other than to help themselves.

However, the concept of a 'risk society' (cf. Beck, 1992) raises important questions about the possible nature of citizen participation and agency within *all* phases of DRM. Importantly, this includes preparation for quicker recovery and adaptation to future extreme weather events. Interest in increased community agency for community resilience goes beyond floods to other hazards, with strong potential for learning across risks. For example, Vallance and Carlton (2015, p27), in reviewing earthquake hazards in New Zealand, described communities as "first to respond, last to leave", "recognising that they need to have 'roles and resilience' across New Zealand's '4Rs'" (Readiness, Reduction, Response, and Recovery).

8.3 Civic agency, community capital, and capacities

The ability of civil society to learn for resilience in managing flood risk is dependent upon local socio-economic, socio-cultural, and demographic characteristics of communities. In framing civic agency, it is helpful to differentiate terms that are notoriously difficult to define: capitals (human and social), alongside differences between 'community' (Section 1.4 of Chapter 1) and 'community capital'. The concept of capitals or assets aligns with an economic model for creating wealth; this idea of different types of capital is widely used in understanding sustainability (e.g., in Forum for the Future's Five Capitals model[3] with human and social capitals).

8.3.1 Differentiating human and social capital

Human capital as an asset consists of the knowledge, skills, and health of individuals – invested in, and accumulated, in their lives (World Bank[4]). It varies with demographic status and is linked to family, educational, and work experiences. Such capital can be applied to achieving personal or collective goals. In contrast, collective social capital is challenging to define, embracing both personal and public good. In simple terms, it can be considered as the "connections among people and organizations or the social glue that makes things happen" (Mattos, 2015, p2), representing "both the source and product of relational interactions which occur within public space" (Carlton and Vallance, 2017, p831). Szreter and Woolcock (2004) distinguished different types of social capital – bonding, bridging, and linking capital – that combine as social capital (Table 8.1).

Ideas of social resilience (Adger, 2000; Section 1.4 of Chapter 1) are related to theories of social capital that emphasise the importance of "social networks, reciprocity, and interpersonal trust", which "allow individuals and groups to accomplish greater things than they could by their isolated efforts" (Patterson et al., 2010, p210). Aldrich et al. (2018) emphasised strong links between these networks, community cohesion, and community resilience. While significant efforts have been made to differentiate 'social capital' and the idea of 'community' (Chapter 1), these concepts are frequently conflated (Colclough and Sitaraman, 2005, p474).

8.3.2 Community capital

A sustainable and successful community is characterised by economic security, social inclusivity, and ecological health (Flora and Flora, 2013; Pitzer and Streeter, 2015). However, the extent to which any community can exercise self-determination in practice is likely to depend on the capital resources ("human, cultural

TABLE 8.1 Defining Different Types of Social Capital (*Szreter and Woolcock, 2004 p654/655; $Smith, 2000–2009, np; permission: Michael Woolcock)

Type	Definition*	Example$
Bonding	"trusting and co-operative relations between members of a network who *see themselves as being similar*, in terms of their shared social identity" (p654/5)	For example, "among family members, close friends and neighbours"
Bridging	"comprises relations of respect and mutuality *between people who know that they are not alike* in some socio-demographic (or social identity) sense (differing by age, ethnic group, class, etc)" (p655)	For example, "more distant friends, associates and colleagues"
Linking	"norms of respect and networks of trusting relationships between people *who are interacting across explicit, formal or institutionalized power or authority gradients* in society" (p655)	For example, those "entirely outside of the community"

TABLE 8.2 Flora and Flora's (2013) Community Capitals Framework (reproduced from Pitzer and Streeter, 2015, Table 1, p359 with permission)

Natural capital	Elements of nature present in a community including land and water resources, weather, and biodiversity
Cultural capital	Values and perspectives of community members that play a major role in self-efficacy in affecting community change
Human capital	Education, skills, health, and self-esteem of community members
Social capital	Trust, collaboration, and shared vision among community members
Political capital	Ability of groups or communities to influence policy and ensure that policies are implemented accordingly
Financial capital	Monetary resources
Built capital	Facilities that contribute to infrastructural capacity of communities

and social") it can access (Flora and Flora, 2013; McCrea et al., 2014, p275). The concept of 'community capital' can be defined as: "the sum of assets including relationships in a community and the value that accrues from these" (Parsfield et al., 2015, p8). They observed (p21) that community capital "is essential for wellbeing and social inclusion" providing wide-ranging co-benefits; however, "it doesn't naturally distribute equitably". The Community Capitals Framework, devised by Flora and Flora (2013), defined aspects of the 'community ecosystem' in terms of seven community capitals: financial, built, social, human, cultural, natural, political, and human. This framework aimed to "map the strategies and impact different capitals are playing in a community's wellbeing"[5] (Table 8.2).

Community capital reflects the resources that can be drawn upon through, and by, existing residents at a place or new people moving into an area (Deeming et al., 2018). So, community capital is not static and can be grown or developed; it can also be latent in certain settings. Parsfield et al. (2015, p21) articulated their "theory of change" for growing community capital — "achieved through efforts to understand, involve, and connect people within communities". This means that those aspiring to increase community capital,

> ... often local public service professionals and policymakers, but also individuals, community groups, charities or businesses – should first seek to understand the specific context within which they are operating and map the assets and social networks that currently exist.

Considered from an environmental risk perspective, a research imperative is to establish how community capitals can support local climate resilience. Kais and Islam (2016) proposed that in a specific community, resilience dimensions impact positively on community capitals while climate change dimensions, like increased variability and extremes, overall impact negatively. The net result of this interaction indicates a particular community's resilience to climate change (Figure 8.1).

Through a literature synthesis, they identified characteristics of a resilient or 'viable' community when encountering slow or rapidly occurring system changes.

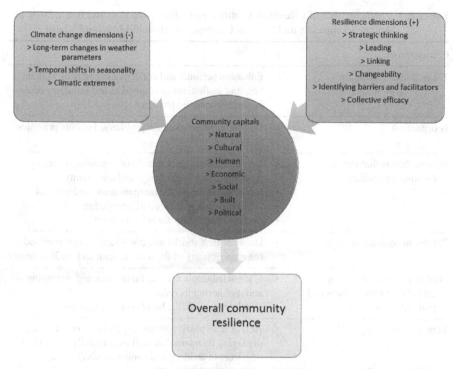

FIGURE 8.1 A conceptual model for climate change impacts, community resilience, and community capitals (Kais and Islam, 2016, adapted from McCrea et al., 2014, with permission Md Saidul Islam)

These characteristics include its organisation, ability to adapt, and ability to connect across scales (individual, community, and regional). Their listing (Table 8.3) can usefully form the basis for local dialogue about how to build community resilience in FRM.

8.3.3 Defining civic agency and citizenship

A working definition of 'civic agency' involves a level of self-organisation in mobilising the capital of civil society agents or groups, their capacities, capabilities, and networks for a particular purpose or effect. Capacities, as an alternative to capitals, are

all the strengths, attributes and resources available within a community, organization or society to manage and reduce disaster risks and strengthen resilience.

UNDRR (ud)[6]; Cretney (2014)

Kuhlicke et al. (2011) provided a typology of social capacities involving knowledge, motivational, network, economic, institutional, and procedural capacities.

Another important but contested concept is 'citizenship' (Chapter 1). Frequently used descriptors include 'responsible', 'active', 'effective', and 'good' citizenship,

TABLE 8.3 Characteristics of a Resilient Community (Edited from Kais and Islam, 2016, p9; Author's Emphasis on Different Capitals; permission: Md Saidul Islam)

Characteristic	Detail
Takes intentional action	• Enhances personal and collective capacity of its members and institutions to respond to and influence course of social/economic change
Is organized	• Has capacity to recognise problems, institute priorities, and act
Fosters factors that enhance community resilience	• Improves community members' capabilities, i.e., by learning to live with change and uncertainty • Nurturing diversity for reorganisation and renewal • Combines different kinds of knowledge • Creates opportunity for self-organisation
Adapts to constant changes	• Does not treat shocks and disturbances only as episodic but regards many of them as constant and gradual threats
Builds its resilience through cumulative mechanisms and pathways over time	• Is knowledgeable and skilful in assessing, managing, and monitoring its risks • Can learn new skills; build on past experiences
Is multi-scalar	• Acts at individual, community, and regional levels, deploying its internal as well as externally networked resources in tackling and coping to adversaries
Assists its members to navigate to resources	• Also negotiates for resources they need
Is relatively autonomous and self-sufficient in relation to economic decision-making	• Has wider economic diversities with a broader range of employment options, income, and financial services (*economic capital*) • Is flexible and resourceful and has capacity to accept uncertainty/ respond proactively to change
Is rich in community capitals	• Includes *economic, social, built, political, and environmental capitals*
Is capable of clearly identifying its barriers and facilitators	• i.e., pre-disaster vulnerabilities, social class, mistrust, race, and ethnicity, gender • i.e., access to community resources, local community civic and faith-based groups, and *bonding–bridging–linking social capitals*
Is connected to external actors (linking *social capital*)	• Includes family friends, religious groups, and government, who deliver wider supportive environment and supply goods and services when needed
Has physical infrastructures and services (*built capital*)	• Includes resilient housing, transport, and power, water, and sanitation systems • Has ability to retain, repair, and renovate these
Can manage its natural assets (*environmental capital*)	• Recognises their value and has ability to protect, enhance, and maintain them

linked to agendas concerned with including *all* citizens. 'Effective' citizenship requires citizen empowerment, agency, and willingness to take responsibility. Horst et al. (2020) explored notions of the 'good citizen' and norms of participation and belonging in Oslo, Norway. They argued for a re-conceptualisation that

> acknowledges present-day spaces of participation as both public and private, and which acknowledges scales of belonging that go beyond and below a narrowly defined national community.
>
> *Horst et al. (2020, p76)*

However described, relations between conception and practices of citizenship at individual, local, and national levels are critical. Personal stories of participation for active citizenship can provide insights into "enablers and benefits, and the barriers and tensions" (Brodie et al., 2011, p71).

8.3.4 Ecological and hydro-citizenship

Growing interest exists in the research and practice of environmental and ecological (or eco-) citizenship and human connection with nature (e.g., Eden, 1993; Dobson, 2003, 2007). Such citizenship is non-contractual and not politically affiliated, and bridges public and private domains. This territory explores the relationships between democracy, citizenship, and their meeting with environmental challenges, inequalities and environmental injustices, including climate change (Lorenzoni et al., 2007). Working in FRM, Nye et al. (2011) used the term "flood risk citizenship" to describe the citizen's role within transition towards a more civic model of policymaking and delivery. This "more collaborative form of engagement" for citizen empowerment has implications for the roles of FRM organisations. Nye et al. (2011, p294) envisaged this as

> both a natural outcome of the different drivers for change, and a necessary tool for tackling the multidimensionality and complexity of flood and coastal risk management.

However, bringing water risk and citizenship together has significant challenges. There is a wider perceived disconnect between citizens and water in the developed world. The character of modern urban water management, with its major legacy of historic engineering, means that traditionally infrastructure has been deliberately hidden underground. Linton (2010, p18) argued "We have left all responsibility for maintaining relationships with water to others". Growing academic and practitioner interest in hydrocitizenship is a response to this, overlaying a burgeoning literature on characteristics of effective citizenship with local water management. Such framing of hydrocitizenship can be conceived as a double mirror, looking at water relations anew through the lens of citizenship and vice versa. McEwen et al. (2020) navigated different interpretations of hydrocitizenship, integrating personal water relationships, with care for people, local place, and environment. Importantly, this involves engaging with water holistically – in resource stewardship and developing resilience to risk.

8.4 Community participation

The act of participation is "to take part in an event or activity" (Cambridge Dictionary). Sense of community (Chapter 2) can be important in participation, with the need to recognise differing experiences of place-based community as both lived and formally organised associations (Dinnie and Fischer, 2020). Hence, community participation needs visioning as both a process and outcome. This section focuses on participation in general and in context of FRM.

In understanding what constitutes meaningful community participation, strong potential exists to learn from research and practice in other domains about how to encourage participation across diverse groups (e.g., in regional development or public health). Burns et al. (2004, p2), working in urban regeneration, emphasised the importance of people's opportunity to be involved in decision-making "about things that affect their lives", while recognising "sometimes people do not want to be involved". They proffered (p2/3) several reasons for why community participation is essential, including "improved democratic and service accountability", the contribution of community knowledge and experience to increase effectiveness, enhanced social cohesion, policy relevance and ownership for communities, and "skills and networks needed to tackle social exclusion".

8.4.1 Models and language of participation

Several models and definitions of participation exist (Bell and Reed, 2021), and community roles vary. An influential model or way of thinking is Arnstein's eight-rung ladder of participation (Arnstein, 1969; Figure 8.2) developed in urban planning.

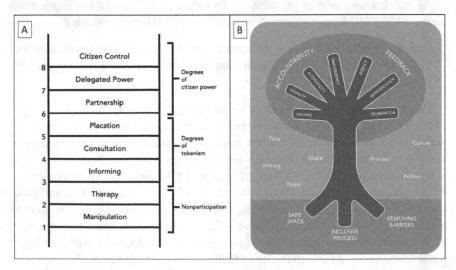

FIGURE 8.2 Participation models and metaphors: (A) Eight rungs on a ladder of citizen participation (Arnstein, 1969) © American Planning Association, Chicago, IL. reprinted by permission of Taylor & Francis Ltd, http://www.tandfonline.com on behalf of American Planning Association, Chicago, IL; (B) the Tree of Participation (ToP) (Bell and Reed, 2021, Figure 2; permission: Karen Bell)

These rungs range from 'lower rung' 'non-participation', through nominal or 'tokenistic engagements' and partnership in decision-making, to 'higher rung' community-led (bottom-up) approaches with goals of 'citizen control'. Schively Slotterback and Lauria (2019) provided a valuable 50-year review of this influential concept.

However, while this visual metaphor is useful in provoking discussion about different practices with local stakeholders, the principle of Arnstein's ladder is much debated, including criticism that its conceptualisation of participation as power is partial and simplistic (e.g., Tritter and McCallum, 2006). This ignores the existence of different relevant forms of knowledge and expertise and lacks recognition that for some, participation may itself be a goal. Tritter and McCallum (2006, p165) perceived Arnstein's ladder as

a linear, hierarchical model of involvement ... [that] fails to capture the dynamic and evolutionary nature of user involvement. Nor does it recognize the agency of users who may seek different methods of involvement in relation to different issues and at different times.

Nevertheless, the principle of who holds the power and resources determining the nature and process of participation still stands, and hence whether participation is meaningful. Other models of participation have since been proffered (e.g., Reed et al.'s, 2018, 'Wheel of Participation'). Bell and Reed (2021) proposed "a new conceptual model of emancipatory, inclusive and empowering participatory decision-making – the 'Tree of Participation (ToP)'" (Figure 8.2B). This non-hierarchical model suggested 12 factors that integrate for inclusive and effective participatory processes (e.g., "safe space", "removing barriers") and seven contextual factors (e.g., transparency, equality) that contribute to the process.

Differences in the language of participation occur including 'community-based' and 'community-led' working, paralleling FRM (Chapters 2 and 10), with potential for cross-sectoral learning. For example, the Child Resilience Alliance,[7] working in community-led child protection, observed

"'Community-based' and 'community-led' processes are often equated. However, there is in fact a world of difference between the two approaches ... as the former are top-down whereas the latter are bottom-up."

The International Association for Public Participation's Spectrum of Public Participation[8] – 'Inform, Consult, Involve, Collaborate, Empower' – was designed to assist with selecting the level of participation that defines the public's role in any public participation process. Alternatively, for some organisations, participation is couched as one engagement method. For example, Australian Institute for Disaster Resilience (2013), in their community engagement model, positioned participation within a wheel of possibilities (information, consultation, collaboration, and empowerment), defined by purpose and context.

Arguably full participation requires a 'community-led' approach – a "for us, by us" culture, posing questions about what such participation looks like in practice. Its principles, involving social inclusion of marginalised groups and local ownership, apply across different risk and resource domains (e.g., water management; rural development; forestry design; planning and public health) and hence are debated in varied professional and cultural settings.

8.4.2 Participation processes and selecting the right tools

In local FRM, citizen participation can potentially involve engaging with an increasingly diverse set of stakeholders, notably statutory organisations like local government, environmental regulators, and water supply providers. It can also involve citizen involvement in wide-ranging community activities (e.g., local governance, volunteering, mutual aid and activism) within the DRM spiral (Chapter 1), working with and through community-based organisations, businesses, and non-government organisations (NGOs) involved in community development and resilience building. Both academic (researchers, teachers, and students) and business sectors (banks, insurance, and corporate) work locally in, and with, communities (e.g., UK's 'Business in the Community' promoting responsible business[9]). Other stakeholders include socially engaged artists and professional gatekeepers to local cultural heritage in galleries, libraries, archives, and museums (Chapters 5 and 10). All bring different skills, knowledge, and values to local participatory processes in building community flood resilience and may have hybrid roles through living within the community.

With active participation and co-working with other stakeholders comes learning that builds human and social capital. Collins and Ison (2009, p369) argued for significant paradigm and epistemic boundaries between information provision *and* consultation, and consultation *and* participation, with an important third boundary between participation *and* social learning to "improve complex situations" (Chapters 4 and 9). This requires different theories of knowledge and practices; for example, citizens and local governments can both benefit from participation in flood risk governance (Wachinger et al., 2013). They observed that the former gain knowledge and personal agency, and the latter, lay knowledge and tailoring of measures. Such ongoing participation has co-benefits, for example, in building trust in messengers for during acute flood events.

Determining meaningful and effective participation processes involves asking six key questions (Table 8.4; Krishnaswamy, 2012). In determining the 'who' and 'how', public participants are unlikely to be homogeneous; it is, therefore, necessary to incorporate citizens' inputs into decision-making in more or less structured ways which can be targeted to access different voices and perspectives. This requires acknowledgement of the diverse methods or tools described by different organisations as 'participatory'. Purposes and deliberative processes to be facilitated are important in tool selection (e.g., Horita and Koizumi, 2009; World Bank Group, 2014). Awareness of

TABLE 8.4 Prompt Questions about Public Participation (Australian Department of Sustainability and Environment, 2005 with permission; Edited from Krishnaswamy, 2012, p5)

WHY?	Situation that calls for, or has produced need for, public participation
WHAT?	Objectives or desired outcomes
WHO?	Profile of potential participants (their interests, experiences, values, etc.)
HOW?	Approach, tools, and methodology to be used
WHEN?	Timeframe for participation
WHERE?	Site or sites for participation

the "continuum of participation" from "nominal" (e.g., information exchange) to "full" (e.g., co-management) improves transparency of process and aids practitioners in selecting appropriate tools (Krishnaswamy, 2012, p3).

Many government organisations promote awareness of participatory methods linked to specific objectives (e.g., NZ, the USA). For example, the American Environmental Protection Agency's (EPA) International Cooperation website[10] differentiates tools by their objectives to "inform the public", "generate and obtain input", and "for consensus building and agreement seeking", building up from 'lower rung Arnstein'. Participatory methods promoting dialogue include citizen juries,[11] World Cafes,[12] scenario-building, serious games, Participatory GIS (PGIS), and social engaged art (Chapter 10). Practical Action, an NGO working in community settings in Less Economically Developed Countries (LEDCs), proposed various participatory tools as options, including drama and storytelling, transect walks, Venn diagrams, and ranking (Pasteur, 2011). However, any participatory method is not inherently good; it is important to understand how participation is experienced by groups in different socio-cultural settings. Evaluation of the use and co-benefits of participatory methods in specific contexts informs co-learning by communities and agencies about 'what works', and hence enhancement strategies.

Strong potential exists for participatory FRM to learn from philosophies and systemic connections made within participatory processes in other settings (e.g., development studies). For example, Participatory Learning and Action (PLA) is a philosophy encapsulating participatory processes that embed critical reflection, and analysis and engender collective action by local people (INTRAC, 2017). PLA, through facilitation,

aims to bring about change, such processes focus on learning by all participants, valuing diversity, supporting group interactions and addressing the importance of context.

The Institute of Development Studies proposed seven key principles for PLA as ways of learning and empowering "people to imagine a different world" (Table 8.5; Chapter 9). Notably, the facilitator's behaviour, attitudes, and reflection are critical to inclusive PLA processes.

TABLE 8.5 Key Principles of Participatory Learning and Action (PLA)[a] (Edited from Institute of Development Studies) (original source: https://www.participatorymethods. org/page/about-participatory-methods. Used with permission of Participation Research Cluster, Institute of Development Studies, University of Sussex)

Principle	Explanation
The right to participate	• That all people have a right to play a part in shaping decisions that affect their lives sounds obvious but is not easy to achieve • Maximising participation of the less powerful is a key feature of PLA
Hearing unheard voices	• Using PMs involves seeking out unheard voices and creating safe spaces that allow them to be heard. Often people who have least say in decisions about their lives are most affected by using PMs
Seeking local knowledge and diversity	• Local people have their own expert knowledge of their community. This should be the starting point for outsiders using PMs to work with them • Also crucial to recognise that there are always different perspectives and realities within communities; every individual brings their own unique experiences and interpretations
Reversing learning	• PMs are about letting go of preconceptions to learn from wisdom of community members. This means being prepared to unlearn what has already been learned
Using diverse methods	• Using a range of PMs helps draw in as many people as possible to undertake learning and analysis on an equal basis
Handing over the stick (or pen, or chalk)	• […] Involves those considered 'expert' – or powerful, or of higher status – sitting back, keeping quiet, and allowing space for others to participate • Thinking about relationships between more and less powerful people, and what those relationships imply for who can speak and who cannot, is important in using PMs
Attitude and behaviour change	• Changing attitudes and behaviours of the powerful is a vital aspect of participatory practice (Chambers, 1997)

[a] https://www.participatorymethods.org/page/about-participatory-methods.

8.4.3 Equitable citizen participation and inclusivity in communities

A key concern in community participation is stereotyping of those who might have the capital to input into participatory decision-making in DRM. Different sectors of communities have different resources, and hence opportunities and capacities to participate, leading to issues of exclusion (e.g., in European FRM; Begg, 2018). Perceptual dichotomies (navigating 'vulnerability' and 'strength') can also pervade understandings – both within communities and statutory agencies – of who might participate (Chapter 10; see Sendai Framework 36). One example is in conceiving older citizens as vulnerable, or as healthy elders with knowledge (local, indigenous) gained through experience and survival that represents a valuable resource to inform local community action. For example, Cohen et al. (2016, p1) in Israel found

> a significant rise in community resiliency scores in the age groups of 61–75 years as compared with younger age bands, suggesting that older people in good health may contribute positively to building community resiliency for crisis.

Another example is whether disabled citizens are perceived as vulnerable or possessing in-built strength, expertise, and resilience gained through dealing with daily challenges that could aid wider community learning. For example, Harrington et al. (2023) explored how gender and disability intersect to influence women's experiences of flood recovery. A third is in considering young people as vulnerable, or with valuable 'pester power', influencing their families and communities to adopt adaptive actions. In contrast, fit, young adults may be perceived as resilient but can be vulnerable if they are unaware of risk and do nothing to prepare.

8.4.4 Evaluating and auditing community participation

Evaluation is essential in any meaningful participatory process, with different assessment frameworks available. Table 8.6, adapted from Beckley et al. (2005), usefully classifies criteria and indicators to evaluate participation tools into three core elements: breadth (range of values), depth (quality), and outcomes (shared goals).

TABLE 8.6 Core Evaluation Criteria and Indicators of Successful Participation Tools (Adapted from Beckley et al., 2005[a], p21 with permission; Explanation[b] from Krishnaswamy, 2012)

Core elements[a]	Explanation[b]	Criteria and indicators[a]
Breadth	Addresses degree to which process adequately incorporates broad range of public values into decision-making process	*Representation* – Incorporate wide range of public values *Accessibility* – Be available to all public interests *Renewal* – Allow for new participants over time *Anonymity* – Protect participants' identities when necessary
Depth	Measures quality of public participation; addresses levels of exchange between participants in participatory process	*Listening & Dialogue* – Foster two-way flow of information *Flexibility* – Be flexible in scope *Deliberation* – Provide opportunities for frank and open discussion *Transparency & Credibility* – Promote and make available in a clearly understandable form, independent input from scientific and other value-based sources *Relationship Building* – Promote positive personal and institutional relationships
Outcomes	Relates to goals of participatory process – how well process met shared vision or goals identified by participants	*Relevance* – Influence decision-making process *Effectiveness* – Improve quality of decisions *Mutual learning* – Contribute to all participants' knowledge *Reciprocity* – Reward or provide incentives *Cost-effectiveness* – Output or outcome cost-effective relative to inputs

Other reports have offered assessment frameworks – with key questions and indicators – for auditing the quality of community participation (e.g., in regeneration; Burns and Taylor, 2000; Burns et al., 2004). For example, the Active Partners[13] framework, developed by COGS (Community and Organisations: Growth and Support, 2000), focused on core dimensions of influence, inclusivity, communication, and capacity in community participation for benchmarking against.

Systematic evaluation of community participation in all its forms is essential. For example, Maskrey et al. (2019, p4) proposed one transferable framework in participatory FRM that comprises four evaluation elements – context, process, substantive outcomes and social outcomes, with attention to "determinants of effectiveness" before and during participation and "dimensions of effectiveness" after the process. Co-evaluation (formative and summative) of participatory processes is also essential for co-learning to inform future practice.

8.5 Community participation in disaster risk management

Building on this review, this section now explores community participation within community-focused DRM[14] (Chapter 2). Critical is who leads the DRM and for what scale of event or disaster. For example, in Local-level Disaster Risk Management (LLDRM), local governments lead the process, with engaged communities, other actors and participatory mechanisms in place (Maskrey, 2011; UNDRR, 2019). In contrast, McLennan (2018) reports on "collective co-production" (Chapter 4) and power-sharing between citizens and government in a community-based disaster risk management (CBDRM) project for bushfire preparedness – 'Be Ready Warrandyte' – in an Australian neighbourhood. A key concern is establishing when and where CBDRM and community-led disaster risk Management (CLDRM) might be appropriate. Tailored community-focused FRM requires multi-stakeholder understanding of local socio-economic and cultural contexts (e.g., risk perception and tolerance; Klijn et al., 2021), alongside scientific determination of risk.

Potential exists to learn from principles and practical guidance for community involvement in DRR within different national settings, including between LEDCs and MEDCs. For example, Šakić Trogrlić et al. (2018), in researching different stakeholder perspectives within CBFRM in Malawi, highlighted challenges of lack of localised ownership and sustainability of processes.

8.5.1 Community-based DRM: Principles and characteristics

Here, participation of community members, as "the main actors and propellers" drawing on local knowledge, is important (Victoria, 2003, p1). The principles and benefits of CBDRM activities and programmes within LEDCs have been articulated in many sources (e.g., Oxfam/ADPC, 2014, Box 8.2), with inclusive participation as essential. Many principles have transferability to community-focused DRM in MEDCs.

BOX 8.2   CBDRM PRINCIPLES (EDITED FROM OXFAM/ADPC, 2014, P23; WITH PERMISSION: ADPC)

- People's participation is essential to make use efficiently of the talent, knowledge, and capabilities of the public.
- Priority is given to the most vulnerable people – both their identification and better care of their needs before, during, and after disasters.
- Interventions are based on Vulnerability and Capacity Assessment (VCA).
- Different perceptions of risk are recognised, with local folk wisdom combined with modern scientific research. This approach caters for different perceptions of disaster risk, offering a comprehensive solution to the problem.
- Risk reduction measures are community specific. While localised primarily, they are integrated with, and are streamlined in, overall national DRR strategies.
- CBDRM activities are managed/operated/sustained by communities themselves. Although initially communities need external expertise and resource help, in the long-term, processes are streamlined in their culture.

Challenges for community-focused, Participatory Disaster Risk Assessment (PDRA) include multi-stakeholder understanding of what it is, degrees of community empowerment, types of knowledge generated and transferability of processes (Pelling, 2007). As Pelling (2007, p377) reflected, "Participation is not a panacea, but it does offer a range of opportunities for progressive policy making". Clarity of purpose among multiple stakeholders is essential to reap its benefits.

Victoria (2003, p2) proposed eight principles for CBDRM activities and programmes derived from practice in the Philippines, but which have transferability. In addition to being "participatory", these include being "responsive", "integrated", "proactive", "comprehensive", "multi-sectoral and multi-disciplinary", "empowering", and "developmental". Such ways of working can act as both targets and performance indicators. A more recent synthesis of learning across studies and cultural contexts by Zwi et al. (2018) yielded valuable and transferable insights into what works within CBDRM. Their systematic review of 27 studies identified key mechanisms and contextual inhibitors of CBDRM in low- and middle-income (LMIC) settings. Key mechanisms were "Integrated knowledges", "Expressed empowerment", "Actioned agency", and "Resilient livelihoods" alongside "Gender and social equity promotion" and "Technological innovation and communication" (p3–5). Linking mechanisms were also important. Inhibitors included "failure to identify local knowledge and to integrate it in disaster risk management"; "marginalisation and failure to take seriously the advice of community elders";

and "excluding communities from the decision process". Zwi et al. (2018, p5) concluded that CBDRM programmes have potential

> to contribute to reducing risk and vulnerability, and may contribute to enhancing resilience and the capacity of affected groups, thus mitigating the long-term economic and social impact of disasters.

However, achieving positive outcomes depends on enabling mechanisms and controlling inhibitors within programme design.

8.5.2 Community-led DRM: Principles and characteristics

CRS/CAFOD and Caritas Australia (Turnbull and Moriniere, 2017) proposed three principles that underpin their approach to CLDRM: inclusion of *all* social groups within the community; leadership of the process *by* the community; and promotion of accountability by *all* involved. They emphasised multiple benefits to ownership and equity:

> These principles not only reflect our commitment to humanitarian values and priorities; we know from experience that they also generate impacts beyond risk reduction. The approach leads to a process and results ... owned by the community and ... likely to be sustained and contribute to equality and equity.
>
> *p6*

Strong potential exists to learn from CLDRM in other cultural and hazard contexts (Cobbing et al., 2023). For example, Kenney et al. (2015) undertook community-based, participatory partnership research into characteristics of New Zealand Māori communities after earthquakes which meant they worked effectively to facilitate community recovery and resilience. Their modelled attributes involved intersections of Māori knowledge, values and principles, and cultural practices (Figure 8.3). Kenney et al. (2015, p11) reflected on the implications of such embedded cultural practices – 'Tikanga' – as "the physical manifestation of Māori knowledge and values" for community resilience. Community history and experience played critical roles.

> Traditional environmental risk mitigation practices such as land mapping and settlement fortifications protected communities by preventing land slippage from episodic flooding as well as ensuring that settlements were developed on stable bedrock. Coastal marae (community centres) were situated so inhabitants could identify early indicators for tsunami and/or king tides and respond accordingly. Inland settlements were located in proximity to rivers to facilitate food security, with secondary sites established as flood evacuation centres.

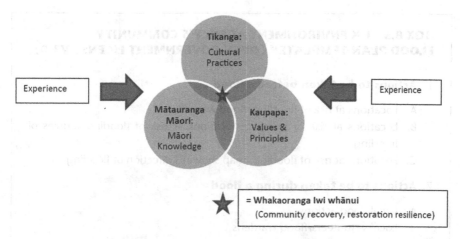

FIGURE 8.3   Model conceptualising traditional Maori approach to disaster risk reduction (Kenney and Phibbs, 2015, Figure 1; Permission: the authors; Licence: Elsevier)

This poses questions about what intermittent or transient communities in MEDC settings may have lost in terms of past intergenerational knowledge for resilience (Chapter 4), and how re(connections) can be brokered (Chapter 5).

8.6   Participation in resilience planning in practice: From household to community

A complex interplay exists between the risk perception of diverse groups (Chapter 3) and their participation in community-embedded mitigation strategies that build from individual to collective measures (Chapter 7). Community agency in preparedness actions and participation in different types of volunteering have potential to step-change local adaptation (e.g., in community emergency planning; 'mutual aid' involving reciprocity and cooperation; community flood groups (CFGs) involving warning or activism; and citizen science activities; Chapter 10).

In terms of emergency planning, community participation can support collective working (e.g., at street level like 'neighbourhood watch') to plan for practical action, building up from household-level planning for flood response. Individual citizen empowerment can be important in mobilising such collaborative working. For example, UK Environment Agency provides a community flood plan template to record intended collective actions before, during, and after a flood (Box 8.3). The chapter now considers some specific forms of citizen and community participation and agency in FRM.

8.6.1   Civic agency in volunteering: Care of communities and place

Defining citizen volunteering – or 'activity by free choice' – is a long-standing challenge (Carson, 2000; Forrest et al., 2023). Western definitions can pervade,

---

BOX 8.3   UK ENVIRONMENT AGENCY'S COMMUNITY FLOOD PLAN TEMPLATE[15] (OPEN GOVERNMENT LICENSE V3.0)

1. **Actions to be taken before a flood**

   A. Locations at risk of flooding: flood warnings
   B. Locations at risk of flooding: locations at risk of flooding/sources of flooding
   C. Locations at risk of flooding: map showing direction of flooding

2. **Actions to be taken during a flood**

   A. Local flood actions
   B. Local volunteers/flood wardens
   C. Important telephone numbers
   D. Available resources
   E. Arrangements between authorities
   F. Vulnerable residents, properties, and locations

3. **After a flood**

   A. Reputable contractors

---

with pleas for less narrowness and more inclusivity in activities encompassed (e.g., Petriwskyj and Warburton, 2007). In practice, NVCO, an English Charity championing voluntary action, defines volunteering as:

> any activity that involves spending time, unpaid, doing something that aims to benefit the environment or someone (individuals or groups) other than, or in addition to, close relatives.[16]

In social science, volunteering is often discussed in relation to "prosocial behavior and altruism; charity and philanthropy behavior; social responsibility; democratic behavior and active citizenship; [and] community development" (Andronic, 2014, p457). Freedom of choice intersects with degree of formality and flexibility. Importantly, volunteers may not recognise themselves as such (Forrest et al., 2023).

Propensity for participation in volunteering and mutual aid is influenced by cultural norms and religion, as well as individual, community, institutional, and governance factors. Within the FRM spiral, volunteering can take varied forms including local flood champions organising CFGs, flood wardens being the 'eyes and ears on the ground', and river stewards observing obstructions and clearing local watercourses. The same citizens may use their agency for increasing wider community resilience to other extreme weather risks and emergencies (e.g., heat, snow).

In the UK, the environmental regulator provides training to flood wardens in monitoring river levels (Forrest et al., 2019) and reporting flood risks. However, warden schemes can fall into abeyance in extended periods without floods.

Some countries have well-established, formal state-organised and trained programmes of voluntary service that allow mobilisation of volunteers (e.g., Czech Republic Government, 2002: "assisting at floods, environmental, or humanitarian disasters"). Dostál (2020) researched the revealed value of volunteering in the Czech Republic; while financial values can be quantified for government, co-benefits also exist for the health and well-being of the citizens participating. As Aoki (2016, p112) articulated, volunteering "encourages positive emotions, such as happiness and a sense of security, which serves to reduce stressors and thereby improve the health status of an individual". However, challenges associated with volunteering for local climate resilience include environmental justice issues, and who has time and resources to volunteer (Forrest et al., 2023).

8.6.2 Spontaneous volunteering

Spontaneous or transient volunteers (SVs) can be defined as:

> Individuals who are unaffiliated with existing official response organisations yet, without extensive preplanning, are motivated to provide unpaid support to the response and/or recovery to an emergency.
>
> *Shaw et al. (2015, p6); UK Cabinet Office (2019)*

Such volunteering can occur both during floods and in supporting recovery (see Paciarotti et al.'s, 2018, review). Examples include volunteers from Muslim Hands travelling to provide support to flood-impacted communities during the 2014 Somerset floods, UK, and young people – 'the Mud Angels' – arriving in Florence to help rescue art masterpieces in after the 1966 Arno floods.[17]

Harris et al. (2017) found motivations for SV included empathy, previous personal flood experience, or imagining such experience, more generalised commitment to the community, and social pressure in involvement. However, tensions in SV exist including perceptual and procedural issues. Harris et al. (2017, p353) proposed a new theory – the "involvement/exclusion" paradox – to explain common emergency situations with "helpers wanting to be involved, juxtaposed with pressures for managers to exclude them". This complex tensioned space provides both opportunities and challenges, posing questions about the scope and desirability of increased volunteer involvement during floods.

Coordination of SV is necessary to capitalise effectively on opportunities – for official emergency responders to be aware of SV's distinctive contribution, and to anticipate their likely arrival in local emergency planning (Rivera and Wood, 2016; Harris et al., 2017; Paciarotti et al., 2018). This includes development of training materials for specific roles and tasks. Shi et al. (2018), working in China, also emphasised the importance of improving policies, regulations, coordination mechanisms, and volunteer training and support. Foreseeing possible patterns of SVs is also essential. Stewart et al. (2014) explored spatiotemporal patterns of volunteering and the role of social networks in community response after severe flooding (2009) impacted Fargo, South Dakota, USA. They found that both social network density and diversity were related to different types of volunteer behaviour.

8.6.3 Participation in community activism: Role of flood groups

Volunteering in FRM can take other forms. The notion of 'associations' has been used to describe small groups galvanised to a particular goal and applied to collective local FRM. Hemming (2011) in *How Small Groups Achieve Big Things* focused on the 'power of giving',[18]

> Groups that last longer consist of members who make an equal contribution, creating fellowship, camaraderie and value. With everyone giving, people can achieve more together than they can on their own.
>
> *Hemming (2011, np)*

He described a positive case study of a CFG active within a rural flood-impacted village on the River Severn, UK. This CFG, possessing significant human capital, was able to work in partnership with the environmental regulator to secure funding for higher structural flood protection. However, CFGs can have more variable human resources to draw on.

Some national settings have seen increases in CFGs as models of local flood volunteering (Forrest et al., 2023) but less so in others. In England, numbers affiliated with the National Flood Forum (NFF), a UK NGO providing flood victim support, have grown since 2000 to over 300. CFGs are also called voluntary flood groups or flood action groups (Forrest et al., 2017). Their roles and status can sit within wider development of local community activism or campaign groups that possess multiple roles and benefits (Figure 8.4).

However, community-initiated group development may not be a simple process. NFF has developed a two-stage support process for CFGs to develop their knowledge, skills, and dispositions for effective multi-agency working with statutory FRM authorities (Chapter 9). McEwen et al. (2018) evaluated the development of CFGs in contrasting community settings, emphasising importance of both human capital and recent flood experience. For example, reasons for CFG participation in more socially disadvantaged communities included gaining employability skills. They identified 'The 6Ss' Framework (Scoping, Situating, Solidarity, Sustainability, Scaffolding, and Sensitive Supporting) for effective participatory CFG building processes in challenged settings with variable human capital and limited flood experience (Table 8.7).

FIGURE 8.4   Multiple benefits of community flood groups (source: The Flood Hub, Newground https://thefloodhub.co.uk/community/; with permission)

TABLE 8.7 'The 6Ss Framework' for Supporting Flood Group Development (adapted from McEwen et al., 2018)

Concern	Explanation
SCOPING	Importance of early baselining of flood groups to identify prior flood experience and existing local capital
SITUATING	Early attention to developing 'local' facilitators who help build networks, maintain connections, and ensure progression
SOLIDARITY	Potential to connect developing groups with less/more capital and with/without flood experience.
SUSTAINABILITY	Early planning for continuance of participatory processes and 'what happens afterwards'
SCAFFOLDING	Recognition that shifts from external to internal facilitation may not be achievable in medium or even longer-term with some groups
SENSITIVE SUPPORTING	Facilitator role is critical – judging support needs in formative stages and how/when facilitator withdraws for group sustainability

The relational territory between CFGs and statutory agencies in FRM can be complicated to navigate. Geaves and Penning-Rowsell (2015, p440) identified two broad categories of relationship: 'contractual' ("a level of protection provided by the authority in exchange for taxes or similar support") and 'collaborative' ("public knowledge, social and financial resources are equal and complementary to those of authority and seeking 'collective security'"). Conflicts can occur when the relationship and degree of reciprocity are implicit, and one or both groups misconstrue the commitment (Chapter 10). However, whatever the model, the ability of groups to access resources was a success factor.

The status and activity of local CFGs can be important in active remembering within communities. Indeed, flood memory can act as the 'grit' that initiates and sustains CFGs. Factors affecting longevity of CFGs along the lower River Severn, UK include: major flood events experienced; personalities and local politics; extent of trust and working relationships with FRM agencies; and levels of support from NGOs. Several different models exist for local CFGs: rarely are groups established to deal with future flooding; more frequently, they are established by flood-affected citizens during or immediately after floods. Some persist only until their immediate goals are met; others have persistence, particularly when integrated with core community development or resilience activities. As Vallance and Carlton (2015, p27) observed, in relation to community groups,

> different types of communities have the potential to weave disaster readiness, response and recovery, and risk reduction into their core business, and therefore represent a valuable – though often underestimated and poorly understood – resource.

CFGs can provide inter-related benefits (e.g., for uptake of property-level resilience; Chapter 7). Dittrich et al. (2016), in Scotland, found positive correlation between the functioning of CFGs and household adoption of FRM measures. They reported,

> positive adoption effects for flood warnings, floodgates and to an extent for insurance, and a positive correlation with increased confidence of implementing and belief in the effectiveness of the measures
>
> *(p471)*

They concluded that statutory agencies supporting CFGs can be cost-effective in promoting household-level measures.

8.6.4 Volunteering as citizen science in environmental monitoring

Another burgeoning and impactful form of participation is in citizen science (CS) activities (Goodchild, 2007; Wehn et al., 2015; Hecker et al., 2018). This territory intersects with 'volunteer monitoring', 'citizen observatories', and 'citizen-sensing' (Assumpção et al., 2018). Buytaert et al. (2014, p1) provided an inclusive definition of CS as "the participation of the general public (i.e., non-scientists) in the generation of new knowledge" with

> members of the public [intentionally] engaging in authentic scientific investigations: asking questions, collecting or processing data, and/or interpreting results.

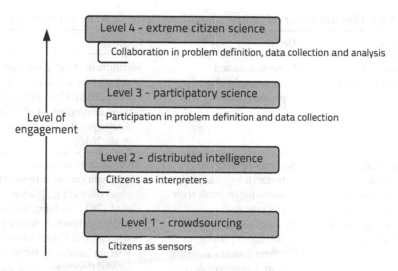

FIGURE 8.5   Levels of participation and engagement in citizen science projects (Assumpcão et al., 2018, Figure 1; adapted from Haklay, 2013)

Citizen science is systematic in approach and rigorous, but with multiple meanings and diverse practices in citizen involvement. The concept, process, and practice of CS have developed from models that involved citizens working with academic researchers in collecting monitored field data (e.g., in Earthwatch[19]) to more autonomous authentic monitoring of the local environment. One initiative – Extreme Citizen Science (ExCiteS)[20] – explicitly links emergent technology and place-based, citizen-led knowledge creation with aspiration for transformation – as

a situated, bottom-up practice that takes into account local needs, practices and culture and works with broad networks of people to design and build new devices and knowledge creation processes that can transform the world.

Assumpção et al. (2018) differentiated CS in terms of levels of engagement, public participation, and citizen role within the process (Figure 8.5). Identification of success factors for CS is critical. San Llorente Capdevila et al. (2020), in context of water quality monitoring, found attributes of citizens (e.g., knowledge and experience, and socio-economic character), attributes of institutions (e.g., types, funding), and their interactions with citizens (e.g., communication, feedback) were important, with motivation and support as key factors. A key question is how to involve citizens with less scientific and human capital in CS and identify training needs.

While much previous hydrological monitoring by CS has focused on rainfall and water quality (Buytaert et al.'s, 2014, Table 1), there is growing involvement of citizen data in mapping, monitoring, and recording floods, both over time at-a-site and through crowdsourcing (Table 8.8; Chapter 10).

TABLE 8.8 Different Models of Citizen Science Data Gathering with FRM Examples

Concepts	Detail	Examples
Volunteered geographic data	Flood videos and photographs shared or posted online	Scientific data extracted from video of flash flood (Lewis and Park, 2018). Can improve post-flood determination of river discharges (Le Boursicaud et al., 2016)
Urban living laboratories (involves social sensing in 'real-time'; citizens as sensors)	Stakeholder-focused research approaches, operating in small river catchments "a set of sensing and data collection paradigms where data are collected from humans or devices on their behalf" (Wang et al., 2015, p1)	River catchments as living labs with community-based observation (e.g., Starkey et al., 2017 studying surface water catchment response). Also project FloodCitiSense[a] – living laboratory for urban pluvial flooding
Community-based monitoring	Citizen monitoring of local water and environmental risk and resource	Community rainfall data incorporated to improve flood early warning systems for surface water flooding (e.g., Abon et al., 2012; Gharesifard et al., 2019)
Citizen observatories (location-based crowdsourcing)	"An open and shared information system dedicated to the collection of data on the environment and natural resources, using ICT, and the volunteer participation of individuals in data collection" (Montargil and Santos, 2017, p1)	Citizen observatories help citizens actively engage in flood monitoring (e.g., Assumpção et al., 2018)

[a] https://jpi-urbaneurope.eu/project/floodcitisense/.

While citizens do not need to collaborate with FRM agencies in their scientific data gathering, it can be mutually beneficial when they do. Local knowledge can observe how complex floods play out at 1:1 (Chapters 2 and 4), with some communities taking on monitoring of smaller water courses. This includes establishing local factors that interact to cause or intensify local flooding – collecting measurements of rainfall, water levels, velocities and flow patterns during floods, and observing location of blockages, efficacy of drainage structures, and impacts of land use/management under different flow conditions.

Data scarcity and spatiotemporal uncertainty have always been problems in specialist flood modelling. CS data properties relate to its spatial and temporal coverage, uncertainty and volume, influencing its ease of integration to improve modelling (Mazzoleni et al., 2017). There is demonstrable value in using community-based CS observations within specialist catchment modelling and characterisation, with strong potential for collaborative working between communities and scientists (e.g., Goodrich et al., 2020 in pluvial flood risk modelling). Such data provision through lived knowledge (e.g., captured in videos and photography on mobile phones) can ground truth specialist modelling during floods and proffer alternative scenarios, with potential for integrating local and scientific knowledges (e.g., Thapa et al., 2019; Chapter 4).

A key question is how CS observations can be integrated with new technological advances in measuring floods in their locale for improved FRM and evidence for catchment management. Mazumdar et al. (2016) reported how "citizen observatories" had been created through low cost, static and portable sending and transfer devices to allow the public to collaborate positively with authorities and other organisations during and outside emergencies (Chapter 10). For example, the EU 'WeSenseIT'[21] project, working in the UK, Netherlands, and Italy, explored mechanisms for how communities can participate in monitoring their water courses for both floods and droughts through ICT-enabled citizen observatories during different phases of the FRM spiral (Wehn et al., 2015; Table 8.8). Such mechanisms can be intended/explicit with citizen training or unintended/implicit.

Attention also needs to be given to hard and soft elements (i.e., both hardware and techniques to harness "citizens' "collective intelligence" – "information, experience, and knowledge embodied") within a participatory process (Wehn et al., 2015, p227). This poses important questions about reciprocity – whether participation in CS makes citizens more resilient. While a growing literature reflects on the advantages and disadvantages of citizen data in flood risk monitoring, less evidence exists on benefits to citizens in inclusive CS. Key aspects involve CS's role in empowering citizens and communities – building their science capital and risk awareness (Bonney et al., 2015); increasing wider engagement (Starkey et al., 2017); and building potential for participation in decision-making (Wehn et al., 2015). The social sciences and humanities have important but currently underplayed roles in exploring this territory in terms of citizen behaviour and agency (Tauginienė et al., 2020).

8.6.5 Widening conception of citizen participation in knowledge sharing

Citizen agency as researcher-observer is not just about contributing 'hard' science data ('big data') but also recognising value in less rigorous 'soft' and deeper 'thick' data in FRM. 'Soft' data may be "descriptive or qualitative and can be

TABLE 8.9 Approaches to Gathering and Sharing Soft and Thick Citizen Data with FRM Examples

Approaches	Detail	Examples
Public Participatory GIS (PP GIS)	P-GIS data (past, present, and future) should be described as 'qualitative data' as "based on people's knowledge, opinions and perceptions" (p13)	Community maps for past, present, and future FRM (Forrester and Cinderby, 2014)
Community mapping Participatory GIS (P GIS) (S)		Integration of community sketch mapping as data for flood vulnerability assessment (Brandt et al., 2020)
Participatory modelling (S/T)	"Purposeful learning process for action that engages the implicit and explicit knowledge of stakeholders to create formalised and shared representation(s) of reality"[a]	Design and implementation of participatory modelling approach to identify intervention options for local FRM (Maskrey et al., 2016)
Digital storytelling (T)	Garnering stories of flood preparedness as local knowledge for resilience	Bank of flood stories for sharing within/between communities (Holmes and McEwen, 2020)

[a] https://participatorymodeling.org/
S = soft; T = thick

used to help interpret hard data" (APA Dictionary of Psychology[22]). 'Thick' data is captured by anthropologists, human geographers, and sociologists, using qualitative, ethnographic research methods, to "provide more nuanced and contextualized information" that reveals people's stories, worldviews, and emotions (Wang, 2016; Hong et al., 2022, p2). Table 8.9 provides examples of such data gathering within FRM. Inevitable issues in soft data quality exist when used for flood risk planning purposes. For example, Forrester and Cinderby (2014, p13) highlighted the potential fuzziness of citizen data in PGIS:

> In some situations, the information can be very precise – for example marking the exact location of a drainage outflow that gets clogged (which may not even be present in official data) – but equally likely (and sometimes on the same map) quite vague – such as where a new riverine planting to slow flood waters should be concentrated.

Citizens contributing thick data to local FRM can also be about sharing personal stories (e.g., of resilience) as lay knowledge or 'citizen as researcher' investigating local flood heritage (e.g., oral flood histories; Chapter 5). As Veer et al. (2016, p321) emphasised

> The importance of allowing individuals to be able to tell their stories and the significance of having spaces and places where shared experiences can be fostered, even when the recipient of the story is unknown.

Importantly, such participatory spaces and places for social learning can be both traditional face-to-face and online (Chapter 9).

8.7 Conclusions

This chapter has systematically explored how and why *meaningful* participation of citizens and communities in diverse ways is critical to effective community-focused FRM. There is a need for multi-stakeholder recognition of its conditions, and the valuable social and community capital that communities bring to CBDRM and CLDRM, scaling up from individual human capital. Here, Flora and Flora's (2013) Community Capitals Framework is valuable for mapping and understanding a community's resilience in advance of engagement and co-working in all models of community-focused FRM.

Citizen participation in FRM can take increasingly diverse forms throughout the DRM spiral, moving out from acute events. This includes activities such as volunteering that might not be formally conceived as 'participation' by both citizens and FRM agencies. Potential for transformation exists in integrating citizen agency (e.g., in citizen science) with new technology, bringing local and scientific knowledge together in local DRM (Chapter 10). However, challenges exist in funding inclusive adoption of technology after trials of participatory pilot projects end. Numerous benefits of community participation in CBDRM are cited by researchers and practitioners, including building on local coping strategies and capacities, promoting ownership of FRM strategies, building confidence and empowerment, strengthening community cohesion and mobilising resources, both internal and external to the community. However, community participation and agency can also bring tensions over the understanding of responsibilities in FRM.

A key question is how to increase community participation and agency in DRM in settings where expectation of a key role of the state still pervades. This is even more critical in context of increasing flood risk and exposure due to climate change and urban development. Postings on ReliefWeb,[23] a knowledge exchange portal provided by the UN's Office for Coordination of Humanitarian Affairs, testify to the valuable potential for mutual learning about 'what works' from exchanging case studies of community participation in CBDRM within different cultural, socio-economic, and governance settings. This occurs particularly where the state has more limited roles, and where community agency is a long-standing part of living with water. Importantly, valuable opportunities exist to integrate effective community participation within decision-making for local anticipatory governance of risk (Chapter 10).

Notes

1 https://www.europarl.europa.eu/factsheets/en/sheet/7/the-principle-of-subsidiarity.
2 Cook et al. (2013, p754) highlight the "persistence of 'normal catchment management' despite the 'participatory turn'".
3 https://www.forumforthefuture.org/the-five-capitals.
4 https://www.worldbank.org/en/publication/human-capital/brief/the-human-capital-project-frequently-asked-questions#HCP2.
5 https://www.canr.msu.edu/news/what_are_community_capitals.
6 https://www.undrr.org/terminology/capacity.
7 https://communityledcp.org/guide/chapter-2/2-2-what-is-a-community-led-approach-to-child-protection.
8 https://cdn.ymaws.com/www.iap2.org/resource/resmgr/pillars/Spectrum_8.5x11_Print.pdf.
9 https://www.bitc.org.uk/.
10 https://www.epa.gov/international-cooperation/public-participation-guide-tools.
11 https://catchmentbasedapproach.org/learn/citizen-juries/.
12 www.theworldcafe.com.
13 http://cogs.uk.net/uploads/File/active%20partners(1).pdf.
14 A range of terms are now used, including CBDRM (Šakić Trogrlić et al., 2018 in Malawi); LLDRM (Maskrey 2011); participatory DRM (e.g., Samaddar et al., 2017 in India) and CLDRM; CLFRM (e.g., McLean et al., 2015 in Scotland).
15 https://www.gov.uk/government/publications/community-flood-plan-template.
16 https://www.ncvo.org.uk/policy-and-research/volunteering-policy.
17 https://www.historytoday.com/history-matters/florence's-mud-angels.
18 http://henryhemming.com/books/together.
19 Earthwatch.org.
20 https://www.geog.ucl.ac.uk/research/research-centres/excites.
21 https://cordis.europa.eu/project/id/308429.
22 https://dictionary.apa.org/soft-data.
23 https://reliefweb.int/.

References

Abon, C.C., David, C.P.C. and Tabios, G.Q. III (2012) Community-based monitoring for flood early warning system. *Disaster Prevention and Management*, 21(1), 85–96. https://doi.org/10.1108/09653561211202728

Adger, N. (2000) Social and ecological resilience: Are they related? *Progress in Human Geography*, 24(3), 347–364. https://doi.org/10.1191/030913200701540465

Albert, A. and Passmore, E. (2008) *Public Value and Participation: A Literature Review for the Scottish Government*. Scottish Government Social Research, Edinburgh. https://www.gov.scot/binaries/content/documents/govscot/publications/research-and-analysis/2008/03/public-value-participation-literature-review-scottish-government/documents/0057753-pdf/0057753-pdf/govscotdocument/0057753.pdf

Aldrich, D.P., Meyer, M.A. and Page-Tan, C.M. (2018) Social capital and natural hazards governance. *Oxford Research Encyclopedia* of Natural Hazard Science https//doi. org/10.1093/acrefore/9780199389407.013.254

Andronic, R. (2014) Definition of volunteering in social sciences. *International Conference of Scientific Paper AFASES*, Brasov, 22–24 May 2014, 457–460.

Aoki, Y. (2016) Donating time to charity: Working for nothing? *Oxford Economic Papers*, 69, 97–117.

Arnstein, S.R. (1969) A ladder of citizen participation. *Journal of the American Institute of Planners*, 35(4), 216–224. https://doi.org/10.1080/01944366908977225

Assumpção, T.H., Popescu, I., Jonoski, A. and Solomatine, D.P. (2018) Citizen observations contributing to flood modelling: Opportunities and challenges. *Hydrology and Earth System Sciences*, 22, 1473–1489. https://doi.org/10.5194/hess-22-1473-2018

Australia Department of Sustainability and Environment (2005) The Engagement planning workbook, Book 2: Effective engagement. Government of Victoria, Australia. http://www.dse.vic.gov.au/__data/assets/pdf_file/0020/105824/Book_2_-_The_Engagement_Planning_Workbook.pdf (accessed March 2012).

Australian Institute for Disaster Resilience (2013) *National strategy for disaster resilience: Community engagement framework*. Handbook 6. https://knowledge.aidr.org.au/media/1761/handbook-6-national-strategy-for-disaster-resilience-kh-final.pdf

Bath, J. and Wakerman, J. (2015) Impact of community participation in primary health care: What is the evidence? *Australian Journal of Primary Health*, 21(1), 2–8. https://doi.org/10.1071/PY12164

Beck, U. (1992) *Risk Society: Towards a New Modernity*. SAGE, London and New York.

Beckley, T., Parkins, J. and Sheppard, S. (2005) *Public Participation in Sustainable Forest Management: A Reference Guide*. Sustainable Forest Management Network, Edmonton. 55pp.

Begg, C. (2018) Power, responsibility and justice: A review of local stakeholder participation in European flood risk management. *Local Environment*, 23, 383–397. https://doi.org/10.1080/13549839.2017.1422119

Bell, K. and Reed, M. (2021) The tree of participation: A new model for inclusive decision-making. *Community Development Journal*, bsab018. https://doi.org/10.1093/cdj/bsab018

Bonney, R., Phillips, T., Ballard, H. and Enck, J. (2015) Can citizen·science enhance public understanding of science? *Public Understanding of Science*, 25. https://doi.org/10.1177/0963662515607406

Brandt, K., Graham, L., Hawthorne, T. et al. (2020) Integrating sketch mapping and hot spot analysis to enhance capacity for community-level flood and disaster risk management. *The Geography Journal*, 186, 198–212. https://doi.org/10.1111/geoj.12330

Brodie, E., Hughes, T., Jochum, V., Miller, S., Ockenden, N. and Warburton, D. (2011) Pathways through participation: What creates and sustains active citizenship? Available at: http://www.sharedpractice.org.uk/Downloads/Pathways_final_report.pdf. Accessed 30 October 2015.

Burns, D., Heywood, F., Taylor, M., Wilde, P. and Wilson, M. (2004) *Making Community Participation Meaningful: A Handbook for Development and Assessment*. Joseph Rowntree Foundation, The Policy Press, Bristol.

Burns, D. and Taylor, M. (2000) *Auditing Community Participation: An Assessment Handbook*. The Policy Press, Bristol/York.

Buytaert, W., Zulkafli, Z., Grainger, S. et al. (2014) Citizen science in hydrology and water resources: Opportunities for knowledge generation, ecosystem service management, and sustainable development. *Frontiers in Earth Science*, 2(26). https://doi.org/10.3389/feart.2014.00026

Carlton, S. and Vallance, S. (2017) The commons of the tragedy: Temporary use and social capital in Christchurch's earthquake-damaged Central City. *Social Forces*, 96(2), 831–850. https://doi.org/10.1093/sf/sox064

Carson, E.D. (2000) On defining and measuring volunteering in the United States and abroad. *Law and Contemporary Problems*, 62, 67–71. https://scholarship.law.duke.edu/cgi/viewcontent.cgi?article=1147&context=lcp

Challies, E., Newig, J., Thaler, T., Kochskämper, E. and Levin-Keitel, M. (2016) Participatory and collaborative governance for sustainable flood risk management: An emerging research agenda. *Environmental Science & Policy*, 55(2), 275–280.

Chambers, R. (1997) *Whose Reality Counts? Putting the First Last*. Practical Action Publishing, Rugby, UK.

Cobbing, P., Waller, E. and McEwen, L.J. (2023) The role of civil society in extreme events through a narrative reflection of pathways and long-term relationships. *Journal of Extreme Events*. https://doi.org/10.1142/S2345737622500038

COGS (Community and Organisations: Growth and Support) (2000) *Active Partners: Benchmarking Community Participation in Regeneration*. Yorkshire Forward, Yorkshire.

Cohen, O., Geva, D., Lahad, M. et al. (2016) Community resilience throughout The lifespan – The potential contribution of healthy elders. *PLoS ONE*, 11(2), e0148125. https://doi.org/10.1371/journal.pone.0148125

Colclough, G. and Sitaraman, B. (2005) Community and social capital: What is the difference? *Sociological Inquiry*, 75, 474–496. https://doi.org/10.1111/j.1475-682X.2005.00133.x

Collins, K. and Ison, R. (2009) Jumping off Arnstein's ladder: Social learning as a new policy paradigm for climate change adaptation. *Environmental Policy and Governance*, 19, 358–373.

Cook, B.R., Kesby, M., Fazey, I. and Spray, C. (2013) The persistence of 'normal' catchment management despite the participatory turn: Exploring the power effects of competing frames of reference. *Social Studies of Science*, 43, 754–779. https://doi.org/10.1177/0306312713478670

Cretney, R. (2014) Resilience for whom? Emerging critical geographies of socio-ecological resilience. *Geography Compass*, 8(9), 627–640.

Czech Republic Government (2002) *Act on Volunteer Service*. 198/2002 Coll.

Deeming, H., Davis, B., Fordham, M. and Taylor, S. (2018) River and surface water flooding in Northern England: The civil protection-social protection nexus. In H. Deeming, M. Fordham, C. Kuhlicke, L. Pedoth, S. Schneiderbauer, and C. Shreve (eds.) *Framing Community Disaster Resilience*. Wiley-Blackwell, Chichester.

Dinnie, E. and Fischer, A. (2020) The trouble with community: How 'Sense of Community' influences participation in formal, community-led organisations and rural governance. *Sociologia Ruralis*, 60, 243–259. https://doi.org/10.1111/soru.12273

Dittrich, R., Wreford, A., Butler, A. et al. (2016) The impact of flood action groups on the uptake of flood management measures. *Climatic Change*, 138, 471–489. https://doi.org/10.1007/s10584-016-1752-8

Dobson, A. (2003) *Citizenship and the Environment*. Oxford University Press, Oxford.

Dobson, A. (2007) Environmental citizenship: Towards sustainable development. *Sustainable Development*, 15, 276–285. https://doi.org/10.1002/sd.344

Dostál, J. (2020) Revealed value of volunteering: A volunteer centre network. *Annals of Public and Cooperative Economics*, 91(2), 319–345. https://doi.org/10.1111/apce.12271

Eden, S. (1993) Individual environmental responsibility and its role in public environmentalism. *Environment Planning A*, 25, 1743–1758.

Facer, K. and Enright, B. (2016) *Creating Living Knowledge: The Connected Communities Programme, Community University Relationships and the Participatory Turn in the Production of Knowledge*. University of Bristol/AHRC Connected, Bristol.

Flora, C. and Flora, J. (2013) *Rural Communities: Legacy and Change* (4th edn). Westview Press, Boulder.

Forrest, S., Dostál, J. and McEwen, L.J. (2023) The future of volunteering in extreme weather events: Critical reflections on key challenges and opportunities for climate resilience. *Journal of Extreme Events*. https://doi.org/10.1142/S2345737623410038

Forrest, S., Trell, E.-M. and Woltjer, J. (2017) Flood groups in England: Governance arrangements and contribution to flood resilience. In E.-M. Trell, B. Restemeyer, M.M. Bakema and B. van Hoven (eds.) *Governing for Resilience in Vulnerable Places*. Routledge, Oxon, pp92–115. https://doi.org/10.4324/9781315103761

Forrest, S., Trell, E.M. and Woltjer, J. (2019) Civil society contributions to local level flood resilience: Before, during and after the 2015 boxing day floods in the Upper Calder Valley. *Transactions of the Institute of British Geographers*, 44, 422–436. https://doi-org.ezproxy.uwe.ac.uk/10.1111/tran.12279

Forrester, J. and Cinderby, S. (2014) *Guide to using Community Mapping and Participatory-GIS*. Prepared as part of Managing Borderlands project and funded by the Rural Economy and Land Use (RELU) programme, Economic & Social and Natural Environment Research Council.

Geaves, L.H. and Penning-Rowsell, E.C. (2015) 'Contractual' and 'cooperative' civic engagement: The emergence and roles of 'flood action groups' in England and Wales. *Ambio*, 44, 440–451.

Gharesifard, M., Wehn, U. and van der Zaaga, P. (2019) Context matters: A baseline analysis of contextual realities for two community-based monitoring initiatives of water and environment in Europe and Africa. *Journal of Hydrology*, 579, 124144.

Goodchild, M.F. (2007) Citizens as sensors: The world of volunteered geography. *GeoJournal*, 69, 211–221. https://doi.org/10.1007/s10708-007-9111-y

Goodrich, K.A., Basolo, V., Feldman, D.L. et al. (2020) Addressing pluvial flash flooding through community-based collaborative research in Tijuana, Mexico. *Water*, 12(5), 1257. https://doi.org/10.3390/w12051257

Haklay, M. (2013) Citizen science and volunteered geographic information: Overview and typology of participation. In D. Sui, S. Elwood and M. Goodchild (eds.) *Crowdsourcing Geographic Knowledge: Volunteered Geographic Information (VGI) in Theory and Practice*. Springer, Netherlands, pp105–122.

Harrington, E., Bell, K., McEwen, L. and Everett, G. (2023) Is there room on the broom for a crip? Disabled women as experts in disaster planning. *Journal of Extreme Events*. https://doi.org/10.1142/S234573762350001X

Harris, M., Shaw, D., Scully, J., Smith, C.M. and Hieke, G. (2017) The Involvement/ Exclusion paradox of spontaneous volunteering: New lessons and theory from winter flood episodes in England. *Nonprofit and Voluntary Sector Quarterly*, 46(2), 352–371.

Hecker, S., Haklay, M., Bowser, A., Makuch, Z., Vogel, J. and Bonn, A. (2018) *Citizen Science: Innovation in Open Science, Society and Policy*. UCL Press, London. https://doi.org/10.14324/111.9781787352339

Hemming, H. (2011) *Together: How Small Groups Achieve Big Things*. John Murray, London.

Holmes, A. and McEwen, L.J. (2020) How to exchange stories of local flood resilience from flood rich areas to the flooded areas of the future. *Environmental Communication*, 14(5), 597–613. https://doi.org/10.1080/17524032.2019.1697325

Hong, A., Baker, L., Prieto Curiel, R., Duminy, J., Buswala, B., Guan, C. and Ravindranath, D. (2022) Reconciling big data and thick data to advance the new urban science and smart city governance. *Journal of Urban Affairs*. https://doi.org/10.1080/07352166. 2021.2021085

Horita, M. and Koizumi, H. (eds.) (2009) *Innovations in Collaborative Urban Regeneration.* Springer, Switzerland.

Horst, C., Bivand Erdal, M. and Jdid, N. (2020) The "good citizen": Asserting and contesting norms of participation and belonging in Oslo. *Ethnic and Racial Studies*, 43(16), 76–95. https://doi.org/10.1080/01419870.2019.1671599

INTRAC (2017) Participatory learning and action (PLA). https://www.intrac.org/resources/ participatory-learning-action/participatory-learning-and-action/

Kais, S.M. and Islam, M.S. (2016) Community capitals as community resilience to climate change: Conceptual connections. *International Journal of Environmental Research and Public Health*, 13(12), 1211. https://doi.org/10.3390/ijerph13121211

Kenney, C.M. and Phibbs, S. (2015) A Māori love story: Community-led disaster management in response to the Ōtautahi (Christchurch) earthquakes as a framework for action. *International Journal of Disaster Risk Reduction*, 14, 46–55. https://doi.org/10.1016/ j.ijdrr.2014.12.010

Kenney, C.M., Phibbs, S.R., Paton, D., Reid, J. and Johnston, D.M. (2015) Community-led disaster risk management: A Maori response to Otautahi (Christchurch) earthquakes. *Australasian Journal of Disaster and Trauma Studies*, 9(1), 9–20.

Klijn, F., Marchand, M., Meijer, K., van der Most, H. and Staparu, D. (2021) Tailored flood risk management: Accounting for socio-economic and cultural differences when designing strategies. *Water Security*, 12, 100084. https://doi.org/10.1016/j.wasec.2021.100084

Kotus, J. and Sowada, T. (2017) Behavioural model of collaborative urban management: Extending the concept of Arnstein's ladder. *Cities*, 65, 78–86. https://doi.org/10.1016/j. cities.2017.02.009

Krishnaswamy, A. (2012) Strategies and tools for effective public participation. *Journal of Ecosystems and Management*, 13(2), 1–13. https://doi.org/10.22230/jem.2012v13n2a124

Kuhlicke, C., Steinführer, A., Begg, C. et al. (2011) Perspectives on social capacity building for natural hazards: Outlining an emerging field of research and practice in Europe. *Environmental Science and Policy*, 14, 804–814. https://doi.org/10.1016/j.envsci.2011.05.001

Landström, C. (2020) *Environmental Participation: Practices Engaging the Public with Science and Governance.* Palgrave Macmillan, Switzerland.

Le Boursicaud, R., Pénard, L., Hauet, A., Thollet, F. and Le Coz, J. (2016) Gauging extreme floods on YouTube: Application of LSPIV to home movies for the post-event determination of stream discharges. *Hydrological Process*, 30, 90–105. https://doi.org/10.1002/ hyp.10532

Lewis, Q.W. and Park, E. (2018) Volunteered geographic videos in physical geography: Data mining from YouTube. *Annals of the American Association of Geographers*, 108(1), 52–70. https://doi.org/10.1080/24694452.2017.1343658

Linton, J. (2010) *What Is Water? The History of Modern Abstraction.* UBC Press, Vancouver.

Lorenzoni, I., Nicholson-Cole, S. and Whitmarsh, L. (2007) Barriers perceived to engaging with climate change among the UK public and their policy implications. *Global Environmental Change*, 17(3–4), 445–459.

Maskrey, A. (2011) Revisiting community-based disaster risk management. *Environmental Hazards*, 10(1), 42–52. https://doi.org/10.3763/ehaz.2011.0005

Maskrey, S. A., Mount, N. J., Thorne, C. R. and Dryden, I. (2016) Participatory modelling for stakeholder involvement in the development of flood risk management intervention options. *Environmental Modelling & Software*, 82, 275–294. https://doi.org/10.1016/j.envsoft.2016.04.027.

Maskrey, S.A., Priest, S. and Mount, N.J. (2019) Towards evaluation criteria in participatory flood risk management. *Journal of Flood Risk Management*, 12, e12462. https://doi.org/10.1111/jfr3.12462

Mattos, D. (2015) Community capitals framework as a measure of community development. *Cornhusker Economics*, 811. https://digitalcommons.unl.edu/agecon_cornhusker/811

Mazumdar, S., LanFranchi, V., Ireson, N. et al. (2016) Citizen observatories for effective earth observations: The WeSenseIt approach. *Environmental Scientist*, 56–61. https://www.the-ies.org/resources/they-walk-among-us-rise

Mazzoleni, M., Verlaan, M., Alfonso, L., Monego, M., Norbiato, D., Ferri, M. and Solomatine, D. (2017) Can assimilation of crowdsourced data in hydrological modelling improve flood prediction? *Hydrology and Earth System Sciences*, 21, 839–861. https://doi.org/10.5194/hess-21-839-2017

McCrea, R., Walton, A. and Leonard, R. (2014) A conceptual framework for investigating community wellbeing and resilience. *Rural Society*, 23(3), 270–282. https://doi.org/10.1080/10371656.2014.11082070

McEwen, L., Gorell Barnes, L., Phillips, K. and Biggs, I. (2020) Reweaving urban water-community relations: Creative, participatory river "daylighting" and local hydrocitizenship. *Transactions of the Institute of British Geographers*, 45, 779–801. https://doi.org/10.1111/tran.12375

McEwen, L.J., Holmes, A., Quinn, N. and Cobbing, P. (2018) 'Learning for resilience': Developing community capital through action groups in lower socio-economic flood risk settings. *International Journal of Disaster Risk Reduction*, 27, 329–342. https://doi.org/10.1016/j.ijdrr.2017.10.018

McEwen, L.J., Leichenko, R., Garde Hansen, J. and Ball, T. (2023) CASCADE-NET: Increasing civil society's capacity to deal with changing extreme weather risk: Negotiating dichotomies in theory and practice. *Journal of Extreme Events*. Journal of Extreme Events, 9(2). http://doi.org/10.1142/S2345737623300016

McLean, L., Beevers, L., Waylen, K., Wright, G. and Wilkinson, M. (2015) Learning from community led flood risk management. *CREW report CD2014-12*. https://www.crew.ac.uk/sites/www.crew.ac.uk/files/sites/default/files/publication/CREW_COS_full-report.pdf

McLennan, B.J. (2018) Conditions for effective coproduction in community-led disaster risk management. *Voluntas*, 31, 316–332. https://doi.org/10.1007/s11266-018-9957-2

Montargil, F. and Santos, V. (2017) Citizen observatories: concept, opportunities and communication with citizens in the first EU experiences. In Paulin, A. A., Anthopoulos, L. G., and Reddick, C. G. (eds.) *Beyond Bureaucracy: Towards Sustainable Governance Informatisation.* Springer International Publishing, Cham. 167–184.

Norris, E. and McLean, S. (2011) *The Civic Commons: A model for social action.* http://www.thersa.org/__data/assets/pdf_file/0003/385518/RSA-Civic-Commons-Final.pdf.

Nye, M., Tapsell, S. and Twigger-Ross, C. (2011) New social directions in UK flood risk management: Moving towards flood risk citizenship? *Journal of Flood Risk Management*, 4(4), 288–297. https://doi.org/10.1111/j.1753-318X.2011.01114.x

Oxfam/ADPC (2014) *Community-Based Disaster Risk Management for Sindh Province, Pakistan.* https://www.adpc.net/igo/category/ID790/doc/2015-nRIu4Y-ADPC-publication_CBDRMHandbookSindhPRINTER.pdf

Paciarotti, C., Cesaroni, A. and Bevilacqua, M. (2018) The management of spontaneous volunteers: A successful model from a flood emergency in Italy. *International Journal of Disaster Risk Reduction*, 31, 260–274.

Parsfield, M., Morris, D., Bol, M., Knapp, M., Park, A., Yoshioka, M. and Marcus, G. (2015) (eds.) *Community Capital the Value of Connected Communities.* Connected Communities Report. https://www.thersa.org/globalassets/pdfs/reports/rsaj3718-connected-communities-report_web.pdf

Pasteur, K. (2011) *From Vulnerability to Resilience: A Framework for Analysis and Action to Build Community Resilience.* Practical Action Publishing, Rugby UK.

Patterson, O., Weil, F. and Patel, K. (2010) The role of community in disaster response: Conceptual models. *Population Research and Policy Review*, 29, 127–141. https://doi.org/10.1007/s11113-009-9133-x

Pelling, M. (2007) Learning from others: The scope and challenges for participatory disaster risk assessment. *Disasters*, 31(4), 373–385.

Petriwskyj, A.M. and Warburton, J. (2007) Redefining volunteering for the global context: A measurement matrix for researchers. *Australian Journal on Volunteering*, 12(1), 7–13. https://search.informit.org/doi/10.3316/ielapa.840215979693516

Pitzer, K.A. and Streeter, C.L. (2015) Mapping community capitals: A potential tool for social work. *Advances in Social Work*, 16, 358–371. https://doi.org/10.18060/17470

Priest, S., Suykens, C., Van Rijswick, H.F.M.W. et al. (2016) Societal resilience: Lessons from the implementation of the floods directive in six European countries. *Ecology and Society*, 21(4), 50. https://doi.org/10.5751/ES-08913-210450

Reed, M. S., Vella, S., Challies, E. *et al.* (2018) A theory of participation: What makes stakeholder and public engagement in environmentalmanagement work? *Restoration Ecology*, 26, S7–S17. https://doi.org/10.1111/rec.12541

Rivera, J.D. and Wood, Z.D. (2016) Disaster relief volunteerism: Evaluating cities' planning for the usage and management of spontaneous volunteers. *Journal of Emergency Management*, 14(2), 127–138.

Šakić Trogrlić, R., Wright, G.B., Adeloye, A.J., Duncan, M.J. and Mwale, F. (2018) Taking stock of community-based flood risk management in Malawi: Different stakeholders, different perspectives. *Environmental Hazards*, 17(2), 107–127. https://doi.org/10.1080/17477891.2017.1381582

Samaddar, S., Okada, N., Choi, J. et al. (2017) What constitutes successful participatory disaster risk management? Insights from post-earthquake reconstruction work in rural Gujarat, India. *Natural Hazards*, 85, 111–138. https://doi.org/10.1007/s11069-016-2564-x

San Llorente Capdevila, A., Kokimovaa, A., Sinha Ray, S. and Avellánb, T. (2020) Success factors for citizen science projects in water quality monitoring. *Science of the Total Environment*, 728, 137843.

Schively Slotterback, C. and Lauria, M. (2019) Building a foundation for public engagement in planning. *Journal of the American Planning Association*, 85(3), 183–187. https://doi.org/10.1080/01944363.2019.1616985

Shaw, D., Smith, C.M., Heike, G., Harris, M. and Scully, J. (2015) *Spontaneous volunteers: Involving citizens in the response and recovery to flood emergencies.* Final report FD2666 for UK DEFRA, July 2015.

Shi, M., Xu, W., Gao, L. et al. (2018) Emergency volunteering willingness and participation: A crosssectional survey of residents in northern China. *BMJ Open*, 8, e020218. https://doi.org/10.1136/bmjopen-2017-020218

Smith, M. K. (2000-2009) 'Social capital', The encyclopedia of pedagogy and informal education. https://infed.org/mobi/social-capital/. Retrieved: 12/11/2023

Starkey, E., Parkin, G., Birkinshaw, S., Large, A., Quinn, P. and Gibson, C. (2017) Demonstrating the value of community-based ('citizen science') observations for catchment modelling and characterisation. *Journal of Hydrology*, 548, 801–817.

Stewart, K., Glanville, J.L. and Bennett, D.A. (2014) Exploring spatiotemporal and social network factors in community response to a major flood disaster. *The Professional Geographer*, 66(3), 421–435. https://doi.org/10.1080/00330124.2013.799995

Szreter, S. and Woolcock, M. (2004) Health by association? Social capital, social theory and the political economy of public health. *International Journal of Epidemiology*, 33(4), 650–667.

Tauginienė, L., Butkevičienė, E., Vohland, K. et al. (2020) Citizen science in the social sciences and humanities: The power of interdisciplinarity. *Palgrave Communications*, 6, 89. https://doi.org/10.1057/s41599-020-0471-y

Thapa, A., Bradford, L., Strickert, G., Yu, X., Johnston, A. and Watson-Daniels, K. (2019) "Garbage in, Garbage Out" does not hold true for indigenous community flood extent modeling in the Prairie Pothole region. *Water*, 11(12), 2486. http://dx.doi.org/10.3390/w11122486

Tritter, J.Q. and McCallum, A. (2006) The snakes and ladders of user involvement: Moving beyond Arnstein. *Health Policy*, 76, 156–168.

Turnbull, M. and Moriniere, L.C. (2017) *Guide to Facilitating Community-Led Disaster Risk Management*. CRS, CAFOD and Caritas Australia.

UK Cabinet Office (2008) *The Pitt Review: Lessons learned from the 2007 floods*. http://www.cabinetoffice.gov.uk/thepittreview.aspx. Accessed 1 May 2013

UK Cabinet Office (2019) *Planning the coordination of spontaneous volunteers in emergencies*. June 2019. https://assets.publishing.service.gov.uk/government/uploads/system/uploads/attachment_data/file/828201/20190722-Planning-the-coordination-of-spontaneous-volunteers-in-emergencies_Final.pdf

UNDRR (2015) *Sendai Framework for Disaster Risk Reduction (2015–2030)*. https://www.undrr.org/publication/sendai-framework-disaster-risk-reduction-2015-2030

UNDRR (2019) *Words into action guidelines: Implementation guide for local disaster risk reduction and resilience strategies*. https://www.undrr.org/publication/words-action-guidelines-implementation-guide-local-disaster-risk-reduction-and

Vallance, S. and Carlton, S. (2015) First to respond, last to leave: Communities' roles across the '4Rs'. *International Journal of Disaster Risk Reduction*, 14(1), 27–36.

Veer, E., Ozanne, L.K. and Hall, M. (2016) Sharing cathartic stories online: The internet as a means of expression following a crisis event. *Journal of Consumer Behaviour*, 15, 314–324.

Victoria, L. (2003) Community-based disaster management in the Philippines: Making a difference in people's lives. *Philippine Sociological Review*, 51, 65–80. http://www.jstor.org/stable/44243073

Vinke-de Kruijf, J., Kuks, S.M.M. and Augustijn, D.C.M. (2015) Governance in support of integrated flood risk management? The case of Romania. *Environmental Development*, 16, 104–118. https://doi.org/10.1016/j.envdev.2015.04.003

Wachinger, G., Renn, O., Begg, C. and Kuhlicke, C. (2013) The risk perception paradox – Implications for governance and communication of natural hazards. *Risk Anal*, 33(6), 1049–1065.

Wang, T. (2016) Why Big Data needs Thick Data. *Ethnography Matters*. https://medium. com/ethnography-matters/why-big-data-needs-thick-data-b4b3e75e3d7

Wang, D., Abdelzaher, T., and Kaplan, L. (2015) Chapter 1 – A new information age. In D. Wang, T. Abdelzaher and L. Kaplan (eds.) *Social Sensing*. Morgan Kaufmann, Burlington, pp1–11. https://doi.org/10.1016/B978-0-12-800867-6.00001-7

Wehn, U., Rusca, M., Evers, J. and Lanfranchi, V. (2015) Participation in flood risk management and the potential of citizen observatories: A governance analysis. *Environmental Science and Policy*, 48, 225–236.

World Bank Group (2014) Strategic Framework for Mainstreaming Citizen Engagement in World Bank Group Operations. Washington, DC. © World Bank. https://openknowledge. worldbank.org/handle/10986/21113. License: CC BY 3.0 IGO.

Zwi, A.B., Spurway, K., Marincowitz, R., Ranmuthugala, G., Hobday, K. and Thompson, L. (2018) *Do CBDRM Initiatives Impact on the Social and Economic Costs of Disasters? If So, How, Why, When and in What Way(s)?* EPPI-Centre, London.

9

COMMUNITY LEARNING FOR FLOOD RESILIENCE

Strategies and pitfalls

9.1 Introduction

Flood education is essential within integrated and sustainable flood risk management (FRM) strategies so that individuals and communities are able and empowered to take responsibility in local flood risk decision-making (Figure 2.4). This chapter explores community flood education and learning for resilience within the wider setting of disaster education, drawing on various theories and practices. The objective of any education is learning not teaching; here 'flood education' is used inclusively to embrace all learning. Indeed, in FRM, the language has shifted from 'education' to 'learning' (Dufty, 2020). Increasingly, the need for learning is recognised as a 'response modification mitigation' option when implementing key non-structural or behavioural interventions to increase community resilience (Chapter 7). This is a significant step forward, reflecting growing awareness that flood management is not solely about structural measures and technology (Chapter 2). Different types of communities need to have, or acquire knowledge, skills, values, and attitudes as capital and capabilities to be active participants in processes of local resilience building (Chapter 8).

The UNDRR Sendai Framework for Disaster Risk Reduction (2015–2030, 24m) states the international imperative

> To promote national strategies to strengthen public education and awareness in disaster risk reduction, including disaster risk information and knowledge, through campaigns, social media and community mobilization, taking into account specific audiences and their needs.

Historically, flood education strategies and programmes have tended to be designed by government agencies, with emphasis on passive transfer or one-way broadcast of information and linear learning (Chapter 6). This differs from promoting two-way

DOI: 10.4324/9781315666914-9

knowledge exchange or 'co-generation' with distinct groups and communities for social learning. While the importance of flood education and learning is now better acknowledged, less awareness exists about what can be learnt from research on effective pedagogies in the framing of ideas and ways of working to inform policy and practice that addresses wicked problems (Chapter 2). This includes how different people (by age, demographics, culture, etc.) learn about risk and adaptation for flood resilience, and the differences between acquiring knowledge and action. There are also important debates about how individual learning for resilience scales up to building community resilience and a society that 'learns for resilience'. This academic interdisciplinary backdrop involves education, psychology, and sociology. This chapter asks:

- How can community flood education best be framed? What can be learnt from research and practice from pedagogies for sustainability, resilience, citizenship, and wider environmental education?
- What might community flood education comprise, and what does 'learning for resilience' mean in terms of knowledge, skills, values, and attitudes?
- What is the potential of different pedagogies in disaster risk management education within both formal learning and informal or social learning in community or organisational settings?
- What are the opportunities and pitfalls in engaging communities in learning for resilience?

9.2 Defining 'flood education'

'Flood education' can be defined as learning that builds individual human and/ or collective community capital for flood resilience (Chapter 8). Both individuals and communities can be construed as learners (e.g., as in learning communities). Flood education itself can be framed in a nested thematic hierarchy of learning, set within an understanding of the flood-drought continuum (risk of water excess and deficit), water education (including water quality and ecology as well as quantity; e.g., UNESCO[1]), broader risk and resilience in 'disaster education' (e.g., Torani et al., 2019), and environmental education (e.g., Krasny et al., 2010; Keong, 2021).

Critical questions include 'why', 'for whom', 'what', 'where', 'when', and 'how' flood education might take place, and 'for whom'. These require critical exploration as the field of flood education develops. Brief reflections below are expanded later in the chapter.

9.2.1 Why is flood education needed?

Learning is a critical element in resilience building, and any system of learning is more resilient if it is capable of adaptation and transformable (Voss and Wagner, 2010). So, how can education support transitions in FRM from "fighting against water" to "living with water" (van Herk et al., 2015, p559)? In specific flood-risk communities, there can be continua of both experience ('no flood experience' to 'many events experienced') and social capital ('high' to 'low') that intersect in different ways (McEwen et al., 2018). Well-recognised differences also exist between experience, risk awareness, knowledge, action, and behaviour change to mitigate losses (e.g., Cologna et al., 2017; see refuted Knowledge-Action-Behaviour theory; Chapter 3). This poses questions: how is learning matched to learner characteristics like extent of flood experience or science capital, and how does knowledge become 'understanding' and 'actionable'? Different types of learning are required for flood preparedness versus effective response during acute events and recovery. Preparedness and adaptive actions requiring prior learning for efficacy include, for example, producing local community-level plans and implementing property-level structural interventions (PLP) (Chapter 7).

9.2.2 Who might be learning?

A critical concern is *who* is doing the learning – individuals or groups, collectives or communities – and the nature of the learning relationships therein. Varied groups for learning in communities include residents (including of special land uses like hospitals and caravan parks), businesses, schools/universities, as well as specific vulnerable groups. This involves concern for both individual learners, and how learners and learning might amass in flood risk settings as collective capital. Key groups have potential roles in cascading learning among peers and across generations (e.g., school children and students; Williams et al., 2017).

Another issue is the government priority given to learning for communities and training professionals. For example, Chadderton (2015), in examining disaster education and preparedness for national emergencies in Germany, emphasised the importance of who is educated, envisaging disaster education as an emerging study area within 'lifelong education'. However, she indicated that in Germany, education is "not generally extended to the general public, rather confined to trained experts, decentralised, localised and exclusive" (Chadderton, 2015, p589).

9.2.3 What should be the focus of learning?

Experiential learning (a cycle of experiencing, reflecting, thinking, and acting[2]; cf. Kolb, 1984) from 'living through floods' needs assimilation and sharing at individual and community levels. This supports shifts from coping to adaptive strategies that help maintain livelihoods during and after floods. Learning for transformation (Section 9.3), involving the questioning of personal feelings, perspectives, beliefs and assumptions, might be the goal. Relationships between flood memory, lay knowledge (experiential; intergenerational), and resilience need to be developed working in, and with, communities (McEwen et al., 2016). Here again, the concept of 'actionable knowledge' (Antonacopolou, 2008) – knowledge with potential for embedding in practice – is important (Chapter 4). This concerns how latent knowledge and skills learnt in different ways are rehearsed and put into action by individuals and groups to build flood-resilient communities.

9.2.4 Where could learning take place and who are the facilitators?

Formal settings for flood education include: schools, colleges, and universities; through training and continuing professional development (CPD) in statutory organisations; and in business continuity planning. In contrast, informal learning occurs through personal experiential and social learning (e.g., through 'flood buddies' or community flood groups). NGOs often work as facilitators in this hybrid space. Flood learning can therefore potentially be facilitated by an increasingly diverse range of individuals and organisations: between individuals within a community, between communities, between communities and statutory FRM organisations or NGOs, or between academia and other groups in FRM (communities, organisations). There are questions about the relative merits of different settings and facilitators for learning for distinct social and cultural groups, and how 'what works' can be strategically interwoven in the policy and practice of community flood education.

9.2.5 When should learning take place?

Identifying timings for learning is critical throughout different phases within the Disaster Risk Management (DRM) spiral (preparedness, response, recovery, mitigation; Chapter 1), *and* across the human life course (Schuller, 2017). Major learning opportunities can present after floods (Kuang and Liao, 2020). Temporal relationships between collective learning and preparedness actions need unravelling, along with personal timelines for learning in childhood as young citizens, and through adulthood in lifelong learning. This sits within, and is enabled by, the 'master concept' and wider culture of a 'learning society' that continues to learn (Field, 2010), committed to active citizenship, equality, and individual well-being.

9.2.6 How should learning take place?

Learning theories selected will determine the nature of learning activities in disaster risk reduction (DRR) (Dufty, 2013; Kitagawa, 2021). For example, active learning strategies recognise the value of different teacher roles – whether a 'sage on the stage' (knowledge holder) or 'guide on the side' (facilitator). The latter develops activities that encourage learners to be active and engaged within learning processes, while in the former, with knowledge transmission and transfer, the learner predominantly listens. Flood education also needs to accommodate potentially wide-ranging demographics in a community, including differing levels of prior learning, cultural learning preferences, and dominant learning styles. The latter is exemplified by

Fleming and Mills' (1992) model of four sensory modalities (VARK) – Visual (spatial), Auditory, Read/write, and Kinaesthetic (physical). One learning style (e.g., written word) is unlikely to fit all learners within a community. Learning styles and preferences also vary in different socio-economic and cultural settings (e.g., visual approaches where first languages differ in migrant communities).

The 'How' includes concern for the value of social (interpersonal) learning styles in communities. Bandura (1971) developed social learning theory, recognising that "new patterns of behavior can be acquired through direct experience or by observing the behavior of others" (p3). This poses questions about what effective social learning for community flood resilience could look like.

9.3 Pedagogies for flood education

Flood education has potential to learn from other forms of community education and different pedagogies. The term 'pedagogy' is used inclusively for learning strategies, however, there is potential to broaden the scope to include andragogy (methods and practices for teaching adult learners) and heutagogy (strategies that promote autonomy in self-directed or self-determined learning). The latter recognises a knowledge-based future where "knowing how to learn" is a fundamental skill (Blaschke, 2012, p59) to be able to "cope and flourish".[3] Figure 9.1 highlights their differences

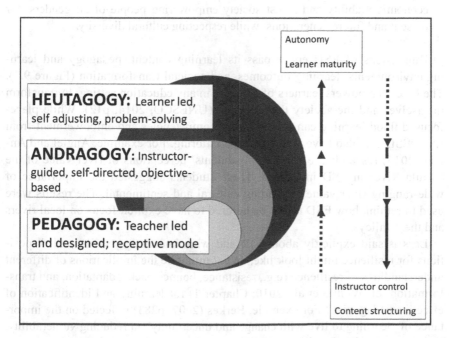

FIGURE 9.1 Differences between pedagogy, andragogy and heutagogy (adapted with permission from figure by Simon Whittemore, 2021)

in terms of learner-instructor competencies and approaches. All have application in conceiving community flood education as a process of lifelong learning.

9.3.1 Sustainability education and learning for resilience

Several related theoretical contexts provide useful background for framing flood education – including Education for Sustainable Development (ESD) (more recently Learning for Sustainability, LfS), Learning for resilience (LfR), and 'Citizenship Education' (or 'education for effective citizenship'). An extensive literature proposes individual knowledge, skills, attitudes, and values required as learning outcomes in ESD (e.g., UNESCO, 2014, 2020). These include holistic, systems and critical thinking, the ability to imagine future scenarios and make collaborative decisions. Such learning involves affect – emotional learning, and questioning of value systems, worldviews and behaviours. Previous shifts in the sustainability research landscape (Tilbury and Cooke, 2005) have parallels in how LfS is conceived (i.e., learning that informs cf. learning that transforms).

The UNESCO (2020, p8) road map aims to implement its global action programme on ESD.

> ESD empowers learners with knowledge, skills, values and attitudes to take informed decisions and make responsible actions for environmental integrity, economic viability and a just society empowering people of all genders, for present and future generations, while respecting cultural diversity.

Dimensions of ESD encompass its learning content, pedagogy and learning environments, learning outcomes, and societal transformation (Figure 9.2). The latter "empowers learners of any age, in any education setting, to transform themselves and the society they live in" (UNESCO, 2014, p12). While place-focused flood learning can draw its core competencies and empowerment from ESD, ESD can also have water-focused tailoring. For example, Angiel and Angiel (2015) researched older school students' perceptions of river value of the Vistula River in ESD in Poland. These students suggested the importance of wide-ranging river values including cultural and sentimental. The results were used to rethink how ESD might be tailored to have explicit focus on local rivers and their valleys.

Less is said explicitly about LfR and what knowledge, skills, and dispositions for resilience might look like.[4] What might be the implications of different understandings of resilience (e.g., resistance, bounce back, adaptation, and transformation; cf. Whittle et al., 2010; Chapter 1) for learning and identification of effective pedagogies? For example, Berkes (2007, p283) reflected on the importance of "learning to live with change and uncertainty" in reducing vulnerability. He argued that local resilience building needs to involve expanding the types of knowledges used within learning and problem-solving. This requires accessing

Learning content: Integrating critical issues, such as climate change, biodiversity, disaster risk reduction [DRR], and sustainable consumption and production [SCP], into the curriculum.

Pedagogy and learning environments: Designing teaching and learning in an interactive, learner-centred way that enables exploratory, action-oriented and transformative learning. Rethinking learning environments – physical as well as virtual and online – to inspire learners to act for sustainability.

Learning outcomes: Stimulating learning and promoting core competencies, such as critical and systemic thinking, collaborative decision-making, and taking responsibility for present and future generations.

Societal transformation: Empowering learners of any age, in any education setting, to transform themselves and the society they live in.

- Enabling a transition to greener economies and societies.
  - Equipping learners with skills for 'green jobs'.
  - Motivating people to adopt sustainable lifestyles.
- Empowering people to be 'global citizens' who engage and assume active roles, both locally and globally, to face and to resolve global challenges and ultimately to become proactive contributors to creating a more just, peaceful, tolerant, inclusive, secure and sustainable world.

FIGURE 9.2 Dimensions of ESD (UNESCO, 2014, p12, reproduced with permission)

both lay knowledges (e.g., experiential and intergenerational; Chapter 4) and specialist scientific and technical knowledge. Resilience in learning can also be construed in other ways. The concept of the 'resilient adult learner' (Chapman Hoult, 2013) involves being able to deal with change with grit and durability, as well as having adaptive and problem-solving skills to deal with being 'knocked back' by events and learning more.

LfR through formal or social learning can be considered through different frames (e.g., social, emotional and psychological, and adaptive resilience and community capital; Chapter 8). Dufty (2008), working in Australia, noted increasing adaptive capacity as a key ongoing element in 'new' community flood education throughout the DRM spiral. Importantly, developing LfR also needs to consider actual learning processes (e.g., roles of participation and experience). Significant value can also exist in intergenerational LfR through exchanges between elders and children in communities, alongside capitalising on different strategies and settings for doing this (Chapters 4 and 5).

9.3.2 Citizenship education

The concept of effective 'citizenship education' – for individual efficacy, civic agency, and 'empowering for change' – also has resonance in how flood education is framed (Nye et al., 2011; McEwen et al., 2020; Chapter 8). Topics relevant to LfR include respect for diversity, participation, democratic processes, and the role of civil society (International Institute for Educational Planning, 2015).

9.3.3 Other framings for learning for resilience

Useful learning theories and approaches include significant learning, levels of learning, transformative learning, and inter-and trans-disciplinary learning.

9.3.3.1 Significant learning

Engaging with different taxonomies of learning can also help frame flood education. This includes Fink's (2003) *Taxonomy of Significant Learning* developed for higher education but with wider applicability. It includes learning about self and others and developing new feelings, interests, and values (Figure 9.3). Feelings and values are important influences on behaviour in FRM. Fink (2003) emphasised that each type of learning is interactive, stimulating and supporting other kinds of learning.

9.3.3.2 Levels of learning

Linear learning involves acquiring factual knowledge and skills. However, also important in LfR is the ability to distinguish between different levels of learning, defined by Argyris (1976) as single- and double-loop learning (DLL). Lower-level, single-loop learning involves developing better ways of doing things (e.g., improving a skill) within an aspect of knowledge. Here goal divergence and adaptation errors are recognised and corrected without questioning underlying beliefs and values. In contrast, setting up potential for DLL or 'learning for improvement' is a higher-level process triggering additional learning. DLL encourages people to think critically and in-depth on their theories and assumptions about underlying causes and beliefs in determining goals in their decision-making. This may involve cognitive change – challenging and re-evaluating a worldview or examining then redeveloping a problem. Drawing on literature evidence, Thomson et al. (2014, p1184) argued that promoting DLL requires FRM and its stakeholders "to be proactive, seek new knowledge, be creative, question, and be holistic when making future based decisions". In contrast, longer-term, triple-loop learning is about values, attitudes, norms, and worldviews, and enabling personal change and transformation (Tosey et al., 2012).

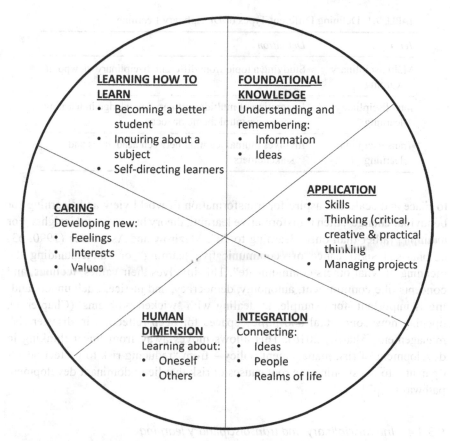

FIGURE 9.3 A taxonomy of significant learning (Fink, 2005; reproduced with permission)

9.3.3.3 *Transformative learning or 'learning that promotes change'*

Another useful theoretical framing for LfR is Transformative Learning Theory, proposed by Mezirow and Associates (1990) and Mezirow (1991). He described this as:

> learning that transforms problematic frames of reference – sets of fixed assumptions and expectations (habits of mind, meaning perspectives, mindsets) – to make them more inclusive, discriminating, open, reflective, and emotionally able to change.

> *Mezirow (2003, p58/59)*

Here, the roles of, and relationships between, the learner, facilitator or teacher and learning settings are crucial, as is space for critical reflection. As LfR tailored

TABLE 9.1 Defining Different Types of Disciplinary Learning

Term	Definition
Multidisciplinary learning	Studying a topic from different disciplinary viewpoints
Interdisciplinary learning	Integration of multidisciplinary knowledge in learning across a central theme or focus
Transdisciplinary learning	Involves mutual learning between academics and stakeholders

to place and scale can aspire for transformation of world view and 'learning for behaviour change', then transformative learning theory brings useful insights. For example, rather than only "learning to do", Mezirow and Associates (1990, p3) emphasised significance of "communicative learning" or "understanding the meaning of what others communicate". This involves their values, feelings, and concepts like commitment, autonomy, democracy, and justice. Such understanding is important, for example, in dealing with 'wicked problems' (Chapter 1), opening new conceptual and policy spaces for deep reflection in disaster risk management (Sharpe, 2016). This allows moves away from linear thinking in development of risk management policy – from "reducing risk to protect development – to questioning the root causes of risk that lie in dominant development pathways" (p213).

9.3.3.4   Interdisciplinary and transdisciplinary learning

Interdisciplinary (IDL) and transdisciplinary learning (TL) both have important value in flood education; Table 9.1 distinguishes between terms. IDL[5] focuses on the connection of concepts (here around 'resilience') and synthesis of ideas across disciplines, which can promote greater creativity (Weller and Appleby, 2021). Notably learning outcomes for IDL and ESD also map well against each other (McEwen et al., 2009). Establishing a balance between generalism and specialism in learning is critical expertise for those delivering community-focused FRM. Ideas around TL, exchanging knowledge and skills between academic researchers and practice (e.g., in sustainable water futures; Schneider and Rist, 2014), and enabling community-based learning (CBL) (e.g., for local students in flood risk settings), are also relevant.

9.4   Different settings for flood-related learning within communities

Three complementary settings for learning about floods are now highlighted – social learning, CBL, and child-centred learning in building resilient communities.

9.4.1   Social learning for resilience – Informal settings

Social learning is important within flood education and wider water resource and environmental management. "In a social learning system, competence is historically and socially defined" (Wenger, 2000, p226). Varied conceptual and contested understandings of social learning exist in the literature. Reed et al. (2010, p1) proposed that its processes:

> must (1) demonstrate that a change in understanding has taken place in the individuals involved; (2) demonstrate that this change goes beyond the individual and becomes situated in wider social units or communities of practice; and (3) occur through social interactions and processes between actors within a social network.

Social learning is frequently treated as a designed intervention in processes facilitated or even initiated by a third party (Johannessen and Hahn, 2013). However, to be successful, social learning within communities needs to be more organic and free flow, emphasising mutual learning. Its value (Benson et al., 2016) is in gaining new knowledge to enable change, linked to the development of actionable knowledge for resilience. The currency of social learning systems is

> collegiality, reciprocity, expertise, contributions to the practice, and negotiating a learning agenda; not affiliation to an institution, assigned authority or commitment to a predefined deliverable.
>
> *Wenger (2000, p243)*

An example of a good social learning system in FRM is where community flood groups function effectively. As communities of practice with identities shaped by the participation of their members, they combine what Wenger (2000, p229) described as "joint enterprise, mutuality, and a shared repertoire of communal resources".

Managing local risk for resilience requires ongoing social learning to increase adaptive capabilities within communities. However, Pahl-Wostl et al. (2008) also emphasised the growing importance of social learning across sectors within policy development and its implementation in wider water resource governance and sustainability science.

> The emphasis is now moving from the need to simply 'know more' and deploying even more information to policy and expert circles to developing adaptive cross-sectoral capacities and new types of knowledge to respond adequately to the changing dynamics of social–ecological systems in concrete contexts of action.
>
> *Pahl-Wostl (2008, p1)*

9.4.2 Community-based learning

While the focus of social LfR lies within the actual communities at risk, CBL (termed 'service learning' in the USA) can be construed as people, often tertiary students in formal education, capitalising on opportunities for co-learning within community settings. It can be defined as:

> A form of experiential education where learning emerges through a cycle of action and reflection where students work through a process of applying what they are learning to community problems and, at the same time, reflecting upon their experience as they seek to achieve real objectives for the community and deeper understanding and skills for themselves.
>
> *adapted from Eyler and Giles (1999); National Coordinating Centre for Public Engagement NCCPE (2017)*

Increasingly important in LfR are local communities' relationships with their secondary and tertiary education establishments, and mutual understanding of potential for CBL (Mason O'Connor and McEwen, 2020). CBL brings together engaging 'community', however defined, with learning through various instructional strategies (Mason O'Connor and McEwen, 2012). For example, CU-Engage[6] (University of Colorado Boulder) emphasises "its reciprocal and mutually beneficial partnerships between instructors, students, and community groups" and its goal "to address community-identified needs and ultimately create positive social change".

Many civic universities are developing experiential CBL in their curricula, recognising that students learn skills, values, and confidence that equip them for effective citizenship and community participation. Importantly, CBL also has benefits to community groups and NGOs – often underfunded; understaffed – bringing time and knowledge to co-investigating local issues like flood risk (Klein et al., 2011). Mason O'Connor and McEwen (2020) outlined ten principles of real-world learning that link to environmental justice agendas, including reciprocity, equity (social justice), and valuing diversity. Teachers in tertiary education, possessing hybrid knowledge (specialist and lay) and personal community connections, can bridge between local education institutions and their communities in addressing local risks. Supporting teachers' professional development in CBL maximises co-benefits for students and communities (Mason O'Connor and McEwen, 2012).

9.4.3 Child-centred learning

United Nations (1989) 'Convention on the Rights of the Child' emphasised the importance of engaging, empowering, and mobilising youth through education to prepare and respond to disasters. While the Sendai Framework emphasises participation of children and young people in DRR, this can be limited by adult perceptions. CUIDAR (2018, p9) highlighted

adult imaginaries or prejudices about childhood e.g., where children and young people are seen as a homogeneous, passive and intrinsically vulnerable group.

They noted that issues of "ageism/adultism" can pervade in identification of "specific issues or topics that should not be discussed with children because these are difficult, harmful or complicated".

However, attention needs to be given to children's voice, role, and responsibility during disasters (e.g., as in the Disaster Management curriculum in Bangladeshi schools). Back et al. (2009) proposed that investing in child-centred DRR is strategically important because learning and practising while young helps embed changed behaviours that can be integrated into adult life. However, the Australian Institute for Disaster Resilience (AIDR, 2021, p16) warns:

There is a perception that 'by educating young people, we will educate the community'. Whilst DRE for young people will assist in building community disaster resilience through family and other social interactions and learning, it should not be the only means of DRE [Disaster Risk Education] in a community.

More case studies are needed that explore the effectiveness of child-centred flood learning activities in schools; see, for example, Williams et al.'s (2017) research on a 'treasure box' (flood grab bag) idea shared in primary schools within flood risk settings in England. Critical is linking design and evaluation of learning interventions to a child's st(age).

The report *Disaster Education* (2007), by Building Research Institute (BRI) and National Graduate Institute for Policy Studies (GRIPS), audited the status of disaster education strategies for different countries globally at that time. Each national curriculum – and specific environmental hazards including floods, hurricanes, cyclones and droughts – was appraised for its coverage of the cause and nature of disasters, effects of disasters, lessons from past disasters, disaster risk reduction/mitigation, preparedness, response-rescue and relief, reconstruction and rehabilitation and the role of community/institutions. They found large diversity in coverage, emphasis, and family/community connection of curricula. For example, Sinha et al. (2007, p4) emphasised the distinctive importance of family and community connection within Costa Rica's curriculum which contained:

a clear picture of reality as something that can be transformed in order to reduce risk and natural disasters. Conceptual and scientific components integrated with research by students and teachers and involvement at the family and community levels, so that civil society can play a role in the process.

While disaster education – dealing with national risks – forms part of the formal national curriculum in some Less Economically Developed Countries (LEDCs), traditionally this has occurred less frequently in More Economically Developed Countries (MEDCs). Here environmental hazards can be 'othered' – perceived as only happening in other countries. Many countries have different school boards, following different curricula; no compulsory requirement existed for these to include disaster management (e.g., in Australia, Bangladesh, India, and the UK; Sinha et al., 2007). Concern for improving this territory is reflected in more recent national guidance; for example, Australian Institute for Disaster Resilience (2021) focused on participation and empowerment in child-centred DRR. Creativity is also needed in identifying disciplinary settings within schools where disaster education is, or could be, located. For example, in the UK, DRM tends to sit within geography but could equally be located within mathematics or creative writing. Design and availability of resource packs on FRM for children, young people, and teachers to support curriculum embedding can fuel this process (e.g., UK Geographical Association's collaboration with the environmental regulator[7]).

9.5 Developing knowledge, skills, and dispositions in learning for flood resilience

When framing and targeting LfR, there is a need to understand three main learning domains: cognitive (knowledge and thinking), psychomotor (skills), and affective (dispositions, attitudes, values, and emotions). In any learning setting, learning strategies and priorities can be envisaged as triangulating between these. A critical decision is where to position learning for different groups, alongside the desired learning outcomes (e.g., increased awareness of risk and responsibility, action, behaviour change). Should the focus be on developing knowledge or skills or resilient dispositions (confidence, empowerment) or all three? Importantly, however, resilience can be learnt in developing character strength, skills, and behaviours that support recovery.[8]

In scoping content for learning for disaster resilience content in 'Education, Communication and Engagement' (ECE), there is a need to appraise skills for different intersecting areas – DRR, emergency management and their connection with individuals and citizens (see Dufty's, 2014b, framework; Figure 9.4). Importantly, developing personal and collective resilience involves "developing thoughts, behaviors, and actions that allow preparation for, and recovery from, traumatic or stressful events".[9] For example, preparedness tasks that may need learning support build from household scale to community-based and community-led risk reduction actions. International Red Cross and Red Crescent Societies (2018) distilled key elements (Box 9.1; Chapter 8, Box 8.4), each requiring different knowledge, skills, and attitudes for successful delivery. So learning to make and

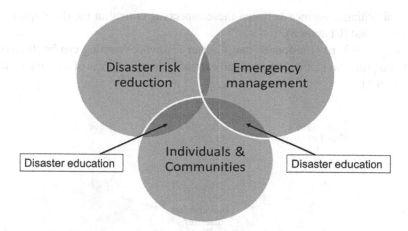

FIGURE 9.4   Scoping a framework for exploring possible Learning for Disaster Resilience (LfDR) content (Dufty, 2014a, Figure 2, permission: Neil Dufty)

carry out better flood action plans – whether household, community, or business continuity – is a key skill, whatever the national setting.

When considering community disaster education and education strategies, Kitagawa (2015, p371), reflecting on the Japanese system for disaster education, observed its "complexity",

> particularly its two-dimensional aspect, namely, 'the science of disasters' on the one hand, and 'life skills for disasters' on the other.

BOX 9.1   ALL-HAZARDS HOUSEHOLD AND FAMILY DISASTER PREVENTION IN A NUTSHELL (SOURCE: INTERNATIONAL RED CROSS AND RED CRESCENT SOCIETIES, 2018, P23)

- Find out what could happen. Stay informed.
- Make a household disaster and emergency plan, considering everyone in your household.
- Reduce structural, non-structural, and environmental risks around your home.
- Learn response skills and practice your plan.
- Prepare response provisions to survive for about a week. Prepare evacuation bags.
- Work together with your workplace, schools, neighbours, and local community to assess your risks, plan to reduce them and prepare to respond.

Establishing a balance between these aspects is critical in the development of human capital (Chapter 8).

Dufty (2013, p14) proposed that disaster resilience learning can be differentiated into principal internal ('host') and external ('hazard') areas of learning content (Figure 9.5).

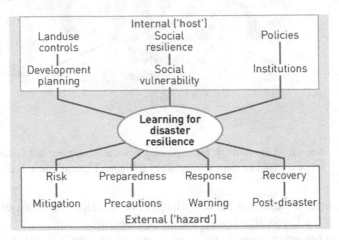

FIGURE 9.5 Main content areas for disaster resilience learning (Dufty, 2013, 2014b, Figure 2, p14, permission: Neil Dufty)

Such learning needs to occur across different phases in the DRM spiral, with individuals, businesses, and communities learning for quick and effective recovery. His proposed content also emphasised the importance of learning about the community itself, including how to reduce social vulnerabilities and connect communities through formation of social capital (Chapter 8). Dufty's (2014a) framework for designing a flood resilience ECE programme in Australia included guiding *principles* (e.g., learner-centred, participatory, cross-hazard); *palettes* from which to choose appropriate learning content and processes; and *filters* (e.g., the nature of risk, learners and community) so programmes are tailored to specific place-based, flood risk communities.

9.5.1 Appropriate flood knowledge

Historically emphasis on knowledge acquisition has focused on specialist (flood) science and understanding key threshold concepts like 'uncertainty' (Chapter 6). However, the role of different knowledges (specialist science, lay, vernacular, intergenerational, hobbyist, citizen science) is increasingly considered in learning for resilience (McEwen and Jones, 2012; Chapters 4 and 8). For example, McEwen (2011) identified starter questions to promote dialogue about local flood science (quantitative, observational) and local exchanges between specialist and lay knowledge (Table 9.2).

9.5.2 Skills

Within DRR, skills for resilience like agency, power, ability to participate, and ability to anticipate are already being explored (Pelling, 2007; Twigg and Bottomley, 2011). This involves establishing what resilient capabilities and skills (e.g., co-working or leadership; 'knowledge as practice'; Chapter 4) might look like as individual or group attributes (Table 9.3).

TABLE 9.2 Starter Questions for Community Flood Science (Modified from McEwen, 2011)

Flood parameter	Indicative questions
Generating conditions	What factors cause and intensify local floods?
Size (or magnitude)	What was the relative size of a given flood (in relation to past floods)?
Frequency	What size are recent floods on this river relative to past floods?
Seasonality	In what seasons (s) do floods occur? Do floods vary in their season of occurrence?
Areas inundated	Which parts of the catchment (upper, middle, lower) are affected by flooding and why?
Flood character	(How) has the speed of flood rise and/or the duration of floods changed over time?
Water quality	What were the water quality issues and why?
Physical impact	Did the flood cause any alterations to the river or its floodplain (e.g., soil erosion, bank erosion, sediment deposition, channels changes)?
Human impact	What are the human impacts and have these changed over time?
Flood memory	What constitutes a large flood on this river?

Participatory active learning activities (individual and collective), involving keeping reflective diaries and digital storytelling about flood experiences, encourage critical reflection as important skill development.

9.5.3 Attitudes and dispositions

Those needed in LfR include agency, confidence, empowerment, flexibility, adaptability, empathy, and willingness to help others (see Fink's, 2003 'significant learning'; Figure 9.3). In considering resilience as a disposition in longer-term living with floods, Soen (2020) highlighted adaptability to floods and other stresses as embedded within medieval peasantry – peasant society, its socio-ecological system, and individual peasant households.

9.5.4 Integration

Examples exist of bringing together development of knowledge, skills, and dispositions – to increase a community's ability to participate in local FRM. For example, the UK's National Flood Forum runs multi-agency groups to develop the knowledge, skills, and dispositions of community flood groups to support their effective working

TABLE 9.3 Examples of Skills Acquisition in Learning for Flood Resilience

Skill	Flood-related example
Having positive relationships	Links to social capital (e.g., flood buddies' and support networks)
Having capacity to make realistic plans	Developing household and community flood plans
Being able to carry out a plan	Implementing flood plan on receipt of warning (e.g., in street level action)
Having good communication skills	Communicating risk and action to others (e.g., in local flood warning; flood wardens)
Having good problem-solving skills	Frequently identified as key skill for resilience – both incremental problem solving and to support a step change in learning and potential behaviours (e.g., how to flood proof a property)
Anticipatory skills	Ability to look ahead (may be affected by socio-economic context); reaction times to a stimulus (like a warning)
Practical construction skills	Knowledge as practice (e.g., knowing how to sandbag properly; how to implement temporary resistance measures)
Critical reflection on own thinking and practice	Linking back to double loop learning; ability to reflect on flood experience

See https://positivepsychology.com/resilience-skills/.

with statutory stakeholders in local FRM (Chapter 8). This involves knowledge (e.g., organisational responsibilities), skills (e.g., cognitive intelligence; brokering, problem-solving, synthesising evidence), and dispositions (e.g., emotional intelligence; empathy, understanding positionality of others, empowerment) to channel energies constructively in brokering longer-term relationships. McEwen et al. (2018), evaluating community flood group development in lower socio-economic settings in Swindon, UK, proposed key skills (e.g., negotiation, listening) and attitudes/dispositions (e.g., confidence, responsibility, adaptability, flexibility) as a starting point for discussions about defining LfR in local flood risk settings.

9.6 Opportunities and barriers to flood education

Development of community flood education strategies therefore needs to focus on factors that encourage learning and secure capabilities for self-organised, adaptive, agile, and flexible development. Concerns include the nature and appropriateness of learning approaches within different phases in the FRM spiral; establishing, and if necessary challenging, the dominant flood narrative and any 'flood myths', language used, ethics, and event scale.

9.6.1 Process and timing

While one-off learning activities may have impact, longitudinal participatory engagements are more likely to lead to positive learning relationships and trust building with FRM professionals, particularly when outside the acute event stage of the FRM spiral. An example of where this approach was taken is in local government working with regional networks of community flood groups in the UK.

9.6.2 Identifying dominant flood narratives

Flood narratives have a role in learning for place responsiveness – both as opportunities and barriers. Narrative emerges as a "'central organising principle' of place and identity and can be viewed as both stories that connect us and stories that make us different" (Cameron, 2003, p99). Critical in FRM is identification of dominant versus counter and sticky narratives (Chapters 3 and 4). As an example of a national dominant narrative in disaster education, Chadderton (2015, p589) argued that it "is unlikely Germany will develop a more inclusive, universal, formalised and high-profile campaign in disaster education in the foreseeable future" due to the contemporary political narrative around Germany as "secure, federal, peaceful and unified". If local narratives of blame and denial dominate in circulation within communities and wider society, these can act as barriers to LfR. Introduction of 'resilience stories' into conversations promulgated by the formal media as counter-narratives (Chapter 4) can offer one way forward.

9.6.3 Language

Technical terms can function as barriers to LfR, with differences in the ways academic, professional, and lay knowledges are articulated and shared (Chapter 6). An example is using the term 'scenarios' – as a stimulus for learning or role-play to promote problem-solving and creative thinking in community engagement. 'What ifs' can provide better ways into collective problem-solving with communities for uncertain futures (e.g., Liguori et al., 2021).

9.6.4 Ethics

Critical ethical considerations affecting those designing LfR include whether to engender fear as a trigger for engagement (Chapters 3 and 6). For example, Bosschaart et al. (2016, p271), in designing learning processes within a flood education programme in the Netherlands, aimed for:

> arousal of moderate levels of fear [that] should prompt experiential and analytical information processing. In this way, understanding of flood risk in the surroundings should prompt students' threat and coping appraisal.

9.6.5 Scale

Valuable LfR can occur across disaster events of different scales. Voss and Wagner (2010, p657) highlighted – in relation to flash flooding – that "learning rarely takes place within an institutional setting that is subjected to *small* disasters, because the stakeholder's focus remains on only one level". They argued for learning from all scales of events. Reasons for learning from smaller scale events included a "more relaxed setting" when damage is "relatively containable" and dealt with collaboratively by local organisations and communities. Changing scale is also important, with some evidence that when people learn from their experience of coping with small-scale stresses and contained impacts, they can gain socio-ecological resilience to larger ones (e.g., Voss and Wagner, 2010; Di Baldassarre et al., 2015; Chaudron, 2016, 2018).

9.7 Pedagogies for place-based flood education: A flood heritage approach?

Different pedagogic approaches to flood education, and resources that promote active place-based learning, provide a valuable learning tool kit for capacity building in DRR. Pedagogic options are also increasing with technological advances. Approaches can be combined to support tailoring of effective learning to specific groups and community settings. This selection requires insight into how risk and resilience information is communicated and processed by different types of learners with preferential learning styles. For example, McEwen (2011) outlined different pedagogic approaches to exploring local flood risk through researching place-based flood histories on the River Severn, UK, including experiential and CBL. The author now proposes a place-based, flood heritage-focused approach to community flood education, combining science and stories, and tied strongly to local risk (Chapter 5). Approaches and resources below can be nested to promote learning:

- *Place-based learning*: Learning emphasises importance of linking specialist and vernacular flood science to the generation and sharing of flood stories as lay knowledge, so they act as mutual stimuli. This starts by researching evidence about the character and impact of historic floods from diverse sources (scientific, archival, oral histories), and then reflecting on this evidence as a learning community. Personal testimonies provide valuable learning resources for intergenerational learning, for example, with local elders reminiscing about floods with young people (Chapter 5). Such memories act as stimuli for discussions about past and future resilience strategies (e.g., older people's memories of 1968 floods in Bristol, UK[10] shared with local school children). Sharing resilience narratives rather than narratives of victimhood can be powerful for social learning. This approach is synergetic with exploring watery sense of place and place attachment, and in promoting river stewardship and hydrocitizenship.

*   *Authentic event-based science*: Learning is explicitly linked to individual and collective interest and concern triggered after severe floods. This approach emphasises importance in reconstructing the 'flood event' in the evolution of community flood science knowledge, viewing flood experience as a chance to learn (e.g., comparison of the March 1947 and July 2007 floods on the lower River Severn, UK, generated under very different conditions).
*   *Active problem-solving*: Learning is shifted from studying flood risk to exploring and solving flood risk problems (cf. Dennison, 2008). Different stimuli for community participation might include establishing differential impacts of contrasting scales of flooding (routine, severe) or exploring the scientific bases for local 'flood myths' – understood as misconceptions or falsehoods. A local case-study approach can encourage community learning through exploring possible solutions to complex, real-world problems (Scholz et al., 2006).
*   *Use of visual stimuli:* Place-based learning is stimulated by visual prompts like local maps or historic photographs. For example, specialist flood risk mapping can be used to trigger dialogue between citizens, community interest groups, government regulators, and other stakeholders in FRM to increase awareness of the strengths and challenges of flood modelling and mapping. Historic sequences of local maps – as boundary objects – act as ways into multi-stakeholder conversations about implications of changes in rivers and their management, riparian zones, and catchment land use (Section 6.5 of Chapter 6). This approach was used to start creative engagements exploring floods, water security, development, and climate change with community participants in Tewkesbury, River Severn, UK.
*   *Reflective experiential learning:* Learning encourages abilities to link, and reflect on, personal and community experiences of antecedent flood conditions (e.g., saturated ground), critical causal characteristics (e.g., interaction of flood types), and local impacts (e.g., exposed activities), during and after extreme floods to develop personal and collective observational flood knowledge. This can be linked to appraising the efficacy of individual coping and adaptive strategies (e.g., PLP) to reduce flood impacts.
*   *'Hooks' to community engagement for learning:* Learning around flood risk can be linked to wider science communication agendas about climate change science, mitigation, and adaptation for all stakeholders. At small scale, McEwen (2011) found that exploring human impacts on local water courses captured public imagination and capitalised on community concern about place. Learning can also be stimulated by empowered flood stories and unravelling 'flood myths' (Chapter 6), with a need to recognise 'grey', as well as 'black' and 'white'. Recurrent flood science misconceptions (e.g., the efficacy of specific management interventions) can dominate within some groups

(McEwen, 2011; Chapters 2 and 6). Additionally, climate change myths can act as hooks to wider conversations about changing extreme weather risks (e.g., Before the Flood's *Top 6 Climate Myths*[11]; WWF's 10 myths about climate change[12]).

- *Future stories and scenario-based learning*: Here the focus is exploring 'What ifs', for example, using design thinking. This involves exploring multiple possible adaptive place-based futures, for example, linked to climate change projections or local urban development options. Facilitator care is needed in brokering desired versus fearsome futures.

- *Community science – arts links in flood science communication:* Links between flood science and socially engaged arts practice provide large potential for participatory learning (Chapters 6 and 10). Rather than using explicit 'hooks' about flood risk, the engagement can be more oblique in considering water, environment, and care of place (e.g., through participatory deep mapping of knowledge and values; McEwen et al., 2020).

9.8   Other learning approaches for flood learning

Other approaches to flood learning involve working across disciplinary and professional domains, drawing on specific pedagogies, science, technology, and the arts. Examples include learning through role-play, serious games, catchment flood models, and the socially engaged arts.

9.8.1   Role-play as a learning strategy

Role-play and inverse debating – or exchanging roles – can be used in event preparation for evolving flood situations in formal and informal learning settings (McEwen et al., 2014). These can involve diverse stakeholders, whether in formal education, in communities or as FRM professionals. For example, in community LfR, Dufty (2014a) cited *Resilientville* – a role-playing activity for professional stakeholders (becoming "neighbourhood stakeholders") that develops participant awareness of short and longer-term benefits of problem-solving at neighbourhood level. Resilientville was developed by the Canadian Community for Dialogue and Deliberation to "build resilient neighbourhoods" (e.g., in Toronto, Canada).

> By working together as a community on issues that present themselves daily, residents develop crucial decision-making skills and relationships that over time strengthen their ability to respond to a wide variety of unforeseen challenges and opportunities.[13]

Such role-play demonstrates that all participants can contribute ideas and resources in recovery, and hence have a role to play.

9.8.2 Game-based learning

Design and use of serious games (or games not simply for entertainment) is a rapidly expanding safe space within flood learning for children and adults (e.g., Khoury et al., 2018; Forrest et al., 2022). Co-design requires interdisciplinary collaborations (e.g., between sociology and engineering). The Red Cross/Red Crescent Climate Centre[14] provides five reasons for using games in learning and dialogue (Box 9.2), recognising that games can be "a good way to communicate complex messages".

BOX 9.2   WHY GAMES? (SOURCE: CLIMATE CENTRE; WITH PERMISSION: RED CROSS RED CRESCENT CLIMATE CENTRE)

Five reasons for using games in learning and dialogue:

- They encourage **active learning** and **active engagement** in dialogues.
- Games allow you to **simplify complex systems**.
- In games, you have to take decisions and **receive feedback** on the result of that decision.
- Games provide opportunities for **reflection, discovery, exploration, and challenge**.
- And .... they are **fun**! Considering that emotions matter in learning – this is also a serious goal.

source: https://www.climatecentre.org/priority_aas/innovation/climate-games/

Game-initiated learning approaches and resources are diverse with low technology (Table 9.4) and technology-enhanced learning (e.g., video or online). Persuasive technologies are where designers seek to influence or change users' attitudes or behaviours for social and environmental benefit (e.g., flood protection computer games for disaster education; Tsai et al., 2015).

Learning opportunities of serious games are being explored across cultures. The Climate Centre provides 28 examples of serious 'games for a new climate'[15] to build community resilience in LEDCs. These can be filtered by duration and topic (including DRR), with information about intended audiences and learning outcomes. For example, 'Before the Storm' is a decision-making game designed to introduce possible actions that can be taken in responding to weather forecasts. Some games are deliberately developed for young children to promote flood preparedness, linked to their potential for influence in households (Chapter 10). For example, UNESCO and the government of Thailand developed a bespoke mobile game application – 'Sai Fah the Flood Fighter'[16] – as a response to the devastating 2011 floods. The storyline involves a young boy avoiding hazards in trying to

TABLE 9.4 Examples of Serious Games Using Low Technology: Focus and Target Groups

Name	Focus/rationale	Target group	Reference
Hazagora – a board game	Raising awareness about geohazards and disaster risk reduction strategies	School students, citizens, scientists, and stakeholders involved in DRM activities	Mossoux et al. (2016)
Ready! (uses cards and dice)	Can use any disaster scenario; most effective using a realistic scenario for participants. Enables focused conversations about disaster preparedness and DRR	Communities	Global Disaster Preparedness Center[a] van den Homberg et al. (2015)
Wheel of Flooding Fortune	Interactive game of risk and probabilities focused on buying PLP to prevent or reduce flood damage. Responses inform software development	Citizens	FATE Flood Awareness Through Engagement (Herriot Watt University, Scotland)[b]
'Play your way to preparedness' – a Flood Resilience Challenge Game	Role-play on selecting FRM measures (competing interests; trade-offs); re-evaluation after flood	Students and FRM stakeholders	University of Waterloo, Canada[c] Bogdan and Cottar (2022)

[a] https://preparecenter.org/resource/ready-game/.
[b] https://www.fifex.co.uk/wheel-of-flooding-fortune.
[c] https://www.frcgame.com/.

reunite with his mother. This is intended to educate children about how to recognise when flooding is likely, and about preparedness actions.

Solinska-Nowak et al. (2018, p1013) provided a review of use of games in DRM and their focus within the DRM spiral, noting the ability of games to engage across groups, and wider emotional and perceptual benefits of serious games.

With a capacity of reaching diverse audiences (embracing adults, children, experts and communities) and of realistically simulating disaster reality, serious games/simulations may assist DRM, especially in the realm of disaster risk awareness raising, identifying hazards, undertaking preventive actions, empathy triggering and perspective-taking.

They also emphasised the importance of evaluating learning outcomes, and researching links between learning during gaming and future preparedness actions.

9.8.3   Learning from catchment-based flood models

Use of physical river catchment models as 'flood demonstrators' – both as physical structures and online – is being trialled to promote social learning and facilitate disaster planning within communities. 'Hands on' catchment models can positively engage communities in problem-solving around FRM options (Vetra-Carvalho, 2018; Table 9.5).

TABLE 9.5 Examples of Interactive Physical Flood Models for Community Learning

Type of model	Objectives	Lead organisation
Natural Flood Management Catchment Model (physical)	Shows how using NFM techniques to store and slow flow of water can help manage flooding	Tweed Forum, UK (river NGO)[a]
Portable hydraulic flume	Demonstrates performance of various coastal defences under different wave regimes	JBA Trust[b]
Augmented reality (AR) sand box	Combines physical sand box with 3D computer visualisation tools which simulate rain and water flow	Reed et al. (2016)

[a] https://www.youtube.com/watch?v=8pPGEINRYW8.
[b] http://www.jbatrust.org/how-we-help/physical-models/hydraulic-flume-free-standing/.

9.8.4   The arts in learning for flood resilience

The socially engaged arts represent a growing area in community risk education, with diverse roles for artists, communities, and other FRM stakeholders (Chapters 6 and 10). Various arts media – theatre/performance, film, storytelling – are being drawn on in multi-sensory learning focused on affective domains. Activities are being trialled within flood-affected communities and in wider public engagement about flood impacts, as ways into discussing differential climate change impacts and environmental justice issues. This includes a bringing together of international and national issues of vulnerability, victims, and inequality (e.g., 'Drowning World' – Photographer Gideon Mendel's 8 Year Project[17] working in Kashmir and the UK).

Several examples exist of plays as output of collaborations between local communities, statutory FRM organisations, and artists (e.g., drama-based teaching groups; Table 9.6). The Scottish environmental regulator (SEPA) has worked with performance theatre practitioners in engaging various groups for learning (see 'Rapid Departure' and 'Keep Calm and Call Floodline' targeted at children) while the play

TABLE 9.6 Flood-Inspired Theatre in Community Learning for Resilience

Title	Character	Learning	Production
"Rapid Departure" – interactive play; immersive comedy Toured Scotland, UK	Audience becomes the 'village locals'	Gave people opportunity to consider how flooding could impact their community and how they can take steps to prepare	Delivered by Right Lines Production and sponsored by Floodline[a]
"Keep Calm and Call Floodline"	Short play for school children	Educates children and young people about SEPA's flood monitoring, dangers of flooding and how to prepare and deal with a flood	In partnership with Theatre&, a drama-based teaching group, SEPA created play with resource pack for teachers[b]
"Downstream[c]" University of Saskatchewan, Canada	Travelling theatre	Combines scientific knowledge from water security research and performance art	Forum theatre with audience participation in decision-making when flooding or drought threatens as story progresses (Strickert and Bradford, 2015)

[a] http://rightlines.net/rightlines/rapid-departure/.
[b] https://www.sepa.org.uk/media/69279/sepateacherresourcepack.pdf.
[c] https://water.usask.ca/news-items/2014_2013/downstream-marries-water-research-and-performance-art-.php.

'Downstream' (Saskatchewan, Canada) developed participatory theatre methods to share water security research with diverse audiences (Strickert and Bradford, 2015).

Another "powerful and effective hazard education tool" is film or video in place-based risk communication. Hicks et al. (2017), in context of volcanic hazards, co-created films in partnership with monitoring scientists and at-risk communities; local DRM agencies ensured meaningful and locally contextualised content and messaging. Importantly, the impact of these films on learning and affect within these communities was evaluated. They found that

> use of local content and actors to share experiences and teach valuable lessons were inspirational. Recognizable faces and spaces helped to convey disaster risk reduction messages. They also motivated audiences to consider ownership of risk and potential actions to reduce risk and strengthen resilience.

Lessons in co-producing place-based, artistic artefacts can have wider transferability to different risk-learning contexts.

Applied storytelling, co-producing short reflective digital stories (2–3 minutes audio along with personally selected imagery) with community members, has also been developed for sharing personal and collective flood resilience stories. This combines processes and outputs that encourage social learning. Research indicates that stories can act as vehicles to share knowledge and affect (emotion) from trusted community members. The 'Sustainable Flood Memories' project (Holmes and McEwen, 2020; Chapter 5) worked with participants with experience of an extreme flood to co-develop stories capturing their memories and learning to inform community flood preparedness.[18] Here storytellers had personal reflections on their experiences to share with people at future flood risk. Aspects of stories found to travel well between communities included senses of empowerment and community cohesion during floods. These digital stories have since been used in training for community engagement by the environmental regulator.

9.9 Community participation in co-development of learning strategies

Design of resources that support LfR needs to combine sound educational design principles with a learner-centred – rather than teacher-centred – ethos. For example, Bosschaart et al. (2016) used a variety of student-directed elements, such as serious games and flood simulations, in designing a flood risk education programme for 15-year-olds in the Netherlands. At worst, learning programmes can be poorly designed and delivered in a relatively ineffective, 'top down' manner. Co-production and participatory design methods (Chapters 4 and 8) can yield significant returns in the development of learning resources, including potential to embed different knowledges and social learning. Integrated participatory Action Research and place-based education, involving collaborations between researchers and communities, can also build resilience knowledges in how to adapt the built environment (Meyer et al., 2018).

Longitudinal working and co-ownership of learning resources with communities are important. The SESAME project co-designed a prototype e-learning tool with local small business owners in a UK town subject to repeat flooding (McEwen et al., 2016). The tool aimed to promote flood resilience, with emphasis on identifying longer-term adaptive strategies (e.g., "be ready to work differently"; "work with customers and suppliers"), in addition to short-term contingency planning. Importantly, the messenger about preparedness options was the business owner, sharing learning experiences through films and digital testimonies.[19]

Building capacities of educators, trainers, and facilitators in flood education is also critical in improving community-focused, disaster risk learning programmes, with increased roles for peer-to-peer learning, teachers, educators in a range of organisations, and training of trusted community facilitators or volunteers.

For example, the national Red Cross and Red Crescent Societies have a long history of providing public education on disaster risks and the steps that households and communities can take to increase their safety and resilience. They carry out this work through various delivery mechanisms including mobilising youth and junior Red Cross Red Crescents for peer education; and training and organising communities through community-based risk reduction actions.

Timing of improvement of learning resources is also important. Frequently adaptation of existing education programmes occurs after stresses caused by extreme weather events, providing important reflective opportunities for changing educators' practices. Dubois and Krasny (2016), in researching how education programmes adjusted in content and organisation after the 'disturbance' of Hurricane Sandy (in 2012), USA, argued for an action research approach that quickly feeds evaluative insights into environmental education practice.

9.10 Conclusions

LfR is increasingly recognised as essential within integrated, community-focused FRM for dealing with residual risk. However, multi-stakeholder awareness of the opportunities and challenges of different pedagogic approaches is critical. Historically top-down transfer and passive approaches to learning have tended to predominate (Chapter 6). However, increasing awareness now exists of the value of different learning theories, active learning, community-based and learner-centred participatory methods and place-based learning. Communities at flood risk are likely to have diverse science capital and starting points for learning. This means that any learning programme needs to be aware of levels of prior education, dominant learning styles, and cultural preferences alongside need for tailoring to place.

Future avenues for investigation include participatory, co-production, and action research processes that involve communities in the design, delivery, and evaluation of their learning programmes and resources. High value exists in developing these to support the knowledge, values, and preparedness actions for floods and other disasters within more marginalised groups. These need to go beyond solely verbal means of communication to increase accessibility and engagement. Media like serious games and socially engaged arts practice offer new opportunities for promoting active learning and local multi-stakeholder dialogue. At national and international scale, the WWW supports sharing of a burgeoning set of e-learning resources that promote learning about flood and DRR across risk and cultural settings, while inclusive, face-to-face, place-tailored interventions can be co-created and delivered locally. Questions should always be asked about the transferability and required attuning of learning resources to distinct socio-cultural contexts, and whether learning approaches and resources have been co-evaluated.

Finally, a cross-hazard risk approach to community learning can have multiple benefits, including economies of scale for managers in integrating education programmes that reinforce similar preparedness behaviours and build community resilience competencies and leadership across hazards.

Notes

1 https://en.unesco.org/themes/water-security/hydrology/water-education.
2 https://experientiallearninginstitute.org/programs/assessments/kolb-experiential-learning-profile/.
3 https://www.skilla.com/en/heutagogy-developing-agile-reflective-lifelong-learners/.
4 https://positivepsychology.com/resilience-skills/.
5 https://rse.org.uk/pillars-lintels-interdisciplinary-learning/.
6 https://www.colorado.edu/cuengage/about-us/what-community-based-learning.
7 https://www.geography.org.uk/teaching-resources/flooding.
8 https://positivepsychology.com/resilience-skills/.
9 https://positivepsychology.com/resilience-skills/.
10 https://www.watercitybristol.org/hidden-rivers-and-daylighting.html.
11 https://www.beforetheflood.com/explore/the-deniers/top-climate-myths/;.
12 https://www.wwf.org.uk/updates/10-myths-about-climate-change.
13 https://resilientneighbourhoods.ca/resilientville-role-playing-game/.
14 https://www.climatecentre.org/priority_areas/innovation/climate-games/.
15 https://www.climatecentre.org/priority_areas/innovation/innovation_tools/climate-games/.
16 https://bangkok.unesco.org/content/sai-fah-flood-fighter.
17 https://gideonmendel.com/; https://floodlist.com/dealing-with-floods/drowning-world-gideon-mendels-8-year-project.
18 https://esrcfloodmemories.wordpress.com/digital-stories/.
19 www.floodresilientbusiness.co.uk.

References

Angiel, J. and Angiel, P.J. (2015) Perception of river value in education for sustainable development (The Vistula River, Poland). *Sustainable Development*, 23, 188–201.

Antonacopolou, E.P. (2008) Actionable knowledge. In R. Clegg and J. Bailey (eds.) *International Encyclopaedia of Organization Studies*. Sage, London

Argyris, C. (1976) *Increasing Leadership Effectiveness*. Wiley, New York.

Australian Institute for Disaster Resilience (2021) *Disaster Resilience Education for Young People Handbook*. Australian Disaster Resilience Handbook Collection. https://knowledge.aidr.org.au/media/8874/aidr-handbook_dreyp_2021.pdf

Back, E., Cameron, C. and Tanner, T. (2009) *Children and Disaster Risk Reduction: Taking Stock and Moving Forward*. Unicef. https://www.preventionweb.net/files/15093_12085 ChildLedDRRTakingStock1.pdf

Bandura, A. (1971) *Social Learning Theory*. General Learning Press, New York.

Benson, D., Lorenzoni, I. and Cook, H. (2016) Evaluating social learning in England flood risk management: An 'individual-community interaction' perspective. *Science & Policy*, 55(2), 326–334.

Berkes, F. (2007) Understanding uncertainty and reducing vulnerability: Lessons from resilience thinking. *Natural Hazards*, 41, 283–295.

Blaschke, L.M. (2012) Heutagogy and lifelong learning: A review of heutagogical practice and self-determined learning. *The International Review of Research in Open and Distance Learning* 13(1), 56–71.

Bogdan, E.A. and Cottar, S. (2022) A serious role-playing game as a pedagogical innovation to strengthen flood resilience. *IEEE Technology and Society Magazine*, 41(3), 98–100. https://doi.org/10.1109/MTS.2022.3197124.

Bosschaart, A., van der Schee, J. and Kuiper, W. (2016) Designing a flood-risk education program in the Netherlands. *Journal of Environmental Education*, 47(4), 271–286. https://doi.org/10.1080/00958964.2015.1130013

Building Research Institute (BRI) and National Graduate Institute for Policy Studies (GRIPS) (2007) *Disaster Education.* https://apadresearch.com/document/disaster-education-building-research-institute-bri-national-graduate-institute-for-policy-studies-grips/

Cameron, J.I. (2003) Educating for place responsiveness: An Australian perspective on ethical practice. *Ethics, Place and Environment,* 6(2), 99–115.

Chadderton, C. (2015) Civil defence pedagogies and narratives of democracy: Disaster education in Germany. *International Journal of Lifelong Education,* 34(5), 589–606. https://doi.org/10.1080/02601370.2015.1073186

Chaudron, G. (2016) After the flood: Lessons learned from small-scale disasters. In E.N. Decker and J.A. Townes (eds.) *Handbook of Research on Disaster Management and Contingency Planning in Modern Libraries* 23pp. https://doi.org/10.4018/978-1-4666-8624-3.ch017

Chaudron, G. (2018) Small disasters seen in sunlight. *International Journal of Disaster Response and Emergency Management,* 1(1). https://doi.org/10.4018/IJDREM.2018010104

Cologna, V., Bark, R.H. and Paavola, J. (2017) Flood risk perceptions and the UK media: Moving beyond "once in a lifetime" to "Be Prepared" reporting. *Climate Risk Management,* 17, 1–10.

CUIDAR Project (2018) *A Child-Centred Disaster Management Framework for Europe.* Lancaster University, Lancaster.

Dennison, W.C. (2008) Environmental problem-solving in coastal ecosystems: A paradigm shift to sustainability. *Estuarine, Coastal and Shelf Science,* 77, 185–196.

Di Baldassarre, G., Viglione, A., Carr, G., Kuil, L., Yan, K., Brandimarte, L. and Bloschl, G. (2015) Debates – perspectives on socio-hydrology: Capturing feedbacks between physical and social processes. *Water Resources Research,* 51. https://doi.org/10.1002/2014WR016416

Dubois, D. and Krasny, M.E. (2016) Educating with resilience in mind: Addressing climate change in post-Sandy New York City. *The Journal of Environmental Education,* 47(4), 255–270. https://doi.org/10.1080/00958964.2016.1167004

Dufty, N. (2008) A new approach to flood education. *The Australian Journal of Emergency Management,* 23(2), 4–8.

Dufty, N. (2013) Towards a learning for disaster resilience approach: exploring content and process. Molino Stewart occasional papers. http://works.bepress.com/neil_dufty/29/

Dufty, N. (2014a) What is flood resilience education? Paper, Floodplain Conference, January 2014. https://www.floodplainconference.com/papers2014/Neil%20Dufty.pdf

Dufty, N. (2014b) Opportunities for disaster resilience learning in the Australian curriculum. *Australian Journal of Emergency Management,* 29, 12–16.

Dufty, N. (2020) *Disaster Education, Communication and Engagement.* Wiley & Sons Online, Hoboken

Eyler, J. and Giles, D.E. Jr. (1999) *Where's the Learning in Service-Learning?* Jossey-Bass, San Francisco.

Field, J. (2010) Lifelong learning. *International Encyclopedia of Education* (3rd edn), pp89–95.

Fink, L.D. (2003) *Creating Significant Learning Experiences: An Integrated Approach to Designing College Courses.* Jossey Bass Higher and Adult Education Series, San Francisco.

Fink, L. D. (2005) *A Self-Directed Guide to Designing Courses for Significant Learning.* Dee Fink and University of Oklahoma, US.

Fleming, N.D. and Mills, C. (1992) *Helping Students Understand How They Learn. The Teaching Professor,* 7(7). Magma Publications, Madison.

Forrest, S.A., Kubíková, M. and Macháč, J. (2022) Serious gaming in flood risk management. *WIREs Water*, 9(4), e1589. https://doi.org/10.1002/wat2.1589

Hicks, A., Armijos, M.T., Barclay, J., Stone, J., Robertson, R. and Cortés, G.P. (2017) Risk communication films: Process, product and potential for improving preparedness and behaviour change. *International Journal of Disaster Risk Reduction*, 23, 138–151. https://doi.org/10.1016/j.ijdrr.2017.04.015

Holmes, A. and McEwen, L. (2020) How to exchange stories of local flood resilience from flood rich areas to the flooded areas of the future. *Environmental Communication* 14(5), 597–613, DOI: 10.1080/17524032.2019.1697325

Chapman Hoult, E. (2013) Resilience in adult learners: Some pedagogical implications. *Journal of Pedagogic Development*, 3, 45–47.

International Federation of Red Cross and Red Crescent Societies (2018) *Public awareness and public education for disaster risk reduction: Key messages.* IFRC, 66pp. https://www.ifrc.org/document/public-awareness-and-public-education-disaster-risk-reduction-key-messages-2nd-edition

International Institute for Educational Planning (2015) *Safety, resilience, and social cohesion: a guide for curriculum developers. Key content: What are the desired learning outcomes?* UNESCO https://unesdoc.unesco.org/ark:/48223/pf0000234814

Johannessen, A. and Hahn, T. (2013) Social learning towards a more adaptive paradigm? Reducing flood risk in Kristianstad municipality, Sweden. *Global Environmental Change*, 23(1), 372–381. https://doi.org/10.1016/j.gloenvcha.2012.07.009

Keong, C.Y. (ed.) (2021) Chapter 6 – The United Nations environmental education initiatives: The green education failure and the way forward. *Global Environmental Sustainability*, 89–349. https://doi.org/10.1016/B978-0-12-822419-9.00006-0

Khoury, M., Gibson, M.J., Savic, D. et al. (2018) A serious game designed to explore and understand the complexities of flood mitigation options in urban-rural catchments. *Water*, 10 (12), 1885. ISSN 2073-4441.

Kitagawa, K. (2015) Continuity and change in disaster education in Japan. *History of Education*, 44(3), 371–390. https://doi.org/10.1080/0046760X.2014.979255

Kitagawa, K. (2021) Disaster risk reduction activities as learning. *Natural Hazards*, 105, 3099–3118. https://doi.org/10.1007/s11069-020-04443-5

Klein, P., Fatima, M., McEwen, L.J., Moser, S., Schmidt, D. and Zupan, S. (2011) Dismantling the ivory tower: Engaging geographers in university-community partnerships. *Journal of Geography in Higher Education*, 35(3), 425–444.

Kolb, D. (1984) *Experiential Learning: Experience as the Source of Learning and Development.* Prentice Hall, London.

Krasny, M.E., Lundholm, C. and Plummer, R. (2010) Resilience in social–ecological systems: The roles of learning and education. *Environmental Education Research*, 16(5–6), 463–474. https://doi.org/10.1080/13504622.2010.505416

Kuang, D. and Liao, K.-H. (2020) Learning from floods: Linking flood experience and flood resilience. *Journal of Environmental Management*, 271, 111025. https://doi.org/10.1016/j.jenvman.2020.111025

Liguori, A., McEwen, L.J., Blake, J. and Wilson, M. (2021) Towards 'Creative Participatory Science': Exploring future scenarios through specialist drought science and community storytelling. *Frontiers in Environmental Science* https://doi.org/10.3389/fenvs.2020.589856

Mason O'Connor, K. and McEwen, L.J. (eds.) (2012) Building staff capacity for curriculum-based community engagement. Staff Educational Development Association (SEDA) Special No 12.

Mason O'Connor, K. and McEwen, L.J. (2020) Real world learning through civic engagement: Principles, pedagogies and practices. In D.A. Morely and M.G. Jamil (eds.) *Applied Pedagogies for Higher Education. Real World Learning and Innovation Across the Curriculum*. Palgrave Macmillan Ltd, London.

McEwen, L., Gorell Barnes, L., Phillips, K. and Biggs, I. (2020) Reweaving urban water-community relations: Creative, participatory river "daylighting" and local hydrocitizenship. *Transactions of the Institute of British Geographers*, 45, 779–801. https://doi.org/10.1111/tran.12375

McEwen, L.J. (2011) Approaches to community flood science engagement: The lower River Severn catchment, UK as case-study. *International Journal of Science in Society*, 2(4), 159–179.

McEwen, L.J., Garde-Hansen, J., Holmes, A., Jones, O. and Krause, F. (2016) Sustainable flood memories, lay knowledges and the development of community resilience to future flood risk. *Transactions of the Institute of British Geographers*, 42(1), 14–28.

McEwen, L.J., Holmes, A., Quinn, N. and Cobbing, P. (2018) 'Learning for resilience': Developing community capital through action groups in lower socio-economic flood risk settings. *International Journal of Disaster Risk Reduction*, 27, 329–342.

McEwen, L.J., Jennings, R., Duck, R. and Roberts, H. (2009) *Masters' Student Experiences of Interdisciplinary Learning*. Published by The Interdisciplinary Teaching and Learning Group, Subject Centre for Languages, Linguistics and Area Studies, Higher Education Academy. https://web-archive.southampton.ac.uk/www.llas.ac.uk/sites/default/files/nodes/3219/interdisciplinary_masters.pdf

McEwen, L.J. and Jones, O. (2012) Building local/lay flood knowledges into community flood resilience planning after the July 2007 floods, Gloucestershire, UK. *Hydrology Research, Special Issue*, 43, 675–688.

McEwen, L.J., Stokes, A., Crawley, K. and Roberts, C.R. (2014) Using role-play for expert science communication with professional stakeholders in flood risk management. *Journal of Geography in Higher Education*, 38, 277–300.

McEwen, L.J., Wragg, A. and Harries, T. (2016) Increasing business resilience to flood risk: Developing an effective e-learning tool to bridge the knowledge gap between policy, practice and business owners. Paper in conference volume: FLOODrisk 2016 3rd European Conference on *Flood Risk Management Innovation, Implementation, Integration*. 18–20 October 2016, Lyon, France.

Meyer, M.A., Hendricks, M., Newman, G.D., Masterson, J.H., Cooper, J.T., Sansom, G., Gharaibeh, N., Horney, J., Berke, P., van Zandt, S. and Cousins, T. (2018) Participatory action research: Tools for disaster resilience education. *International Journal of Disaster Resilience in the Built Environment*, 9(4/5), 402–419. https://doi.org/10.1108/IJDRBE-02-2017-0015

Mezirow, J. (1991) *Transformative Dimensions of Adult Learning*. Jossey-Bass, San Francisco, 247pp

Mezirow, J. (2003) Transformative learning as discourse. *Journal of Transformative Education*, 1(1), 58–63.

Mezirow, J. and Associates (1990) *Fostering Critical Reflection in Adulthood: A Guide to Transformative and Emancipatory Learning*. Jossey-Bass, San Francisco.

Mossoux, S., Delcamp, A., Poppe, S., Michellier, C., Canters, F. and Kervyn, M. (2016) HAZAGORA: Will you survive the next disaster? – A serious game to raise awareness about geohazards and disaster risk reduction. *Natural Hazards and Earth System Sciences* 16, 135–147. https://doi:10.5194/nhess-16-135-2016

National Coordinating Centre for Public Engagement (2017) How to enable community based learning. https://www.publicengagement.ac.uk/sites/default/files/publication/how_to_enable_community_based_learning.pdf (accessed 1 November 2021)

Nye, M., Tapsell, S. and Twigger-Ross, C. (2011) New social directions in UK flood risk management: Moving towards flood risk citizenship? *Journal of Flood Risk Management*, 4, 288–297. https://doi.org/10.1111/j.1753-318X.2011.01114.x

Pahl-Wostl, C., Mostert, E. and Tàbara, D. (2008) The growing importance of social learning in water resources management and sustainability science. *Ecology and Society*, 13(1), 24. [online] URL: http://www.ecologyandsociety.org/vol13/iss1/art24/

Pelling, M. (2007) Learning from others: The scope and challenges for participatory disaster risk assessment. *Disasters*, 31(4), 373–385.

Reed, M.S., Evely, A.C., Cundill, G., Fazey, I., Glass, J. et al. (2010) What is social learning? *Ecology & Society*, 15(4), r1.

Reed, S., Hsi, S., Kreylos, O., Yikilmaz, M.B., Kellogg, L.H., Schladow, S.G., Segale, H. and Chan, L. (2016) Augmented Reality Turns a Sandbox into a Geoscience Lesson. *Eos* 26 July https://eos.org/science-updates/augmented-reality-turns-a-sandbox-into-a-geoscience-lesson

Schneider, F. and Rist, S. (2014) Envisioning sustainable water futures in a transdisciplinary learning process: Combining normative, explorative, and participatory scenario approaches. *Sustainability Science*, 9, 463–481. https://doi.org/10.1007/s11625-013-0232-6

Scholz, R.W., Lang, D.J., Wiek, A., Walter, A.I. and Stauffacher, M. (2006) Transdisciplinary case studies as a means of sustainability learning: Historical framework and theory. *International Journal of Sustainability in Higher Education*, 7(3), 226–251. https://doi.org/10.1108/14676370610677829

Schuller, T. (2017) What are the wider benefits of learning across the life course? *Future of Skills & Lifelong Learning*. Foresight, Government Office for Science.

Sharpe, J. (2016) Understanding and unlocking transformative learning as a method for enabling behaviour change for adaptation and resilience to disaster threats. *International Journal of Disaster Risk Reduction*, 17, 213–219.

Sinha, R., Mahendale, V., Singh, V.K. and Hegde, G. (2007) School education for disaster reduction. In BRI/GRIPS *Disaster Education*. https://www.yumpu.com/en/document/read/24437854/disaster-education-pdf-1068-mb-preventionweb

Soen, T. (2020) Resilience in historical disaster studies: Opportunities and pitfalls. In M. Endress, L. Clemens and B. Rampp (eds.) *Strategies, Dispositions and Resources of Social Resilience: A Dialogue between Medieval Studies and Sociology* (1st edn). Springer, Berlin, pp253–276.

Solinska-Nowak, A., Magnuszewski, P., Curl, M. et al. (2018) An overview of serious games for disaster risk management – Prospects and limitations for informing actions to arrest increasing risk. *International Journal of Disaster Risk Reduction*, 31, 1013–1029. https://doi.org/10.1016/j.ijdrr.2018.09.001

Strickert, G.E. and Bradford, L.E.A. (2015) Of research pings and ping–pong balls: The use of forum theater for engaged water security research. *International Journal of Qualitative Methods*, 14(5), 1609406915621409.

Thomson, C., Mickovski, S. and Orr, C. (2014) Promoting double loop learning in flood risk management in the Scottish context In A.B. Raiden and E. Aboagye-Nimo (eds.) Proceedings of 30th Annual ARCOM Conference, 1–3 September 2014, Portsmouth, UK, Association of Researchers in Construction Management, 1185–1194.

Tilbury, D. and Cooke, K. (2005) *A National Review of Environmental Education and Its Contribution to Sustainability in Australia: Frameworks for Sustainability*. Australian

Government Department of the Environment and Heritage and the Australian Research Institute in Education for Sustainability (ARIES), Canberra.

Torani, S., Majd, P.M., Maroufi, S.S., Dowlati, M. and Sheikhi, R.A. (2019) The importance of education on disasters and emergencies: A review article. *Journal of Education and Health Promotion*, 8, 85. https://doi.org/10.4103/jehp.jehp_262_18

Tosey, P., Visser, M. and Saunders, M.N. (2012) The origins and conceptualizations of 'triple-loop' learning: A critical review. *Management Learning*, 43(3), 291–307. https://doi.org/10.1177/1350507611426239

Tsai, M., Chang, Y., Kao, C. et al. (2015) The effectiveness of a flood protection computer game for disaster education. *Visualization in Engineering*, 3, 9. https://doi.org/10.1186/s40327-015-0021-7

Twigg, J. and Bottomley, H. (2011) Making local partnerships work for disaster risk reduction. *Humanitarian Practice Network*. https://odihpn.org/publication/making-local-partnerships-work-for-disaster-risk-reduction/

UNDRR (2015) *Sendai Framework for Disaster Risk Reduction 2015–2030*. https://www.undrr.org/publication/sendai-framework-disaster-risk-reduction-2015-2030

UNESCO (2014) *UNESCO Roadmap for Implementing the Global Action Programme on Education for Sustainable Development*. UNESCO, France.

UNESCO (2020) *Education for Sustainable Development: A Roadmap*. UNESCO, France. https://unesdoc.unesco.org/ark:/48223/pf0000374802.locale=en

United Nations (1989) *Convention on the Rights of the Child*. https://www.ohchr.org/en/resources/educators/human-rights-education-training/7-convention-rights-child-1989

van den Homberg, M., Cumiskey, L., Oprins, E., van der Hulst, A. and Suarez, P. (2015) Are you Ready! to take early action? Embedding serious gaming into community managed DRR in Bangladesh. *Proceedings of ISCRAM 2015 Conference – Kristiansand, 24–27 May*.

van Herk, S., Rijke, J., Zevenbergen, C., Ashley, R. and Besseling, B. (2015) Adaptive co-management and network learning in the room for the River programme. *Journal of Environmental Planning and Management*, 58(3), 554–575. https://doi.org/10.1080/09640568.2013.873364

Vetra-Carvalho, S. (2018) Playful floods. *Weather and Climate @ Reading*. https://blogs.reading.ac.uk/weather-and-climate-at-reading/2018/playful-floods/

Voss, M. and Wagner, K. (2010) Learning from (small) disasters. *Nat Hazards*, 55, 657–669. https://doi.org/10.1007/S11069-010-9498-5

Weller, M. and Appleby, M. (updated 2021) What are the benefits of interdisciplinary study. OpenLearn. https://www.open.edu/openlearn/education-development/what-are-the-benefits-interdisciplinary-study

Wenger, E.C. (2000) Communities of practice and social learning systems. *Organization*, 7(2), 225–246.

Whittle, R., Medd, W., Deeming, H., Kashefi, E., Mort, M., Twigger Ross, C., Walker, G. and Watson, N. (2010) After the rain – Learning the lessons from flood recovery in Hull, final project report for '*Flood, Vulnerability and Urban Resilience: A Real-Time Study of Local Recovery Following the Floods of June 2007 in Hull.*' Lancaster University, Lancaster.

Williams, S., McEwen, L.J. and Quinn, N. (2017) As the climate changes: Intergenerational action-based learning in relation to flood education? *Journal of Environmental Education*, 48(3), 154–171.

Whittemore, S. (2021) *Heutagogy; Developing agile, Reflective Lifelong Learners*. https://www.researchgate.net/publication/368775564_Heutagogy_developing_agile_reflective_lifelong_learners

10

COMMUNITY RESILIENCE TO FLOOD RISK

Futures, challenges, and opportunities

10.1 Introduction

Within the Anthropocene,[1] settings for future local flood risk management (FRM) are dynamic, complex, and uncertain, with multiple stresses overlaid and unfolding. Shifting externalities that interweave to impact socio-ecological systems in More Economically Developed Countries (MEDCs) include climate, environmental and demographic change, austerity, and the COVID-19 pandemic and its mitigation (Chapters 1 and 2). Policy imperatives for disaster risk reduction (DRR), playing out at nested spatial scales – international, national, regional, and local – require meaningful citizen participation (e.g., UN Sendai Framework for Disaster Risk Reduction, 2015–2030; European Commission's Flood Directive 2007/60/EC; Chapter 8). Effective citizen agency and local community action are critical in decreasing social vulnerability and in building socio-ecological resilience in context of rapid changes in disaster risk resilience practice.

Lozano Nasi et al. (2023) asked "Can we do more than 'bounce back'" when faced with rapidly changing climate risks? Is there potential to thrive rather than just survive (Bullock et al., 2015)? What are opportunities for transformative responses (Leichenko and O'Brien, 2019)? In planning for more risky futures, community resilience needs to move beyond 'coping' with floods to 'adaptation' and 'transformation' (Cutter, 2020). Transformative thinking in DRR requires identifying critical points of leverage and opportunity – and their timing within the disaster risk management spiral (Chapter 1) – to help target interventions that will increase local community resilience. However, conceiving resilience as transforming adversity also needs to attend to systemic social and environmental justice issues, and how to build social capital (Aldrich et al., 2018).

DOI: 10.4324/9781315666914-10

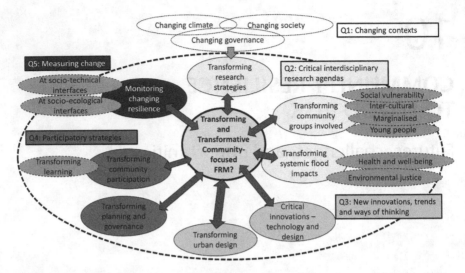

FIGURE 10.1 Some domains of possible transformation in community-focused FRM

Much can already be learnt from existing research and practice into building community resilience within different interdisciplinary, professional, and cultural domains focused on FRM and wider disaster risk management (DRM). A critical concern is whether and how transformative processes and outcomes, as key drivers for future success in developing community-focused FRM, transfer across different socio-ecological settings. This future-facing chapter builds on evidence from the previous eight themed chapters in this book. It asks:

- What are the changing contexts to future community-focused, FRM?
- What are the critical interdisciplinary agendas that need researching to inform building community resilience to future flood risk?
- What innovations, trends, and approaches could transform community participation within community-focused, FRM for longer-term resilience?
- How are such changes in community resilience (transformative or otherwise) meaningfully measured, and who will monitor them?

Figure 10.1 anticipates this territory, juxtaposing areas of current and possible future transformation, while linking back to previous chapters.

10.2 Beyond 'business as usual'

10.2.1 Societal transformation

Given the scale of future societal challenges, exploring the potential for transformation in DRM for equitable and just climate resilience is urgent. This requires a focus that scales up from individual citizens to wider society. Societal transformation can be defined as "deep and sustained, nonlinear systemic change" which generally involves a range of interconnected processes – "cultural, political, technological, economic, social and/or environmental" (Feola, 2015; Patterson et al., 2017; Fazey et al., 2018; Linnér and Wibeck, 2019; Linnér and Wibeck, 2020, p222). Linnér and Wibeck (2020) provided a typology of societal transformation towards sustainability involving interacting system level, pace, and scope at different scales. They recognised four modes of transformation: "quantum leap, convergent, emergent and gradual". They also distinguished between the complementary concepts of transition ("'going across' from one state to another") and transformation ("change in form or shape"). In the context of sustainability, Wibeck et al. (2019, p1) found that "transformative pathways need to splice new structures into the old" when transitioning towards mainstreaming approaches. Similarly, Thaler et al. (2019, p1075), working within FRM, referred to the "period of transformation" during which "'old' and 'new' systems can co-exist, but within the process of change the 'new' will overtake the 'old' leading to a change in society". They emphasised that awareness of opportunities and barriers to transformation is needed at different stages of the policy life cycle – in understanding, planning, and delivering.

So what is the potential for connecting risk management, resilience-building, and sustainability in systemic ways that ensure synergy between community-focused FRM, and the international drive for learning for sustainability to deliver UN Sustainable Development Goals (UNESCO, 2020)? This involves establishing how the trajectory from FRM to community-based FRM (CBFRM) might be developed and supported, in which circumstances communities themselves might lead, and what innovations might be transformative. It is also important to determine transformations in community-focused FRM that involve abrupt change or gradual transitioning to new practices.

10.2.2   Opportunities for systemic approaches

Building community resilience is not just about flood resilience – it encompasses wider systemic resilience to other social shocks, both acute and diffuse. A healthy, resilient community is more likely to be a flood-resilient community (e.g., Patel et al., 2017; Della Bosca et al., 2020; den Broeder et al., 2022). This is not a finite problem. With climate change, there will be new at-risk communities, without previous personal, intergenerational, or archival flood experience. A key concern is therefore how to build new abilities to live with routine and extreme flooding, alongside other stresses.

Chapter 8 identified the importance and implications of linking terminology and ethos in community-focused FRM. Shared language around CBDRM can obscure different ways of doing, reflecting "divergent worldviews, values and experienced histories of their environment" (Heijmans, 2009, p2; Chapters 2 and 8). Heijmans articulated this contested space – whether CBDRM is about "developing technical solutions", "a governance and human rights issue"; "an approach to advance local level decision-making and partnering with local government"; or "a strategy to transform power relations, and to challenge policies and ideologies responsible for generating vulnerability locally" (Heijmans, 2009, p2). She contrasted the "implicit interpretations and worldviews behind the CBDRM tradition" when viewed through a more top-down, internationally promoted "donor-driven" or "home-grown" community lens (Figure 10.2). Navigating these continua is critical in aspiring to societal transformation.

It is therefore timely to re-explore the "promise of a people-centred approach to floods" (Wolff, 2021, p9) and the "frontiers and directions" for targeting improvements (Šakić Trogrlić et al., 2018, p10). This approach requires resilience thinking and bottom-up "renegotiation of existing social contracts between governments' facilitation and private-individual responsibility" (Thaler et al., 2019, p1080). Transformation as a goal requires persistent questioning of the status quo, tackling root causes of inequalities and maximising the co-benefits of actions at different but connected scales. Here, community participation needs to move beyond consultation and information sharing to co-creation of solutions (Chapter 8).

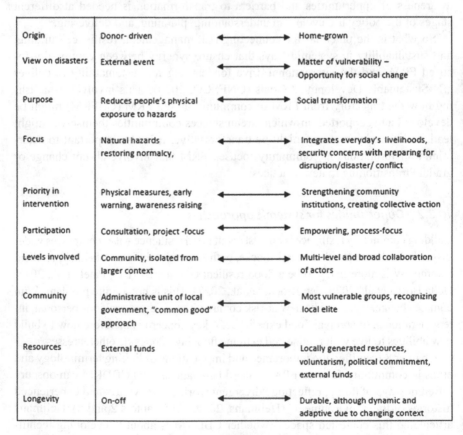

Origin	Donor- driven	←——————→	Home-grown
View on disasters	External event	←——————→	Matter of vulnerability – Opportunity for social change
Purpose	Reduces people's physical exposure to hazards	←——————→	Social transformation
Focus	Natural hazards restoring normalcy,	←——————→	Integrates everyday's livelihoods, security concerns with preparing for disruption/disaster/ conflict
Priority in intervention	Physical measures, early warning, awareness raising	←——————→	Strengthening community institutions, creating collective action
Participation	Consultation, project -focus	←——————→	Empowering, process-focus
Levels involved	Community, isolated from larger context	←——————→	Multi-level and broad collaboration of actors
Community	Administrative unit of local government, "common good" approach	←——————→	Most vulnerable groups, recognizing local elite
Resources	External funds	←——————→	Locally generated resources, voluntarism, political commitment, external funds
Longevity	On-off	←——————→	Durable, although dynamic and adaptive due to changing context

FIGURE 10.2 The nature of CBDRM traditions (interpretations and worldviews) expressed through its primary features on a continuum (Heijmans, 2009, p27; permission: Annelies Heijmans)

10.3 Futures for FRM: Changing risk; changing society

These are dynamic times, with uncertainty, complexity, and change in extreme weather events overlaid spatially and temporally with other systemic risks and socio-economic change. Transformative thinking needs awareness of how existing inequalities are exacerbated by differential impacts from the 'waves' of the pandemic, austerity, climate change, and biodiversity loss (Figure 10.3).

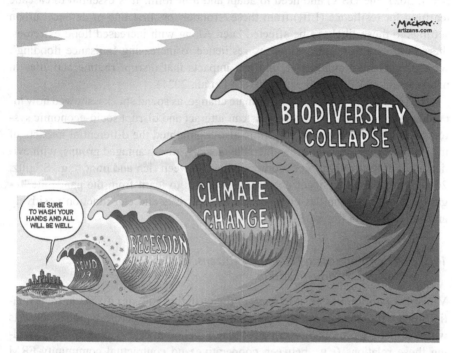

FIGURE 10.3 Looming waves of destruction (copyright Graeme MacKay – licence Artizans. com https://mackaycartoons.net/2020/03/18/wednesday-march-11-2020/)

A key imperative is to integrate equitable and just strategies for DRR, community climate adaptation (CCA) and community-driven development (e.g., as DRR-CCA; Lassa, 2011). This involves (re)interpreting CBDRM as a strategy to transform and challenge.

10.3.1 Changing climate: Future flood risk and other social shocks

FRM represents a major 21st-century challenge that is set to increase with climate change (Alfieri et al., 2017, 2018; IPCC, 2019).

> At 4°C global warming, countries representing more than 70% of the global population and global gross domestic product will face increases in flood risk in excess of 500%. Changes in flood risk are unevenly distributed, with the largest increases in Asia, U.S., and Europe.
>
> *Alfieri et al. (2017, p171)*

Traditionally, research effort has been placed on understanding riverine and coastal flooding. However, floods are increasingly generated by compound, climate-related agents in combination with intensifying factors. Increased urban surface water flooding is arguably most quickly impacted by climate change (Chapter 2). Coastal cities, subject to increased tidal 'sunny day' and storm surge flooding with exposed infrastructure, are particularly vulnerable (e.g., Sandifer and Scott, 2021, the USA) and need to adapt and transform. It is essential to cascade learning for resilience (LfR) from these 'frontier' or 'first line of defence' urban settings to those likely to be affected later. Along with increased flood extremes, there is a need to build community resilience to increasing 'nuisance flooding', with its pervasive cumulative, indirect impacts that disrupt routine activities in urban settings (Moftakhari et al., 2018; Li et al., 2021).

The COVID-19 pandemic and climate change, as social shocks, showed how increasingly severe and uncertain risks can interact and disrupt socio-economic systems (UNDRR, 2022). The pandemic also highlighted the differential impacts of mitigation strategies on more vulnerable, socially disadvantaged groups, with evidence for amplification of economic divides between rich and poor (e.g., Stiglitz, 2020). However, significant potential also exists to learn from the pandemic, regarding how communities engage with hidden risks and can function positively and collectively in times of sustained stress.

10.3.2 Changing society, changing citizens, and moves towards hydrocitizenship

Valuable opportunities exist to explore the changing role of citizens in their relationships with the state within community-focused FRM across national contexts with distinct cultures and governance. Major dichotomies can exist in understanding these relations (e.g., between cooperative and contractual community-FRM agency relationships, Geaves and Penning-Rowsell, 2015; Figure 10.4). It is important to make these relationships explicit in navigating varied engagements undertaken under the banner of participation (McEwen et al., 2023).

For example, the UK National Flood Forum (NFF)'s *Flood Risk Communities' Charter*[2] requires that communities are placed "at the heart of FRM" with "a role for everyone" (Cobbing et al., 2023). Possible citizen responsibilities within flood resilience planning range widely (Chapter 8); Twigger-Ross et al. (2020a) explored how these might vary in gathering evidence, and in design and delivery of interventions, differentiating "knowledge", "campaign", "physically", or "virtually" focused community participation. A key question is how these roles and their resourcing needs might need to change with different types of place-based communities and social disadvantage, and over time with increasing extreme weather stresses.

In identifying how communities might build capital and capacities for resilience, strong potential exists to integrate solutions in LfR across the range of risks at-a-place *and* promote active participatory hydrocitizenship. The latter requires

FIGURE 10.4 Dichotomies in role of civil society in extreme weather adaptation (source: CASCADE-NET)

and stimulates strong links between local and global action. Connecting LfR and hydrocitizenship provides valuable opportunities for creative integration of socio-ecological resilience building, encompassing water as risk and resource. Practically, this may mean engaging with some at-risk and proximal communities holistically about their local water relationships rather than explicitly about risk. Opportunities also exist to co-work with communities through engaging groups that are already water sensitive (e.g., gardeners) in local surface water storage management to 'hold water back' and 'slow the flow' (e.g., Sharp, 2019).

10.4 Researching community resilience to future floods

Aspiring for transformation poses questions about the best approaches for researching the development of community-focused FRM. This book has appraised diverse but intersecting domains within FRM, the value of blurring boundaries and seizing opportunities for interdisciplinary working, with the different thinking and practices that this can bring. Value can lie in creative, non-contiguous interdisciplinarity, bringing disciplines together – natural and social sciences, engineering, law, arts and humanities – to explore wicked problems (Chapter 1; e.g., Sadiq et al., 2019). It is a truism that perception, values, and worldview drive decision-making behaviours rather than facts. This raises the importance of research that explores people's values – empathy, optimism, empowerment, trust – that influence local climate adaptation (e.g., in maintenance of learned behaviours, cf. Sharpe, 2021).

For example, place-based working with socially engaged arts practitioners (Chapters 6 and 9) can promote effective community dialogue and capture stories that reveal local experiences, (environmental) knowledges, values, worldviews, and agency in context of changing and distinctive risky places.

A key imperative for transformation is to build strong relationships between research, policy and practice, with knowledge exchange and mutual learning through, and throughout, research processes. As noted in Chapter 4, the need is not always about more research but also embedding its effective communication, implementation, and evaluation of impact. Research involving academics co-researching in partnership with communities and other stakeholders in local FRM can have high value, despite facing resistance – including potential value-judgements about prestige – within the academy. Such research can bring in different skills and new perspectives on local cultural resilience building (e.g., work with the cultural sector; see "Dialogues on Museum Resilience", Norn et al., 2023; Grace and Sen, 2013; Drubay and Singhal, 2020).

A potentially transformative strategy for place-based resilience involves building longitudinal, community-academic research partnerships with local tertiary education institutions in co-production (Chapter 8). These bring together the concerns, differential capacities, and vulnerabilities within communities with the skill-development needs, time, and agency (curricular; volunteering) offered by students from different disciplines and academics to collaborate on, and help fuel, community-centred projects (Chapter 9).

At the community resilience and innovation nexus, researchers are adopting increasingly inventive ways of involving citizens within research. For example, the concept-approach of 'Living laboratories' (LL) has been used in testing experimental resilience infrastructures for climate adaptation in different national contexts. LLs have a strong spatial element as experimental, collaborative interaction spaces. A good example is their application in smart urbanism, emerging "at the intersection of visions for the future of urban places, new technologies and infrastructures" (Luque-Ayala and Marvin, 2015, p2105). This can involve in-depth, inclusive, participatory processes in knowledge generation. The LL concept is strongly promoted in Europe ("citizen-driven" empowered innovation in European Network of Living Labs ENoLL[3,4]) and includes shifts to bottom-up research, development, and innovation (RDI). Examples include co-design and collaborative implementation of green infrastructure and nature-based solutions. Places where the LL concept has been applied include implementation of green roofs and walls for a climate resilient smart city – incorporating stormwater management – in Liege, Belgium[5] and river restoration for natural flood management (NFM) within the Isar-Plan River Restoration, Munich, Germany (Lupp et al., 2021). Other potentially transformative research design innovations include co-designing Citizen Observatories for gathering crowd-sourced, hydrological data (Chapter 8).

A further research imperative is to secure funding for longitudinal studies which track differential long-term impacts of floods and management measures and interventions in communities, particularly focused on health and well-being (see below) and socio-technical aspects of innovations that support systemic effects. Factors that ratchet down or build up community resilience can be slow to manifest. This chapter now explores some domains of thinking and innovation that have potential to transform citizen participation in community-focused FRM.

10.5 Transforming community roles in DRM

Research into the barriers and enablers affecting perceptions and agency of specific community groups can act as a catalyst for transformative interventions. Pivotal is alertness to the importance, potential, and challenges of place-based community diversity (social, ecological, political) for resilience (Berkes, 2007). This poses questions about who is currently involved, and who else could actively contribute their knowledge, skills, and values to local resilience building. Attention to identifying marginalised or excluded groups is essential.

10.5.1 Transforming social vulnerability: Inequalities and environmental justice

Not all people within communities have similar capital and capacities or resources to prepare for future FRM. Transferring local responsibility for dealing with residual risk without attention to social and spatial justice issues can be a major pitfall of policy (Rufat et al., 2020). Transforming social vulnerability requires attention to engaging marginalised at-risk groups[6] affected by interventions in climate adaptation planning (Alba et al., 2020), and to "fair and inclusive processes" in decision-making (Bell, 2014; Chapter 2), as key concerns in community resilience building. In understanding community complexity, a critical issue is intersectionality – "the interconnected nature of social categorizations such as race, class, and gender as they apply to a given individual or group" (Crenshaw, 1991). This results in "overlapping and interdependent systems of discrimination or disadvantage", playing out in marginalisation or social exclusion.

Such framing can also be transformative for all stakeholders in understanding where and how social vulnerability, and local knowledge for resilience, may connect within place-based communities. Some groups may prove to be more resilient because of their need to be resilient in other aspects of their everyday life (Chapter 8). As Cutter et al. (2014, p56) highlighted, "inherent resilience" is not the opposite of social vulnerability; rather, these constructs can be differentiated both conceptually and empirically.

Need for transformation also exists in pervasive organisational rhetoric and exclusion in practices within community-focused FRM. The motif of 'hard to reach' or 'underserved' requires situational unpicking in community engagement settings about risk (e.g., Good Governance Institute, 2020). Marginalised groups are diverse with exclusion by geography (distance, terrain, or remoteness); transience or nomadic movement (traveller communities, migrants, refugees, homeless); racial minority status; low income or literacy; digital poverty; lack of local language; and disability.

The International Federation of Red Cross and Red Crescent Societies (IFRC, 2018, p4) in "Leaving no one behind" disentangled the language of 'hard to reach' as five types of exclusion:

- "Out of sight: the people we fail to see"
- "Out of reach: the people we cannot get to"
- "Left out of the loop: the people we unintentionally exclude"
- "Out of money: the people we don't prioritise"
- "Out of scope: the people who aren't our problem"

Unpicking 'hard to reach' involves increased understanding of embedded power dynamics by those with power, and a rethinking of who is at the centre of an engagement process (McEwen et al., 2023). The potential exists for FRM to learn from other risk management sectors in navigating this territory, for example research from the voluntary and community sector in social care and health. Regarding the design of inclusive access to services, Flanagan and Hancock (2010) highlighted impacts of negativity from experience and funding issues, but also benefits of respect, trust, participation, and partnership working – all themes which are also important in FRM practice.

10.5.2 Involving children and young people

Conceptions of the societal role of young people are culturally determined. They represent another important group for engagement to transform flood resilience planning, through prioritising their LfR as empowered citizens – now and in the future (Chapter 9). It is increasingly recognised that young people have potential to act as powerful change agents in families and communities (Percy-Smith and Burns, 2013; Winograd, 2016). This positive approach to promoting change in values and adaptive actions is adopted much more in some cultures, with potential for intercultural and cross-sector learning (e.g., Swindle et al., 2020; Williams and McEwen, 2021). However, this cannot become intergenerational delegation of responsibility.

Longer-term, sustainable practice requires strong systemic connections made between building resilience capacities of young people in formal learning and social learning in local communities. Attention is being given to trialling strategies that promote learning for urban climate adaptation, with schools acting as important community 'hubs' in awareness-raising. For example, potential exists to link development of local 'flood champion' groups in schools in flood risk settings with local adult community flood groups (CFGs) for their mutual sustainability (McEwen et al., 2018). Trialling of different approaches to delivering school estates, using co-developed Sustainable Drainage Systems (SuDs), can provide new transformative learning settings, with co-benefits for young people, teachers, and local communities (Chapter 7).

10.5.3 Harnessing cross-cultural learning for resilience

Another underplayed domain of transformation is cross-cultural learning involving communities with strong, embedded watery senses of place that come from routinely living with water and its risks. These may be from predecessor communities within the same locale ('flood heritage'; Chapter 5), or cultural and economic migrants that possess increased human and social capital about how to 'live with floods'. This latter sharing recognises cultural differences in flood risk awareness, connection to flood heritage, adaptive strategies, senses of hydrocitizenship, and local governance arrangements that support community-focused DRM. Given the locally embedded character of community-focused FRM, transformative inter-cultural LfR needs to explore ways of sharing knowledge locally that surmount language barriers.

10.5.4 Developing roles of volunteers and volunteering

A further concern for future community-focused FRM is in transforming volunteering as a critical form of citizen agency and hydrocitizenship within the DRM spiral (Chapter 8). Volunteering during flooding – and other extreme weather events – has many direct and systemic benefits (e.g., its flexibility and speed of response; its integration of local knowledge; its promotion of collective efficacy and well-being; e.g., Ntontis et al., 2021). Transformative participation through volunteering involves establishing what (new) task domains throughout the DRM spiral lend themselves to effective volunteer effort, and identifying existing tasks that could be done better or differently with, or by, volunteers. However, aspiration to transform volunteering in FRM requires awareness of challenges, such as inclusivity and coordination, and dealing with these effectively in policy development (Forrest et al., 2023). Significant co-benefits exist when volunteer community planning groups work to build local climate resilience (e.g., Baxter, 2019) beyond a sole flood focus, and their sustainability.

Flood volunteering can also learn from research into the efficacy of citizen agency in volunteering (spontaneous; mutual aid) during the COVID-19 pandemic (e.g., Mao et al.'s 2020 tool kit). For example, the UK Kruger report (2020) in its 'Levelling up our communities' laid out how positive elements of community agency (e.g., by young people and faith groups) might be harnessed and sustained. Opportunities exist to rethink and develop volunteering systemically and sustainably, building local capacity inclusively for climate resilience and more liveable communities. An example is co-developing volunteering for blue-green recovery, linking recovery of nature and health and well-being (HWB) of local people living by urban floodplain nature reserves (e.g., Tewkesbury Nature Reserve's 'Green Lung' project, UK; https://tewkesburynaturereserve.org.uk/green-lung-project/).

10.6 Transforming systemic flood impacts: Health and well-being for community resilience

Aspiration for transformation also involves identifying and mitigating systemic impacts like flood-related morbidity and mortality. This complex, cross-cutting theme within flood and climate resilience in urban and rural settings (e.g., Longman et al., 2019) requires attention to likely increases in health inequalities (e.g., Lowe et al., 2013; Institute of Health Equity, 2020; Action for Global Health, Beagley et al., 2021). Research evidences shifts in focus from floods impacting physical health during and immediately after floods (e.g., fatalities; susceptibility to disease) to increased awareness of mental HWB impacts (see WHO[7]). Flood experience (particularly if it is repeated), hazard perception, and 'living with increasing risk' have well-established links to mental health issues, with literature reviews affirming systemic causes and impacts (Fernandez et al., 2015; Saulnier et al., 2017; Zhong et al., 2018; French et al., 2019). Recent work recognises longer-term, negative effects (e.g., post-traumatic stress disorder, depression, and anxiety) and health inequalities that emerge. For example, Houston et al. (2021), working in Scotland, found social differentiation in shorter and longer-term flood impacts on communities including health, while Fernandez et al. (2015, p14) noted "the impact of floods on mental health is higher in areas of material deprivation".

Zhong et al. (2018), reviewing research into health impacts of floods and linked secondary hazards, produced a simplified influencing pathway to differentiate short-term and long-term physical and psychological health outcomes after flooding (Figure 10.5). They highlighted wide-ranging risk management and policy factors that influence longer-term mental health outcomes in communities (e.g., relocation, insurance cover, social support, psychological intervention, and health services; Chapter 7). Mitigating pervasive mental HWB impacts has potential to be transformative, scaling up from individual to community. More research – longitudinal and systematic – is needed, focused on "disability, chronic diseases, relocation population, and social interventions after flooding" (Zhong et al., 2018, p165). Research is also required to investigate HWB impacts of different flood types and compound flooding with overlaid or cascading stresses (e.g., COVID-19 pandemic; Chapter 2).

Community engagement for transformation must attend to links between mental HWB and recovery timelines, the role of "flood anniversaries", and implications of active remembering and active forgetting within communities and their sensitive negotiation (McEwen et al., 2016; Chapter 5). Research is also required into impacts of community-embedded, social interventions that support the long-term mental HWB of flood risk communities, so determining what transformative practice might look like. Notably, the key assets within communities for good health and resilience align – the "skills and knowledge, social networks and community organisations" (South, 2015, p5). She outlined community-centred approaches that support positive HWB: the value of strengthening communities, benefits of

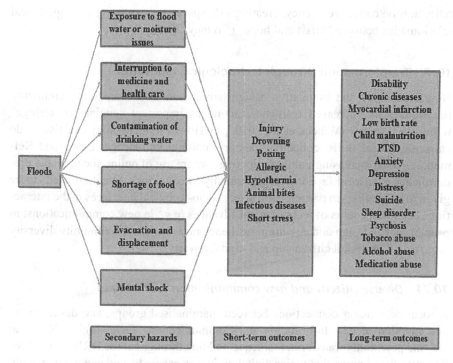

FIGURE 10.5   A simplified influencing pathway of long-term health impacts after flooding (Zhong et al., 2018, Figure 9, p177; reproduced by permission of Elsevier)

volunteer and peer roles, roles of collaborations and partnerships, and access to community resources. Indeed, Walker-Springett et al. (2017, p66) emphasised community connections and interactions as an "important protective element for wellbeing" after floods, while Quinn et al. (2020) also highlighted importance of relational capital and active belonging in self-reported well-being in communities after disasters.

Extending the disciplinary and professional contributions, another potentially transformative realm is the role of embedded arts and artists in supporting communities dealing with trauma, healing and recovery. Artists have been conceived as 'first responders' in healing, with roles for creative and expressive arts alongside health clinicians in offering "a place to keep the chaos" (Grebenau, 2021). While well recognised at the individual level, arts practices (e.g., making, drawing) can also help community members to deal with trauma collectively. Examples include children making pillows for others in their post-disaster 'healing journey' (Global Kids "Project Pillows", connecting Haiti and Baton Rouge, Louisiana; Nemeth and Olivier, 2017), and composite drawing for adults in flood-affected communities as part of psychosocial healing (e.g., Great Wellness Workshops after Great Flood in 2016, Louisiana; Nemeth et al., 2021). Creative practice can also encourage reflection on collective agency, creating safe spaces for exploring entangled local relationships between "crisis and hope" (Cretney, 2018).

10.7 Transformation through technological innovation

New technology and innovative design have potential to transform community functioning for increased resilience, acting as important "engines of change" within civil society (Wibeck et al., 2019, p1). However, technology can also erode capacities (local skills; collective memory; cultural knowledge; Lewis and Kelman, 2010) so increasing vulnerability (e.g., where use of online social media predominates over local face-to-face community interactions). Attention needs to be given to inequalities in participation at key socio-technical interfaces in the interactions between benefits of technological advances (e.g., in new communications; in property-level design and structural resilience measures) and community diversity (e.g., extent of digital citizenship and digital poverty).

10.7.1 Diverse citizens and new communication technologies

A focus on making connections between marginalised groups, and development and adoption of new Information and Communications Technologies (ICT) can aid, and potentially transform, resilience building throughout the DRM spiral. It is, however, well known that major digital divides in access to, and competency with, ICT can occur by age, socio-economic status and in geographically isolated, rural versus urban settings. This sits alongside intergenerational issues of digital 'natives' and 'migrants' in technology adoption. The selection of a specific communications technology, by communities themselves or FRM agencies, has major implications for potential exclusion of marginalised groups. As Reddick et al. (2020, np) pointed out, the digital divide is not exclusively rural/urban but also occurs within cities. In the USA, for example, many low-income and rural households do not have reliable internet access through smartphones or broadband connections (Granicus, ud[8]). However, mobile phone use is more pervasive than smartphone[9] (Pew Research Center, 2019). Use of SMS (Short Message Service) for information dissemination can have a disproportionate impact, compared with other communication channels (Granicus, ud). Importantly, multi-channel communication is essential for inclusivity across groups within communities (Chapter 6).

The role of social media is also rapidly changing, allowing social learning and solidarity within communities beyond the local (Reuter and Kaufhold, 2018). This provides potential for mutual support between proximal and affected communities in a flood risk setting – a spatial relationship that may change with flood severity. Twitter and other Web technologies (e.g., Instagram; Flikr) have already demonstrated potential as learning media for disaster resilience – sharing oral and visual stories – but reflect specific demographic preferences (e.g., by age). Notably, relationships between citizens and FRM agencies through social media can be active or passive – both important in transforming community-focused FRM. In the former, the citizen is aware they are sharing local knowledge with agencies during an unfolding event. For example, the US Federal Emergency Management Agency's (FEMA) app has a "Disaster Reporter" feature[10] permits crowdsourcing and information sharing between FRM actors. Here empowered citizens can contribute images, geo-tagged and mapped in 'real time' during a flood, to assist FEMA in rapidly targeting its relief efforts. In passive models, citizens may not be aware of who they are sharing information with beyond their networks, and how FRM agencies and others are using their data. For example, Silva (2018) reported how government agencies may mine information (e.g., tweets or Flikr tags) about local impacts already publicly shared by citizens, while Tkachenko et al. (2017) highlighted potential for using such evidence for early flood warnings. This data gathering can work alongside a more traditional collection of offline, less immediately garnered, information. However, Haworth et al. (2018) emphasised care in navigating complex issues of community empowerment, privacy, and exclusion.

10.7.2   Valuing big data: Developing citizen science and observatories

Democratising environmental monitoring – increasing communities' roles in their observation of local risk (Chapter 8) – is another domain for transformation. Emerging technologies, big data (its velocity; volume, variety) and use of rapidly emerging ICT can impact local flood awareness and preparedness. This brings together rapid (geo)technological development, participation and citizen science, connecting to wider care of place (Chapter 8). Concerned citizens and communities as observers now have potential to map and monitor floods for themselves as events play out, using their local knowledge. Technologies like mobile mapping and Geographical Information Systems (GIS) are being applied in innovative practice, including information sharing for community preparedness (e.g., Fagg et al., 2015).

Humanitarian aid agencies have developed initiatives like 'Missing Maps[11]' (IFRC, 2018), an "open collaborative project" that uses processes of collecting and sharing local field data to empower vulnerable communities and inform relief efforts of first responders. While this example is for Less Economically Developed Countries (LEDCs), its principles have wider applicability. In the UK, for example, a CFG is trialling free community mapping[12] – overlaying digital, volunteered, flood-related information – on its community-facing website (Cobbing et al., 2023). Potential exists for rapid sharing of learning about 'what works' in local monitoring from one at-risk community to another – by supporting NGOs such as the UK's National Flood Forum.[13] However, local CFGs need to attune to ICT capital in their communities when designing and implementing adaptive measures.

A domain with large potential transformative impact is the development of citizen observatories (CO) and eParticipation within FRM (Chapter 8). Applications include citizen science approaches to community-based flood monitoring for public health (e.g., Wolff, 2021), and the concept of Participatory Early Warning Systems (P-EWS) (Marchezini et al., 2018). Citizens can participate in monitoring their local water courses (levels and obstructions) through mobile apps so raising local risk awareness and enabling knowledge exchange between citizens and statutory FRM agencies (Ferri et al., 2020; Chapter 8). Such activity has multiple benefits; reductions in social vulnerability through CO have been equated to an annual average damage cost avoided (Wehn and Evers, 2015). Further research into improved socio-technical interfacing is, however, needed to determine socio-economic, cultural, and behavioural factors that encourage or inhibit citizens' eParticipation in collecting and sharing local data to permit its mainstreaming.

10.7.3   Valuing thick data: Harnessing the power of stories

Alongside harnessing of big data, transformation in community-focused FRM also needs greater recognition of the value of garnering 'thick' data (Chapter 8). Stories act as interactive experiential memory; listening to stories can stimulate exploration of relationships between people's individual and collective senses of risk, place, identity, and belonging. In LfR, potential exists to capture positive personal and community stories of adaptation and transformation that can help cascade change. There is a role for socially engaged arts in co-producing creative ecological thinking about place-based resilience. Applied storytellers can disrupt existing narratives and coproduce new collective stories – so supporting development of creative ecosystems for resilient societies (Liguori et al., 2023). Value also exists in testing new ways of using new crowd-sourcing technologies to bring together rich narrative insights gained from personal flood experiences (e.g., Garde Hansen's prototype 'Flappy' Developing a Flood Memory App[14]). However, this is not just about *past* place-based resilience stories connecting with flood heritage (Chapter 5), but also about collective visioning of possible adaptive futures that allow communities to thrive rather than just survive. Examples from water risk management include participatory storyboard scenario-ing (visioning) stimulated by scientists' catchment-based drought projections as a method to promote new dialogue (Liguori et al., 2021). This involves brokering imagined futures – both fearsome and desired (Chapter 9).

10.8   Transforming urban design at different scales

Another domain of transformation is urban design, with new approaches to mitigating urban flood and climate risk nested at different scales, from the whole city to neighbourhood level.

10.8.1   Designing water cities – Involving citizens

A growing number of all-city blue-green approaches to flood resilience are developing across different cultural contexts (China; the USA; Chapter 7), with co-benefits involving liveable cities for communities. This human element is known as 'the red (social) imperative' (Wörlen et al., 2016). Such thinking can be creatively aligned with the concept within the arts and humanities of 'water cities' where water and stories circulate (e.g., Water City Bristol[15]), promoting cultural understandings of water in community identity and sense of place. Importantly, such design thinking provides valuable opportunities to engage local citizens about water holistically.

The socio-technical interfaces of blue-green resilience measures need further research. This includes establishing how communities with differing capital and capacities connect with Blue-Green Infrastructure and how to engage all groups inclusively for fair blue-green resilience (O'Donnell and Thorne, 2020). Urban water planning, as a basin and community connector, is promoted within The International Water Association's (IWA) *Principles for Water-Wise Cities* (2016)[16], to facilitate transition to sustainable water management. 'Water-wise communities' represent one level of civil action with citizens driving

> urban planning and design with their understanding of the risks (flooding, scarcity) and opportunities (the liveability outcomes their community values, resource recovery opportunities, reducing dependency on uncertain future resources, and increased well-being). In working collaboratively to achieve these outcomes, water-wise citizens may also adapt their behavior.
>
> *IWA (2016, p5)*

Participation of urban communities as key stakeholders is essential within scaleable innovations that deal with distributed pluvial urban flooding. Potential exists for strengthening community involvement in action-based 'open innovation' (e.g., testing pioneering adaptations like smart green roofs[17] in Belgium; see LL above). Drivers vary but some water-related Living Labs initiatives are citizen-driven, long-term transformative and informal (see Water Europe's Inventory of European Water-Orientated Living Laboratories, WoLLs[18] ud p9). For example, BruSEau (Sensible à l'Eau, p16) shares FRM solutions developed by the inhabitants with local government and water operators in Brussels, Belgium. Cities in recovery can also act as valuable Living Laboratories for resilient urban design involving meaningful community participation (e.g., New Orleans after Hurricane Katrina[19]). Potential co-benefits are large; the Rockefeller Foundation/ARUP's *City Resilience Strategy* (2015) integrated systemic concern for infrastructure and ecosystems, HWB, society and economy, and leadership and strategy, with inclusive approaches to community engagement and ownership.

Transformation can also involve cascading cheaper interventions with creative co-benefits. 'Tactical urbanism' as an approach refers to city, organisational, and/or citizen-led neighbourhood building "using short-term, low-cost, and scalable interventions to catalyze long-term change".[20] For example, the volunteer-led 'Depave' initiative – a Community-Based Approach to Storm Water Management[21] – links community activism, action, and learning in removing local impermeable paving.

Started in Portland, Oregon, USA, this initiative, with its training and open-source resources, has been reproduced in many cities (the USA, Canada, Australia, and the UK).

10.8.2 Socio-technical interfacing: Increasing uptake of property-level protection

Transformation can also occur through new design thinking at household or property scale. A major dilemma for individuals and communities is whether to defend, adapt, or abandon with increased flood risk (Chapter 7). For many, relocation after being flooded is not an option. An area of rapidly emerging and potentially transformative technologies is development of Property Flood Resilience (PFR) measures for new build design, for example, waterproof, rapid construction techniques, and retrofitting, such as water resilient kitchen designs. Designed visual aesthetics, practical ease of use and affordability are key concerns, with innovations needing tailoring for different types of people, floods, properties, and resources. However, transformation is not just about design efficacy (i.e., development of standards) but also concern for behavioural aspects of factors – individual and community – affecting uptake of innovations. For example, NGOs, supporting flood-affected communities in the UK, report major challenges in getting community support for PFR as a mitigation measure (Chapter 7).

Transformation also needs to consider more radical future resilient build, designing properties that will not need flood resilience retrofitting. This is the domain of flood resilient architecture and 'making space for the river' (Barsley, 2020) and involves not only residential property but also resilient critical infrastructure (e.g., schools, hospitals, utilities) that allows communities to continue functioning during floods. Reflecting business entrepreneurialism, trials are underway in rethinking flood resilient building, including "elevated, amphibious and floating" (Piątek and Wojnowska-Hciak, 2020 on the semi-wild Vistula River, Poland). Major challenges are affordability and equity of access. However, the Dutch are one leader in 'development on water', with the Dutch docklands[22] concept of the Dutch Watervalley© aiming to offer "affordable housing on large-scale floating structures".

10.9 Participation: Transformative pathways in governance

Another important domain for transformation is in developing models of participatory local governance that strongly embed community voices and agency in FRM, within settings with complex, uncertain, emergent, and changing risks while also attending to environmental justice issues. This needs to build on underpinning principles and practices of good local governance, making explicit how communities and deliberative democracy sit within these arrangements. Critical are the knowledges underpinning dominant stakeholder discourses and how power is executed and resource distributed (Herath and Wijesekera, 2020, p1). The Organisation for

Economic Co-operation and Development's (OEDC's) 12 *Principles on Water Governance* (2015), highlighting effectiveness, efficiency, and public engagement, have been applied within flood governance to promote dialogue and assessment of performance (OEDC, 2019). Principle 10 is to "Promote stakeholder engagement for informed and outcome orientated contributions to water policy design and implementation" (p83) although community is not explicitly mentioned. This principle highlights concern for issues of under-represented groups and capacity development, and ensuring flexibility in process. For OEDC, good governance is highly contextual, and locally tailored but coordinated between different scales, with responses adaptable to changing circumstances. Importantly, "opening up engagement" needs attention to equality and local power relations, and identification of inclusive pathways for community participation (cf. Triyanti et al., 2020).

Adaptive governance for flood resilience requires adoption of good governance principles tailored to changing community flood risk settings. Exploring flood risk governance that is more resilient (Driessen et al., 2016) and evolving in its systems and processes requires multi-stakeholder recognition that 'not one size fits all'. It also needs attention to characteristics of anticipatory climate governance or "governing in the present to adapt to or shape uncertain futures" (Muiderman et al. 2020, p1). They identified four distinct ways the future might be conceptualised (probable and improbable, plausible, pluralistic, and performative), then mapping these against diverse sets of methods and tools for "action in the present". Such action moved through strategic planning, building adaptive capacities, mobilising diverse actors to interrogating future imaginaries for their political implications (Figure 10.6).

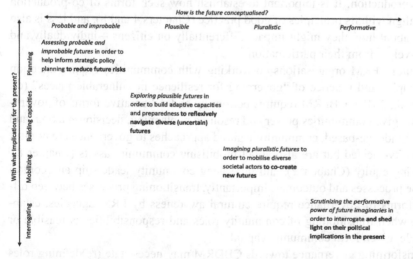

Key: *Italics*: <u>what</u> conception of/engagement with the future; Plain: <u>how</u> these intersect with actions to be taken in the present; **black bold**: <u>why/to what end</u>: the desired ends of engaging in/with anticipatory approaches. (Source: Muiderman et al., 2020; edited to B/W by author)

FIGURE 10.6   Approaches to anticipatory governance: diverse conceptions of the future and actions in the present (Muiderman et al., 2020, Figure 1, p10; reproduced by permission of Elsevier and authors)

So what governance structures facilitate adaptation given uncertainties both in science, and in values and social norms? Hurlbert and Gupta (2016) suggested that uncertainty in science can be dealt with through adaptive management, and uncertainty about social norms through adaptive co-management and participation. A further question is the role of local communities in such participatory processes and their potential for leading, or partnership and co-production (e.g., in revisiting plans; Chapter 4)? For example, Hurlbert and Gupta (2016) proposed anticipatory governance in engaging with people to develop future solutions for multi-hazard risk, highlighting the importance of dealing with floods, droughts and climate change together. Cleaver and Whaley (2018, p1) also argued for "'thicker', contextualized, and power sensitive understandings of how adaptive governance works in practice". Such effort gives prominence to processes of knowledge production that value different types of knowledge.

10.9.1   Transforming community roles in adaptive governance

Roles of community in policymaking, its implementation, and operational management (e.g., around enactment of measures) are generally under studied. Mees et al. (2018), in exploring transitioning to co-productive flood governance (CBFRM) between local authorities and citizens in different national settings, found wide-ranging activities and citizen-government relationships. They identified typologies based on type of interaction, role and type of citizen input and distribution of contributions and benefits, defined as hierarchical, incentivised, and deliberative co-production. It is important to establish how such forms of co-production might align with its principles of good practice (Chapters 4 and 8), and less is also known about how they might impact differentially on citizens – individually and collectively – from their participation.

Statutory FRM organisations co-working with communities require attention to principles and practice of "governing for resilience in vulnerable places" (cf. Trell et al., 2017). CBFRM requires bottom-up and deliberative forms of governance that give communities power and resource in planning, decision-making, and action. Evidence-based, community-centred approaches to governance are not just community-located but are also about mobilising community assets (Chapter 8), promoting equity (Chapter 2), and increasing community leadership and control over the processes and outcomes. Importantly, transitioning processes between different forms of governance require cultural awareness by FRM agencies, establishing when transitioning of community roles and responsibilities is feasible or desirable, given local community capital.

Transforming governance towards CBDRM may necessitate (re)defining roles and responsibilities, requiring statutory DRM agencies to know when to lead and when to step back (Cobbing et al., 2023). Effective communication with communities about how approaches to FRM can be integrated locally, alongside respective roles and commitments, is essential (e.g., NFF's *Flood Risk Communities' Charter*[3]). In CBFRM, "outsiders" (e.g., NGOs) have "supporting and facilitating roles" (Shaw, 2016, p3). Transitioning to CBDRM may require sharing or delegation of responsibility, and building of confidence and empowerment for, and through, participation by community stakeholders. Community-centred approaches involving "deep public engagement" have potential to be transformative; however, transformation at a local scale (Thaler et al., 2019, p1080) requires space and time for bottom-up drivers to function throughout a process from the outset. Processes of transitioning governance arrangements need to value the trialling of innovative arrangements – with reflexivity and transparency through co-evaluation. Benefits of strong community involvement in local risk governance include having different messengers about local risk and adaptation, and communities using their increased knowledge of access to funding to undertake flood-risk measures for themselves.[23]

10.10 Transforming participation

Good practice in meaningful community participation has already been explored (Chapter 8), recognising its importance in research, and policy and practice generally and in community-focused FRM (Chapter 8; Twigger-Ross et al., 2020a). Strong potential exists for these participatory spaces to be synergetic, integrating processes and outcomes that empower citizens. Reflection on how meaningful participation might develop in place-based settings, and potentially become transformative for communities and other stakeholders, requires attention to key questions (Box 10.1).

BOX 10.1 QUESTIONS IN TRANSFORMING COMMUNITY PARTICIPATION

- How can participation be meaningful and potentially transformative for marginalised groups?
- What learning can be gained from other sectors, with strong concern for inclusive participation, like public health or sustainable land management?
- How do we nest community participation in local FRM systemically with that for other risks and wider water relations for resilience and citizenship?
- What could transformative participation look like throughout the DRM spiral, tailored to different places?
- How can arts and humanities disciplines and the cultural sector transform participatory processes?
- How might participation be imaginative and future-facing, feeding effectively into adaptive and transformative governance?

Both context and design of participatory decision-making processes within socio-ecological systems are critical to their outcomes whatever the setting. de Vente et al. (2016, p1) highlighted processes leading to "more beneficial environmental and social outcomes" including:

the legitimate representation of stakeholders; professional facilitation including structured methods for aggregating information and balancing power dynamics among participants; and provision of information and decision-making power to all participants.

They (2016, Table 6, p9) identified five key outcomes identified from participatory processes conducted across 13 study sites internationally:

- "enhanced social networks, collaboration, and trust among participants"
- "learning and knowledge exchange among participants"
- "better problem identification and awareness"
- "consensus, acceptance, and implementation of tailor-made solutions is more likely"
- "more confident and motivated stakeholders".

Maximising potential for transformation through meaningful participation needs to target those likely to benefit most in reducing inequalities. This also requires attuning to different group cultures in communities when compared to statutory workplaces, recognising spaces and places where slow deliberative processes are necessary. Table 10.1 draws together intersecting concerns, principles, and approaches in meaningful participatory decision-making from research and practice, as a framework for planning and delivery.

TABLE 10.1 A Framework for Designing Meaningful Participatory Processes with Potential to Transform (Drawing on Shaw, 2016; IFRC, 2018; de Vente et al., 2016 and Author Experience)

Concern	Principle
Level of participation	• Use approaches that engender meaningful involvement of place-based communities • Think 'higher-rungs' of Arnstein's (1969) ladder of participation
Locate participation	• Be aware of who is at the centre of the process • Encourage systemic thinking
Knowledge	• Value different types of knowledge. Give them space and encourage their interaction • Value hybridity of knowledge, building out from individual level
Learning Ethos	• Focus on social learning through participatory processes • Foster trust among participants. Facilitate repeat engagements outside acute events • Practice active listening as skill for mutual understanding and respect • Plan for dynamic of participatory meetings

(Continued)

TABLE 10.1 (Continued)

Concern	Principle
Equality	• Work to understand social vulnerability • Avoid assuming it • Recognise diversity, inequality, and social disadvantage in participatory settings • Avoid language of 'hard to reach' • Work to include more marginalised voices and vulnerable groups • Pay "particular attention to the people likely to be passed over, such as women, older people and persons with disabilities" (IFRC, 2018, p13)
Process	• Research your participants and adapt process design
Participants	• Understand and appreciate diversity by age, culture, religion, disability, other demographics, and their intersection • Identify pivotal gatekeepers in accessing participation of groups • Actively involve young people as citizens now • Capitalise on potential for inter-cultural learning • Identify different knowledges, values, and worldviews in particular places. These may be unconnected but with potential for connection
Technology	• Capitalise on opportunities for eParticipation • Explore potential for ICT-enabled citizen observatories • Act to mitigate the digital divide. Where possible, engage participants through high- and low-technology versions
Resourcing	• Attend to resources and power development in participation. Match resourcing to devolvement of tasks to community • Adapt locations for participation to participants • Community-embedded settings can work better
Process	• Adapt language of engagement to participants • Be sensitised to technical and academic language as barriers • Identify brokers and translators in communities and FRM organisations • Give attention to different competencies and communication skills
Evaluate	• Co-evaluate participatory processes – formatively and summatively • Ensure capture of multiple voices and perspectives
Legacy	• Attend to sustainability and legacy of interventions • Work to reap value from longitudinal engagement processes • Link to community development and wider local resilience building/ risk management • Integrate with other place-based, participatory processes involving communities

Transformative participation is nimble, inclusive, and reflective in its impacts, linking to "radical forms of social change at the grassroots scale" (Cretney, 2018, p472). Such participation needs to embed social learning, with all participants having potential to build their human capital through its processes. Box 10.2 proposes systemic (non-linear) thinking that increases the likelihood of transformative impacts through participation for communities and other FRM stakeholders.

BOX 10.2 SYSTEMIC THINKING NEEDED FOR TRANSFORMATION THROUGH PARTICIPATION IN FRM

- RECOGNISE importance of communities defining and assessing their own resilience capabilities (cf. Scherzer et al., 2019).
- IDENTIFY existing social capital within communities in form of lay indigenous knowledge (experiential, intergenerational) and capabilities (being resourceful, enterprising, persistent, and cooperating with other community members).
- INVEST in social capital and social infrastructure to support participation for resilience (cf. Aldrich et al., 2018).
- INTEGRATE participatory risk management options in locally tailored ways. Act on opportunities for niche innovation as well as mainstream solutions (cf. Thaler et al., 2019).
- PAY ATTENTION to scale relations, moving up and down from the individual, with 'community' as a critical aggregator of knowledge, interests, agency, etc.
- TARGET full DRM spiral, bringing participation out from acute events.
- DIVERSIFY OUT from statutory interests. Think creatively about possible roles in resilience building with different actors and sectors that work locally.
- RETHINK ways of working within statutory agencies, including funding models for local work.
- AVOID co-producing single-purpose solutions. Identify multiple benefits.
- COLLABORATE with those most at risk but also capitalise on strengths and resources of proximal communities. This involves partnerships with less vulnerable or flood-affected groups.
- SEIZE participatory opportunities to engage communities with flood risk in context of wider extreme weather risk *and* water security for climate resilience.

10.11 Evidencing transformation – Measuring change

Community resilience can increase or decrease over time; it can also be an "illusion" in context of systemic future changes (Chapter 2). Ability to baseline, track, and compare temporal changes in community resilience characteristics (Scherzer et al., 2019) is therefore critical for *all* stakeholders in appraising integrated risk management strategies designed to deliver local change. Continued development of theoretically robust frameworks and tools for measuring disaster resilience that are empirically tested is essential – so matching validity with usability (Schipper and Langstone, 2015; Sharifi, 2016; Keating et al., 2017). This is needed to:

"(a) deepen understanding of the key components of "disaster resilience" to better target resilience-enhancing initiatives, and (b) enhance our ability to benchmark and measure disaster resilience over time, and (c) compare how resilience changes as a result of different capacities, actions and hazards.

Keating et al. (2017, p77)

Resilience is both dynamic process and outcome, with the latter easier to measure. Any measurement also needs to ask, 'by whom, for whom' (Le Dé et al., 2021), linking to different stakeholder roles – now and future – in community-focused DRM. Most resilience indices are created top-down by researchers or consultants rather than bottom-up, working *with* communities. Equally, 'measurement' may not be a priority for finite community resources.

Many assessment tools now exist, with resilience indices developed for spatial units from city-level (e.g., Arup's City Resilience index[24]) to county-level (e.g., BRIC index; see Twigger-Ross et al.'s 2020b review). Table 10.2 provides examples of top-down and bottom-up models. Indices vary in their focus (community resilience; disaster resilience; community flood resilience), data inputs, and intended user. Many indices (e.g., BRIC) use a capitals approach (Chapter 8) or an extension of this. Quantitative metrics tend to use available data in measurement of the construct – 'community resilience'. However, parameters that lend themselves to measurement are not necessarily the most meaningful; moreover, aggregated data for a geographical area will have inevitable limitations in interpretation at community-level. An index, like BRIC, can only ever be:

a static snapshot of inherent resilience, recognizing that the production of resilient community characteristics is dynamic and can vary on an annual, monthly, weekly, daily, or even hourly basis.

Cutter et al. (2014, p66)

Of course, "measuring resilience should in no way replace a vulnerability analysis" (Keating et al., 2017, p88). Social vulnerability indices include SOVI, using aggregated US Census data to explore social vulnerabilities across geographies.

Improvements to resilience can look very different in different communities (Hochrainer-Stigler et al., 2021). A critical question is what aggregate resilience metrics miss at the individual or group level, and how these gaps might be filled. Scherzer et al. (2019, p3) highlighted some limitations of BRIC's approach focusing on a community's existing resilience capacities and not their application or use. They noted that

'community' in the BRIC model is reduced to a locality, side-lining social and relational aspects of community that so often are of critical importance in crises.

As an alternative, Kyne and Aldrich (2020) recently developed a social capital index (SoCI) index focusing on levels of social ties and cohesion within

TABLE 10.2 Examples of Community Resilience Indices with Different Approaches, Scales, and Users

Index	Objective	Approach	Scale	Use/User	Reference
BRIC index (Baseline Resilience Indicators for Communities) (Top-down)	To assess inherent characteristics of a community that contribute to resilience Information developed into six indicators used to compare communities across the USA	Uses capitals approach across six domains (social, economic, community capital, institutional, infrastructural, and environmental)	County level (USA) – serving as proxy for communities	Baselining, monitoring, assessing improvement, comparing counties over time Easy to compute and disaggregate for use in a policy context	Cutter et al. (2014a)
Development of BRIC index (Top-down)	As above	Hierarchical index, using 47 indicators divided into six subdomains, to describe resilience capacities.	Norwegian municipalities	Decision-makers in regional planning	Scherzer et al. (2019)
Australian (Natural) Disaster Resilience Index[25] (Collaborative)	Focus on capacities for resilience in Australian communities Across disaster settings (e.g. bushfires, floods, storms, and earthquakes) Nationally standardised assessment	Focus on coping and adaptive capacity Understanding these capacities; how they differ by place	To help communities, governments and industry work together to cope with, and adapt to, natural hazards	Multi-stakeholder use including communities	Parsons et al. (2016); Patch (2020)
Rural Resilience Index (RRI) (applied disaster resilience assessment index for use in rural/ remote communities) (Bottom-up)	Goals to produce resilience assessment and planning tools that could be used by communities to generate locally relevant data on their current resilience and be able to monitor/enhance their resilience over time	Designed as user-friendly, process-based, qualitative resilience assessment tool Capital based by community domains: human, social, built, economic, natural, governance, and disaster preparedness	Community-centred	By communities (action research project)	Cox and Hamlen (2015)
Analysis of Resilience of Communities to Disasters (ARC-D[26]) toolkit (Bottom-up)	Developed over ten years by international humanitarian and development organisation	Thirty measures and rating guidance: from minimal to full resilience, guidebook, and online platform	Communities; aid agencies	Goal to enable aid organisations to measure community resilience in a way that supports resilience building interventions	Clark-Ginsberg et al. (2020)

communities, using available US census data. Their validity testing supported outcomes comparable to, or better than, capital-based approaches.

10.11.1   Community-implemented approaches to measurement

For community resilience programmes, concerns in measuring resilience are not just about accuracy and validity, but also ease of use, utility, and affordability (Clark-Ginsberg et al., 2020; US National Institute of Standards and Technology, NIST). Robust and easily implemented procedures for community-led resilience measurement are essential and empowering in evaluating success of resilience-building initiatives. Practical Action has developed community resilience measurement approaches to be implemented *by* communities as an informative rather than extractive process. Their V2R (Vulnerability to Resilience) Framework "for analysis and action to build community resilience" (individuals, households, and communities) is based on the Sustainable Livelihoods Framework (Pasteur, 2011). Here, 'usability' is paramount; V2R involves participatory activity with question checklists that aspire to develop local capacity and ensure community ownership. The process helps inform local decision-making and actions (Figure 10.7).

This framework explores what causes community vulnerability and how different factors – "disaster risk reduction, climate change impacts, governance and livelihoods" – interact locally to impact on resilience outcomes.[27] Pasteur also brought together differing scales of analysis – local-scale analysis by communities and collated secondary data – arguing these need to be considered complementary, with differences "explored not ignored" (p69). Importantly, the process is concerned with feeding back findings to the whole community.

This thinking informed Keating et al.'s (2017) development of a framework and tool for measuring community flood resilience, tested in 75 communities in eight countries. They measured 88 potential sources of resilience graded at baseline (initial state) to endline (final state) approximately two years later, verifying sources of resilience empirically. This framework linked several important elements identified above: five capitals (5Cs; human, social, natural, physical, financial) with 4Rs defining quality of community characteristics (redundancy, resourcefulness, rapidity, and robustness), themes, system level, and stages of the DRM spiral. Concerns included issues of thresholds or abrupt changes in indicator parameters selected.

In defining desirable characteristics of a resilience measurement tool, Cutter (2014) suggested that communities seek specific dimensions like openness, goal alignment, and adaptability (Box 10.3). For example, Clark-Ginsberg et al. (2020) in their ARC-D toolkit (Table 10.2) focused on utility and ease of use in a "snapshot" rapid assessment, alongside validity. Their assessment process aimed to empower both communities and field staff to understand and operationalise resilience measures.

In community resilience assessments, there can also be significant value in capturing thick data about lived resilience as a process through stories, observations,

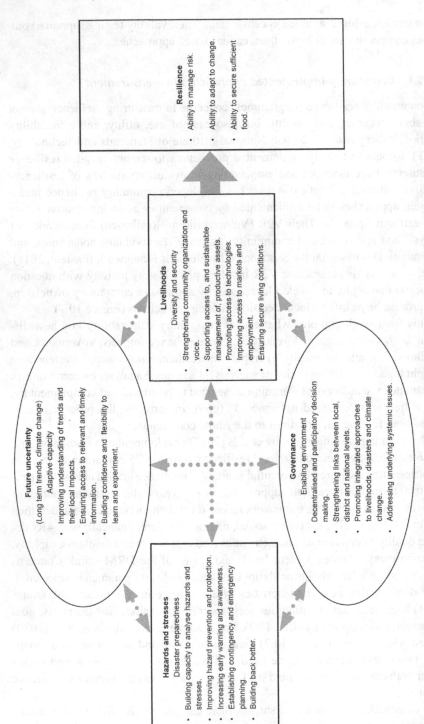

FIGURE 10.7 Practical Action's V2R detailed resilience framework (Pasteur, 2011, Figure 3, p12, permission: Practical Action)

BOX 10.3   WHAT SHOULD A COMMUNITY EXPECT FROM ITS OWN RESILIENCE MEASUREMENT TOOL? (CUTTER, 2014, P12; PERMISSION: SUSAN CUTTER)

- Openness and transparency
- Alignment with the community's goals and vision
- Measurements
  - are simple, well documented, and replicable and address multiple hazards
  - represent community's areal extent, physical (manmade and environmental) characteristics; and composition/diversity of community members
  - are adaptable and scalable to different community sizes, compositions, and changing circumstances

---

and experiences (Scherzer et al., 2019). For example, using qualitative methods at village scale, Amundsen (2012) found six dimensions of community resilience that were activated in processes and activities to respond to current challenges. These were community resources, community networks, institutions and services, people-place connections, active agents, and learning. Conceptualisation from this qualitative data collection can feed back into meaningful development of metrics (e.g., Sherrieb et al., 2010). Finally, Cutter (2014, p17) emphasised balance – important in aspiration for transformation.

Measurement tools cannot create a resilient community, but they can help show the path towards becoming safer, stronger, and more vibrant in the face of unanticipated events.

10.12   Conclusions

Building community resilience to future flood risk sits at a nexus of uncertainties due to dynamic socio-spatial interdependencies and inequalities, and complex system change. Both floods and at-risk communities are becoming more diverse. Paradigm shifts in FRM require radical collective reappraisals of the roles of communities in dealing with residual risk, and in their relationships with statutory agencies throughout the DRM spiral to help build place-based, socio-ecological resilience. This involves identifying potential domains of, and opportunities for, adaptation and transformation that individually, but more likely cumulatively, could significantly increase local resilience. Innovations (e.g., technical, design, governance) and practices (e.g., strong socio-technical interfacing, creative, embedding flood heritage, positively contributing to place-making, with co-benefits) underpin FRM strategies and interventions with potential to be transformative locally.

Aspiration for transformation in FRM requires a strong focus on social infrastructure, attending to critical interfaces in socio-ecological, socio-hydrological, and socio-technical systems, identifying key leverage points for change and conditions for positive systemic effects. It involves co-designing, co-implementing, and co-evaluating individual elements (interventions, approaches, changes in social and organisational norms) and their integration in specific places, working *with* local communities to build back stronger and differently for greater socio-ecological resilience. Positive community experience of such innovations can help stimulate proactive and share-able local preparedness narratives that shift discourse from 'risk' to 'resilience'.

This chapter has drawn together a framework for meaningful participatory processes with potential to transform, identifying key concerns including who is central to a deep engagement process. This recognises that the baton of leadership may change hands (community – other organisations) over time. Concern for developing community-focused FRM must recognise that 'not one size fits all' and attend to 'just' transitioning and transformation processes. Strong potential exists to learn from different risk and cultural settings, and other sectors (e.g., health and social care) that work routinely in practice with diverse and more marginalised communities.

Integrating different knowledges – lay and specialist science – in local decision-making is essential in transformation, taking co-working between communities, statutory FRM agencies and other FRM actors outside the heat of the acute event phase. This involves collaborative experiments in co-production in research and practice. Increasing the involvement of different academic and professional perspectives can also bring valuable new insights to research and practice. Longitudinal, community-based co-working involving communities and students from their local tertiary education institutions, mutually beneficial and empowering, has strong potential for longer-term impact on local resilience building. New collaborations with locally embedded, cultural sectors in developing place-based, flood heritage approaches to community resilience building can also have significant co-benefits.

While this book has by necessity artificially divided domains in community-focused FRM by theme (perception, knowledge, heritage, science communication, mitigation strategies, community action, flood education), holistic and systemic thinking is required to understand local communities and their capitals, and work collaboratively with them to co-create locally owned adaptations. This complex territory is best navigated through longer-term, meaningful participatory processes that build trusted relationships, and which place the needs and aspirations of diverse at-risk and flood-impacted communities at their core.

Notes

1 "The current geological age, viewed as the period during which human activity has been the dominant influence on climate and the environment" (Oxford Languages).
2 https://nationalfloodforum.org.uk/wp-content/uploads/2019/11/Charter_FINAL.pdf.
3 https://enoll.org/.
4 https://issuu.com/enoll/docs/423662117-short-history-of-living-labs-research-an.

5 https://www.smart-city.uliege.be/cms/c_9225506/en/green-roofs-and-walls-solutions-for-making-our-territories-future-proof.
6 https://www.epa.gov/environmentaljustice/learn-about-environmental-justice.
7 https://www.who.int/news-room/fact-sheets/detail/mental-health-strengthening-our-response.
8 https://granicus.com/blog/gov-communicators-need-to-connect-with-isolated-residents-say-researchers.
9 Pew Research Fact sheet: Mobile Available at: https://www.pewresearch.org/internet/fact-sheet/mobile/.
10 https://www.fema.gov/about/news-multimedia/mobile-app-text-messages; https://communities.geoplatform.gov/disasters/fema-disaster-reporter-hurricane-lane/.
11 https://www.missingmaps.org.
12 https://www.floodalleviation.uk/map/.
13 https://nationalfloodforum.org.uk/.
14 https://floodmemoryapp.wordpress.com/ and flappy.warwick.ac.uk.
15 https://www.watercitybristol.org/.
16 https://iwa-network.org/publications/the-iwa-principles-for-water-wise-cities/.
17 https://brigaid.eu/engage-as-an-end-user-2018/living-lab-the-city-of-antwerp-belgium/.
18 https://watereurope.cu/wp-content/uploads/2019/07/Atlas-of-the-EU-Water-Oriented-Living-Labs.pdf.
19 https://www.rockefellerfoundation.org/new-orleans-birth-urban-resilience/.
20 http://tacticalurbanismguide.com/about/.
21 https://floodlist.com/protection/depave-community-based-approach-storm-water-management.
22 Dutchdocklands.com.
23 West Sussex Flood Resilience Community Pathfinder.
24 https://www.arup.com/perspectives/city-resilience-index.
25 https://www.adri.bnhcrc.com.au/.
26 https://resiliencenexus.org/arc_d_toolkit/what-it-is/.
27 https://www.preventionweb.net/news/practical-action-new-climate-and-vulnerability-handbook.

References

Alba, R., Klepp, S. and Bruns, A. (2020) Environmental justice and the politics of climate change adaptation – the case of Venice. *Geographica Helvetica*, 75, 363–368. https://doi.org/10.5194/gh-75-363-2020

Aldrich, D., Meyer, M. and Page-Tan, C. (2018) Social capital and natural hazards governance. *Oxford Research Encyclopedia of Natural Hazard Science*. https://doi.org/10.1093/acrefore/9780199389407.013.254

Alfieri, L., Bisselink, B., Dottori, F., Naumann, G., de Roo, A., Salamon, P., Wyser, K. and Feyen, L. (2017) Global projections of river flood risk in a warmer world. *Earth's Future*, 5, 171–182. https://doi.org/10.1002/2016EF000485.

Alfieri, L., Dottori, F., Betts, R., Salamon, P. and Feyen, L. (2018) Multi-model projections of river flood risk in Europe under global warming. *Climate*, 6(1), 6. https://doi.org/10.3390/cli6010006

Amundsen, H. (2012) Illusions of resilience? An analysis of community responses to change in northern Norway. *Ecology and Society*, 17(4), 46. https://doi.org/10.5751/ES-05142-170446

Arnstein, S.R. (1969) A ladder of citizen participation. *Journal of the American Institute of Planners*, 35(4), 216–224. doi:10.1080/01944366908977225

Barsley, E. (2020) *Retrofitting for Flood Resilience: A Guide to Building and Community Design*. RIBA Publishing, Newcastle upon Tyne.

Baxter, H. (2019) Creating the conditions for community resilience: Aberdeen, Scotland – An example of the role of community planning groups. *International Journal of Disaster Risk Science*, 10, 244–260. https://doi.org/10.1007/s13753-019-0216-y

Beagley, J. et al. (2021) *Health Inequalities and Climate Change: Action for Global Health Position Paper*. Action for Global Health, United Kingdom. https://actionforglobal-health.org.uk/wp-content/uploads/2021/10/Health-Inequalities-and-Climate-Change-Action-for-Global-Health-Position-Paper.pdf

Bell, K. (2014) *Achieving Environmental Justice: A Cross-National Analysis*. Policy Press, University of Bristol, Bristol.

Berkes, F. (2007) Understanding uncertainty and reducing vulnerability: Lessons from resilience thinking. *Natural Hazards*, 41, 283–295.

Bullock, J.A., Haddow, G.D., Haddow, K.S. and Cappola, D.P. (2015) *Living with Climate Change: How Communities Are Surviving and Thriving in a Changing Climate*. CRC Press, Auerbach Publications, Boca Raton.

Clark-Ginsberg, A., McCaul, B., Bremaud, I., Cáceres, G., Mpanje, D., Patel, S. and Patel, R. (2020) Practitioner approaches to measuring community resilience: The analysis of the resilience of communities to disasters toolkit. *International Journal of Disaster Risk Reduction*, 50, 101714. https://doi.org/10.1016/j.ijdrr.2020.101714

Cleaver, F. and Whaley, L. (2018) Understanding process, power, and meaning in adaptive governance: A critical institutional reading. *Ecology and Society*, 23(2), 49. https://doi.org/10.5751/ES-10212-230249

Cobbing, P., Waller, E. and McEwen, L.J. (2023) The role of civil society in extreme events through a narrative reflection of pathways and long-term relationships. *Journal of Extreme Events*. https://doi.org/10.1142/S2345737622500038

Cox, R.S. and Hamlen, M. (2015) Community disaster resilience and the rural resilience index. *American Behavioral Scientist*, 59(2), 220–237. https://doi.org/10.1177/0002764214550297

Crenshaw, K. (1991) Mapping the margins: Intersectionality, identity politics, and violence against women of color. *Stanford Law Review*, 43(6), 1241–1299. https://doi.org/10.2307/1229039

Cretney, R. (2018) "An opportunity to hope and dream": Disaster politics and the emergence of possibility through community-led recovery. *Antipode*, 51(2), 497–516.

Cutter, S.L. (2014) The landscape of resilience measures. Presentation at *Resilient America 562 Roundtable Workshop on Measures of Community Resilience*, September 5. http://sites.nationalacademies.org/cs/groups/pgasite/documents/webpage/pga_152239.pdf

Cutter, S.L. (2020) Community resilience, natural hazards, and climate change: Is the present a prologue to the future? *Norsk Geografisk Tidsskrift – Norwegian Journal of Geography*, 74(3), 200–208. https://doi.org/10.1080/00291951.2019.1692066

Cutter, S.L., Ash, K.D. and Emrich, C.T. (2014) The geographies of community disaster resilience. *Global Environmental Change*, 29, 65–77. https://doi.org/10.1016/j.gloenvcha.2014.08.005

de Vente, J., Reed, M.S., Stringer, L.C., Valente, S. and Newig, J. (2016) How does the context and design of participatory decision making processes affect their outcomes? Evidence from sustainable land management in global drylands. *Ecology and Society*, 21(2), 24. https://doi.org/10.5751/ES-08053-210224

Della Bosca, H., Schlosberg, D. and Craven, L. (2020) Shock and place: Reorienting resilience thinking. *Local Environment*, 25(3), 228–242. https://doi.org/10.1080/13549839.2020.1723510

den Broeder, L., South, J., Rothoff, A. et al. (2022) Community engagement in deprived neighbourhoods during the COVID-19 crisis: Perspectives for more resilient and healthier communities. *Health Promotion International*, 37(2) daab098. https://doi.org/10.1093/heapro/daab098

Driessen, P.P.J., Hegger, D.L.T., Bakker, M.H.N., Van Rijswick, M.H.F.M.W. and Kundzewicz, Z.W. (2016) Toward more resilient flood risk governance. *Ecology and Society*, 21(4), 53. https://doi.org/10.5751/ES-08921-210453

Drubay, D. and Singhal, A. (2020) Dialogue as a framework for systemic change. *Museum Management and Curatorship*, 35(6), 663–670. https://doi.org/10.1080/09647775.2020.1837001

European Commission's Floods Directive (2007/60/EC) https://www.eea.europa.eu/policy-documents/directive-2007-60-ec-of

Fagg, G., Verrucci, E. and Rickles, P. (2015) Exploring new ways of digital engagement: A study on how mobile mapping and applications can contribute to disaster preparedness. Conference paper, GIS Research UK, Leeds, April 2015.

Fazey, I., Moug, P., Allen, S. et al. (2018) Transformation in a changing climate: A research agenda. *Climate and Development*, 10, 197–217. https://doi.org/10.1080/17565529.2017.1301864

Feola, G. (2015) Societal transformation in response to global environmental change: A review of emerging concepts. *Ambio*, 44, 376–390. https://doi.org/10.1007/s13280-014-0582-z

Fernandez, A., Black, J., Jones, M., Wilson, L., Salvador-Carulla, L., Astell-Burt, T. et al. (2015) Flooding and mental health: A systematic mapping review. *PLoS ONE* 10(4), e0119929. https://doi.org/10.1371/journal.pone.0119929

Ferri, M., Wehn, U., See, L., Monego, M. and Fritz, S. (2020) The value of citizen science for flood risk reduction: Cost–benefit analysis of a citizen observatory in the Brenta-Bacchiglione catchment. *Hydrology and Earth System Sciences*, 24, 5781–5798. https://doi.org/10.5194/hess-24-5781-2020

Flanagan, S.M. and Hancock, B. (2010) 'Reaching the hard to reach'–lessons learned from the VCS (voluntary and community sector). A qualitative study. *BMC Health Services Research*, 10, 92. https://doi.org/10.1186/1472-6963-10-92

Forrest, S., Dostál, J. and McEwen, L.J. (2023) The future of volunteering in local climate resilience: Critical reflections on key challenges and opportunities. *Journal of Extreme Events*. https://doi.org/10.1142/S2345737623410038

French, C.E., Waite, T.D., Armstrong, B. et al. (2019) Impact of repeat flooding on mental health and health-related quality of life: A cross-sectional analysis of the English National Study of Flooding and Health. *BMJ Open*, 9, e031562. https://doi.org/10.1136/bmjopen-2019-031562

Geaves, L.H. and Penning-Rowsell, E.C. (2015) 'Contractual' and 'cooperative' civic engagement: The emergence and roles of 'flood action groups' in England and Wales. *Ambio*, 44, 440–451.

Good Governance Institute (2020) COVID-19 22 June 2020 Engaging with the hard-to-reach. https://www.good-governance.org.uk/wp-content/uploads/2020/06/COVID-19-blog-22-06-20.pdf

Grace, D. and Sen, B.A. (2013) Community resilience and the role of the public library. *Library Trends*, 61(3), 513–541.

Granicus (ud) *Serving the Underserved – Accessing Hard to Reach Populations*. https:// granicus.com/blog/serving-the-underserved/

Grebenau, J. (2021) A place to keep the chaos: Using expressive arts to mitigate post-traumatic stress disorder in disaster recovery. Expressive Therapies Capstone Theses. 372. https://digitalcommons.lesley.edu/expressive_theses/372

Haworth, B.T., Bruce, E., Whittaker, J. and Read, R. (2018) The good, the bad, and the uncertain: Contributions of volunteered geographic information to community disaster resilience. *Frontiers in Earth Science*, 6, 183. https://doi.org/10.3389/feart.2018.00183

Heijmans, A. (2009) The social life of community-based disaster risk reduction: Origins, politics and framing. In *Disaster Studies Working Paper 20*, 35. Aon Benfield UCL Hazard Research Center, Boulder.

Herath, H.M.M. and Wijesekera, N.T.S. (2020) Transformation of flood risk management with evolutionary resilience. *E3S Web of Conferences*, 158, 06005. https://doi.org/ 10.1051/e3sconf/202015806005

Hochrainer-Stigler, S., Velev, S., Laurien, F. et al. (2021) Differences in the dynamics of community disaster resilience across the globe. *Scientific Reports*, 11, 17625. https://doi. org/10.1038/s41598-021-96763-0

Houston, D., Werritty, A., Ball, T. and Black, A. (2021) Environmental vulnerability and resilience: Social differentiation in short- and long-term flood impacts. *Transactions of the Institute of British Geographers*, 46, 102–119. https://doi-org.ezproxy.uwe.ac.uk/ 10.1111/tran.12408

Hurlbert, M. and Gupta, A. (2016) Adaptive governance, uncertainty, and risk: Policy framing and responses to climate change, drought, and flood. *Risk Analysis*, 36(2), https://doi. org/10.1111/risa.12510

Institute of Health Equity (2020) *Sustainable Health Equity: Achieving a net-Zero UK*. Health Expert Advisory Group Report for UK Committee on Climate Change. http:// www.instituteofhealthequity.org/resources-reports/uk-climate-change-committee-report

International Federation of Red Cross and Red Crescent Societies (2018) *World disasters report 2018. Leaving no one behind*. https://media.ifrc.org/ifrc/world-disaster-report-2018/.

International Water Association (2016) *The IWA principles for water wise cities*. https:// iwa-network.org/wp-content/uploads/2016/10/IWA_Brochure_Water_Wise_Communi- ties_SCREEN-1.pdf

IPCC (2019) *Climate Change and Land: An IPCC special report on climate change, desertification, land degradation, sustainable land management, food security, and greenhouse gas fluxes in terrestrial ecosystems*. https://www.ipcc.ch/srccl

Keating, A., Campbell, K., Szoenyi, M., McQuistan, C., Nash, D. and Burer, M. (2017) Development and testing of a community flood resilience measurement tool. *Natural Hazards and Earth Systems Sciences*, 17, 77–101.

Kruger, D. (2020) *Levelling up our communities: proposals for a new social covenant*. Report for UK Government. https://www.dannykruger.org.uk/new-social-covenant

Kyne, D. and Aldrich, D.P. (2020) Capturing bonding, bridging, and linking social capital through publicly available data. *Risk, Hazards & Crisis in Public Policy*, 11, 61–86. https://doi.org/10.1002/rhc3.12183

Lassa, J.A. (2011) Achieving resilience through communicating research, policy and practice in DRR and CCA in Indonesia. *Community Adaptation and Risk Reduction Governance Series*. https://www.preventionweb.net/files/23532_23532carrindonesiafirsteditionnov20.pdf

Le Dé, L., Wairama, K., Sath, M. and Petera, A. (2021) Measuring resilience: By whom and for whom? A case study of people-centred resilience indicators in New Zealand. *Disaster Prevention and Management*, 30(4/5), 538–552. https://doi.org/10.1108/DPM-04-2021-0128

Leichenko, R. and O'Brien, K. (2019) *Climate and Society: Transforming the Future*. Polity Press, Cambridge.

Lewis, J. and Kelman, I. (2010) Places, people and perpetuity: Community capacities in ecologies of catastrophe. *ACME: An International E-Journal for Critical Geographies*, 9(2), 191–220.

Liguori, A., Le Rossignol, K., Kraus, S., McEwen, L. and Wilson, M. (2023) Exploring the uses of arts-led community spaces to build resilience: Applied storytelling for successful co-creative work. *Journal of Extreme Events*. https://doi.org//10.1142/S2345737622500075

Liguori, A., McEwen, L.J., Blake, J. and Wilson, M. (2021) Towards 'creative participatory science': Exploring future scenarios through specialist drought science and community storytelling. *Frontiers in Environmental Science*. https://doi.org//10.3389/fenvs.2020.589856

Linnér, B.-O. and Wibeck, V. (2019) *Sustainability Transformations: Drivers and Agents Across Societies*. Cambridge University Press, Cambridge.

Linnér, B.-O. and Wibeck, V. (2020) Conceptualising variations in societal transformations towards sustainability. *Environmental Science and Policy*, 106, 221–227. https://doi.org/10.1016/j.envsci.2020.01.007

Li, S., Wahl, T., Talke, S.A., Jay, D.A., Orton, P.M., Liang, X., Wang, G. and Liu, L. (2021) Evolving tides aggravate nuisance flooding along the U.S. coastline. *Science Advances*, 7(10), eabe2412. https://doi.org/10.1126/sciadv.abe2412

Longman, J.M., Bennett-Levy, J., Matthews, V. et al. (2019) Rationale and methods for a cross-sectional study of mental health and wellbeing following river flooding in rural Australia, using a community-academic partnership approach. *BMC Public Health*, 19, 1255. https://doi.org/10.1186/s12889-019-7501-y

Lowe, D., Ebi, K.L. and Forsberg, B. (2013) Factors increasing vulnerability to health effects before, during and after floods. *International Journal Environmental Research and Public Health*, 10(12), 7015–67. https://doi.org/10.3390/ijerph10127015. PMID: 24336027; PMCID: PMC3881153.

Lozano Nasi, V.L., Jans, L. and Steg, L. (2023) Can we do more than "bounce back"? Transilience in the face of climate change risks. *Journal of Environmental Psychology*, 86, 101947. https://doi.org/10.1016/j.jenvp.2022.101947

Lupp, G., Zingraff-Hamed, A., Huang, J.J., Oen, A. and Pauleit, S. (2021) Living labs – a concept for co-designing nature-based solutions. *Sustainability*, 13, 188. https://doi.org/10.3390/su13010188

Luque-Ayala, A. and Marvin, S. (2015) Developing a critical understanding of smart urbanism? *Urban Studies*, 52. https://doi.org/10.1177/0042098015577319

Mao, G., Fernandes-Jesus, M., Ntontis, E. and Drury, J. (2020) What have we learned so far about COVID-19 volunteering in the UK? A rapid review of the literature. *BMC Public Health*. https://doi.org/10.1186/s12889-021-11390-8

Marchezini, V., Horita, F.E.A., Matsuo, P.M., Trajber, R., Trejo-Rangel, M.A. and Olivato, D. (2018) A review of studies on participatory early warning systems (P-EWS): Pathways to support citizen science initiatives. *Frontiers in Earth Science*, 6. https://doi.org/10.3389/feart.2018.00184

McEwen, L.J., Cornish, F., Leichenko, R., Holmes, A., Guida, K., Burchell, K., Sharpe, J., Everett, G. and Scott, M. (2023) Rebuffing the 'hard to reach' narrative: How to engage diverse groups in participation for resilience. *Journal of Extreme Events*, 9(2).

McEwen, L.J., Garde-Hansen, J., Holmes, A., Jones, O. and Krause, F. (2016) Sustainable flood memories, lay knowledges and the development of community resilience to future flood risk. *Transactions of the Institute of British Geographers*, 42(1), 14–28.

McEwen, L.J., Holmes, A., Quinn, N. and Cobbing, P. (2018) 'Learning for resilience': Developing community capital through action groups in lower socio-economic flood risk settings. *International Journal of Disaster Risk Reduction*, 27, 329–342.

McEwen, L.J., Leichenko, R., Garde-Hansen, J. and Ball, T. (2023) CASCADE-NET: Increasing Civil Society's capacity to deal with changing extreme weather risk - negotiating dichotomies in theory and practice. *Journal of Extreme Events*, 9(2), DOI: 10.1142/S2345737623300016.

Mees, H., Alexander, M., Gralepois, M., Matczak, P. and Mees, H. (2018) Typologies of citizen co-production in flood risk governance. *Environmental Science & Policy*, 89, 330–339. https://doi.org/10.1016/j.envsci.2018.08.011

Moftakhari, H.R., AghaKouchak, A., Sanders, B.F., Allaire, M. and Matthew, R.A. (2018) What is nuisance flooding? Defining and monitoring an emerging challenge. *Water Resources Research*, 54, 4218–4227.

Muiderman, K., Gupta, A., Vervoort, J. and Biermann, F. (2020) Four approaches to anticipatory climate governance: Different conceptions of the future and implications for the present. *WIREs Climate Change*, 11, e673. https://doi.org/10.1002/wcc.673

Nemeth, D.G., Kuriansky, J. and Onishi, Y. (2021) The 3 es of psychosocial recovery after disaster. In K.E. Cherry and A. Gibson (eds) *The Intersection of Trauma and Disaster Behavioral Health*. Springer, Cham. https://doi.org/10.1007/978-3-030-51525-6_18

Nemeth, D.G. and Olivier, T.W. (2017) *Innovative Approaches to Individual and Community Resilience: From Theory to Practice*. Academic Press, Cambridge.

Norn, A., Drubay, D., Debono, S. et al. (2023) Dialogues on museum resilience. In *Museum Activism*. The Woman's Museum in Aarhus and University of Aarhus.

Ntontis, E., Drury, J., Amlôt, R., Rubin, G.J., Williams, R. and Saavedra, P. (2021) Collective resilience in the disaster recovery period: Emergent social identity and observed social support are associated with collective efficacy, well-being, and the provision of social support. *British Journal of Social Psychology*, 60, 1075–1095. https://doi-org.ezproxy.uwe.ac.uk/10.1111/bjso.12434

O'Donnell, E. and Thorne, C. (2020) Chapter 1: Urban flood risk management: The blue–green advantage. In *Blue–Green Cities*. ICE, Thomas Telford Ltd, London, pp1–13. https://doi.org/10.1680/bgc.64195.001

OEDC (2019) Applying the OECD principles on water governance to floods a checklist for action. *OECD Studies on Water.* https://doi.org/10.1787/d5098392-en

Organisation for Economic Co-operation and Development (2015) *OECD Principles on Water Governance*. https://www.oecd.org/cfe/regionaldevelopment/OECD-Principles-on-Water-Governance-en.pdf

Parsons, M., Glavac, S., Hastings, P., Marshall, G., McGregor, J., McNeill, J., Morley, P., Reeve, I. and Stayner, R. (2016) Top-down assessment of disaster resilience: A conceptual framework using coping and adaptive capacities. *International Journal of Disaster Risk Reduction*, 19, 1–11.

Pasteur, K. (2011) *From Vulnerability to Resilience: A Framework for Analysis and Action to Build Community Resilience*. Practical Action Publishing, Rugby UK.

Patch, B. (2020) The new Australian disaster resilience index: A tool for building safer, adaptable communities. *Australian Journal of Emergency Management*, 35(3), 15–17.

Patel, S.S., Rogers, M.B., Amlôt, R. and Rubin, G.J. (2017) What do we mean by 'Community Resilience'? A systematic literature review of how it is defined in the literature. *PLOS Currents Disaster*, 1. https://doi.org/10.1371/currents.dis. db775aff25efc5ac4f0660ad9c9f7db2

Patterson, J., Schulz, K., Vervoort, J., van del Hel, S., Widerberg, O., Adler, C., Hulfbert, M., Anderton, K., Sethi, M. and Barau, A. (2017) Exploring the governance and politics of transformations towards sustainability. *Environmental Innovation and Societal Transitions*, 24, 1–16. https://doi.org/10.1016/j.eist.2016.09.001

Percy-Smith, P. and Burns, D. (2013) Exploring the role of children and young people as agents of change in sustainable community development. *Local Environment*, 18(3), 323–339. https://doi.org/10.1080/13549839.2012.729565

Pew Research Center (2019) Mobile Technology and Home Broadband 2019. June 2019. https://www.pewresearch.org/internet/2019/06/13/mobile-technology-and-home-broadband-2019/

Piątek, Ł and Wojnowska-Heciak, M. (2020) Multicase study comparison of different types of flood-resilient buildings (elevated, amphibious, and floating) at the Vistula River in Warsaw, Poland. *Sustainability*, 12(22), 9725. https://doi.org/10.3390/su12229725

Quinn, T., Adger, W.N., Butler, C. and Walker-Springett, K. (2020) Community resilience and well-being: An exploration of relationality and belonging after disasters. *Annals of the American Association of Geographers*. https://doi.org/10.1080/24694452.2020.1782167

Reddick, C.G., Enriquez, R., Harris, R.J. and Sharma, B. (2020) Determinants of broadband access and affordability: An analysis of a community survey on the digital divide. *Cities (London, England)*, 106, 102904. https://doi.org/10.1016/j.cities.2020.102904

Reuter, C. and Kaufhold, M. (2018) Fifteen years of social media in emergencies: A retrospective review and future directions for crisis informatics. *Journal of Contingencies and Crisis Management*, 26, 41–57. https://doi.org/10.1111/1468-5973.12196

Rufat, S., Fekete, A., Armaş, I., et al. (2020) Swimming alone? Why linking flood risk perception and behavior requires more than "it's the individual, stupid". *WIREs Water*, 7, e1462. https://doi.org/10.1002/wat2.1462

Sadiq, A.-A., Tyler, J. and Noonan, D.S. (2019) A review of community flood risk management studies in the United States. *International Journal of Disaster Risk Reduction*, 41, 101327. https://doi.org/10.1016/j.ijdrr.2019.101327

Šakić Trogrlić, R., Wright, G.B., Adeloye, A.J., Duncan, M.J. and Mwale, F. (2018) Taking stock of community-based flood risk management in Malawi: Different stakeholders, different perspectives. *Environmental Hazards*, 17(2), 107–127. https://doi.org/10.1080/17477891.2017.1381582

Sandifer, P.A. and Scott, G.I. (2021) Coastlines, coastal cities, and climate change: A perspective on urgent research needs in the United States. *Frontiers in Marine Science*, 8, 631986. https://doi.org/10.3389/fmars.2021.631986

Saulnier, D., Brolin Ribacke, K. and Von Schreeb, J. (2017) No calm after the storm: A systematic review of human health following flood and storm disasters. *Prehospital and Disaster Medicine*, 32(5), 568–579. https://doi.org/10.1017/S1049023X17006574

Scherzer, S., Lujala, P. and Ketil Rød, J. (2019) A community resilience index for Norway: An adaptation of the baseline resilience indicators for communities (BRIC). *International Journal of Disaster Risk Reduction*, 36, 101107.

Schipper, E.L.F. and Langstone, L. (2015) A Comparative Overview of Resilience Measurement Frameworks: Analysing Indicators and Approaches, *ODI Working Paper 422*. Overseas Development Institute, London. http://www.odi.org/sites/odi.org.uk/files/odiassets/publications-opinion-files/9754.pdf

Sharifi, A. (2016) A critical review of selected tools for assessing community resilience. *Ecological Indicators*, 69, 629–647.

Sharp, L. (2019) Flood defences simply aren't good enough – here's what needs to be done. *The Conversation* 11. https://theconversation.com/flood-defences-simply-arent-good-enough-heres-what-needs-to-be-done-126781

Sharpe, J. (2021) Learning to trust: Relational spaces and transformative learning for disaster risk reduction across citizen led and professional contexts. *International Journal of Disaster Risk Reduction*, 61, 102354. https://doi.org/10.1016/j.ijdrr.2021.102354

Shaw, R. (2016) Community-Based Disaster Risk Reduction. *Oxford Research Encyclopedias, Natural Hazard Science*. https://doi.org/10.1093/acrefore/9780199389407.013.47

Sherrieb, K., Norris, F.H. and Galea, S. (2010) Measuring capacities for community resilience. *Social Indicators Research*, 99, 227–247. https://doi.org/10.1007/s11205-010-9576-9

Silva, D. (2018) Mobile phones help transform disaster relief. Accessed https://phys.org/news/2018-03-mobile-disaster-relief.html. Retrieved 31 July 2021

South, J. (2015) *A guide to community-centred approaches for health and wellbeing: Full report*. Public Health England. https://www.gov.uk/government/publications/health-and-wellbeing-a-guide-to-community-centred-approaches

Stiglitz, J. (2020) Conquering the Great Divide: The pandemic has laid bare deep divisions, but it's not too late to change course. September 2020. https://www.imf.org/Publications/fandd/issues/2020/09/COVID19-and-global-inequality-joseph-stiglitz.

Swindle, K., McBride, N.M., Staley, A., Phillips, C.A., Rutledge, J.M., Martin, J.R. and Whiteside-Mansell, L. (2020) Pester power: Examining children's influence as an active intervention ingredient. *Journal of Nutrition Education and Behavior*, 52, 801–807. https://doi.org/10.1016/j.jneb.2020.06.002

The Rockefeller Foundation/ARUP (2015) City resilience framework. https://www.rockefellerfoundation.org/wp-content/uploads/City-Resilience-Framework-2015.pdf.

Thaler, T., Attems, M.-S., Bonnefond, M. et al. (2019) Drivers and barriers of adaptation initiatives – How societal transformation affects natural hazard management and risk mitigation in Europe. *Science of the Total Environment*, 650(1), 1073–1082. https://doi.org/10.1016/j.scitotenv.2018.08.306

Tkachenko, N., Jarvis, S. and Procter, R. (2017) Predicting floods with Flickr tags. *PLoS One*, 12(2), e0172870. https://doi.org/10.1371/journal.pone.0172870

Trell, E.-M., Restemeyer, B., Bakema, M.M. and van Hoven, B. (2017) *Governing for Resilience in Vulnerable Places*. Routledge, Abingdon.

Triyanti, A., Hegger, D.L.T. and Driessen, P.P.J. (2020) Water and climate governance in deltas: On the relevance of anticipatory, interactive, and transformative modes of governance. *Water*, 12, 3391. https://doi.org/10.3390/w12123391

Twigger-Ross, C., Sadauskis, R., Orr, P., Jones, R., McCarthy, S., Parker, D., Priest, S. and Simms, J. (2020a) Flood and coastal erosion risk management research and development framework: working with communities: Report. *R&D report for the Environment Agency*. Project: FRS19209/R1.

Twigger-Ross, C., Orr, P., Kolaric, S., Parker, D., Flikweert, J. and Priest, S. (2020b) *Evidence Review of the Concept of Flood Resilience*. Final Report FD2716 May 2020, Department for Environment, Food and Rural Affairs.

UNDRR (2015) *Sendai Framework for Disaster Risk Reduction (2015–2030)* https://www.undrr.org/publication/sendai-framework-disaster-risk-reduction-2015–2030

UNDRR (2022) *Boosting systemic risk governance: Perspectives and insights from understanding natural systems approaches for dealing with disaster and climate risks*. United

Nations Office for Disaster Risk Reduction (UNDRR). https://www.undrr.org/publication/boosting-systemic-risk-governance-perspectives-and-insights-understanding-national

UNESCO (2020) *Education for Sustainable Development: A roadmap*. UNESCO, France. https://unesdoc.unesco.org/ark:/48223/pf0000374802.locale=en

Walker-Springett, K., Butler, C. and Adger, W.N. (2017) Wellbeing in the aftermath of floods. *Health & Place*, 43, 66–74. https://doi.org/10.1016/j.healthplace.2016.11.005

Wehn, U. and Evers, J. (2015) The social innovation potential of ICT-enabled citizen observatories to increase eParticipation in local flood risk management. *Technology in Society*, 42, 187–198. https://doi.org/10.1016/j.techsoc.2015.05.002

Wibeck, V., Linnér, B.-O., Alves, M., Asplund, T., Bohman, A., Boykoff, M.T., Feetham, P.M. et al. (2019) Stories of transformation: A cross-country focus group study on sustainable development and societal change. *Sustainability*, 11(8), 2427. https://doi.org/10.3390/su11082427

Williams, S.-J. and McEwen, L.J. (2021) Learning for resilience as the climate changes: Discussing flooding, adaptation and agency with children. *Environmental Education Research*. https://doi.org/10.1080/13504622.2021.1927992

Winograd, K. (2016) *Education in Times of Environmental Crises: Teaching Children to Be Agents of Change*. Routledge, London.

Wolff, E. (2021) The promise of a "people-centred" approach to floods: Types of participation in the global literature of citizen science and community-based flood risk reduction in the context of the Sendai Framework. *Progress in Disaster Science*, 10, 100171. https://doi.org/10.1016/j.pdisas.2021.100171

Wörlen, M., Wanschura, B., Dreiseitl, H., Noiva, K., Wescoat, J. and Moldaschl, M. (eds.) (2016) *Enhancing Blue-Green Infrastructure and Social Performance in High Density Urban Environments: Summary Document*. Ramboll, Liveable Cities Lab, Überlingen, Germany.

Zhong, S., Yang, L., Toloo, S., Wang, Z., Tong, S., Sun, X., Crompton, D., FitzGerald, G. and Huang, C. (2018) The long-term physical and psychological health impacts of flooding: A systematic mapping. *The Science of the Total Environment*, 626, 165–194.

INDEX

Note: Page references in *italics* denote figures and in **bold** tables.

Ackoff, R.L. 100
actionable knowledge 99, 122–123, 271, 279
action research 296
active citizenship 10, **11**, 239, 250
active forgetting 79, *112*, 139, 155, 314
active problem-solving 289
active remembering 79, *112*, 143–144, 155, 314
adaptation 50, 58–59, 78–79, 105, 198, 209, 222, 270, 276, 296
adaptive governance 321–323
affect heuristic **82**
agency: civil 237–239; in local flood risk management (FRM) 231–260
agricultural land drainage **53**
Alba, R. 50
Aldrich, D.P. 235, 327
Alexander, K.S. 206
Alexander, M. 122
Alfieri, L. 4
Allan, R. 132
andragogy 273, *273*
Anthropocene 303
anticipatory governance 260, 322
applied storytelling 224, 295
archival knowledge **106**
archiving: as community practice 152–156, **153**, **155**; cultural flood archives 136–149; physical flood archives 134–136

Argyris, C. 122–123
Armstrong, A.K. 186
Arnold, M. 118
Arnstein's ladder of participation 240, *240*
arts: in learning for flood resilience 293–295; links in flood science communication 290; practice in communication for flood risk awareness **184**
Atmadja, S.S. 73, 79
attitudes in learning for flood resilience 285
Australian Institute for Disaster Resilience 215, 241
availability heuristic **82**
Aven, T. 175

Babcicky, P. 87
Baker, V. 70
Baker, V.R. 175
Bandura, A. 273
Bang, H.N. 69
Banks, S. 120
Baubion, C. 112
behaviour change: and collective perception 87–89; theories of 88, *89*; tools and policy interventions for 90
Bell, K. 49

Béné, C. 19
Benito, G. 133
biases 80–81; cognitive 80; optimism
        80
big data 108, 257, 317–318
Birkholz, S. 73, 87
Black, Asian and Minority Ethnic (BAME)
        communities 21
Black swan event 175–176, 186
Blaikie, P. 20
'Blue-Green' Cities 212–214
Blue-Green Infrastructure (BGI) 212
blue humanities 132
Blues music 149
Bohensky, E.L. 82
Bonaiuto, M. 78
Bosschaart, A. 287, 295
Bridges of Katrina: Three survivors, one
        interview (Eugene) 140
British Hydrological Society 134
'Broken Levee Blues' (Johnson) 149
Brown, Sterling 146
Buchecker, M. 90
Building Back Better (BBB) 219–221
Building Research Institute (BRI) 281
built capital **236**
Burke, K.L. 167
Burningham, K. 49
Burton, N.C. 69

Canadian Institutes of Health Research
        (CIHR) 114
capacities 234–239
capital: cultural **236**; financial **236**;
        human 235, **236**; natural **236**;
        political **236**; social 235, **235**,
        236
Capstick, S. 75
Carey, M. 137
Carlton, S.J. 72, 186
Carmichael, C. 152
catchment-based flood models 293
catchment-based solutions 202–214;
        'Blue-Green' Cities 212–214;
        communities and buffers/managed
        retreat 205–208; floodplain-based
        approaches 202–205; land and
        water management 208–214;
        sustainable drainage systems
        (SuDS) 211; and their communities
        202–214
'catchment systems thinking' (CaST) 209
Cawood, M. 215

Chadderton, C. 271, 287
Chakraborty, J. 49
Challies, E. 55
child-centred learning 280–282
children and flood risk 312–313
Child Resilience Alliance 241
Christensen, C.M. 9
Chronology of British Hydrological Events
        134
citizen agency 6–7
citizen observatories 317–318
citizens: changing 308–309; participation
        in knowledge sharing 257–259;
        volunteering 249–251; and water
        cities 318–320
citizen science 317–318; in environmental
        monitoring 254–257, *255*, **256**;
        volunteering as 254–257, *255*,
        256
citizenship 10, 237–239; active 10, **11**;
        ecological 239; education 276
civic agency xiv, 10, **11**, 133, 192,
        234–239, 249–251, 276
civil society 10, **11**
Clarke, D. 78
Clark-Ginsberg, A. 329
climate: changing 307–308; future flood
        risk 307–308; social shocks
        307–308
climate change 4, 16; communicating flood
        risk in the context of 186; and
        compound flooding 51; perception
        75–76
'Climategate' 168
coastal flooding 43
Coates, T. 12
cognitive bias 80, **82**
cognitive dissonance (Cd) 80–81
cognitive heuristics 80–81
Cohen, O. 244
Colclough, G. 12
collective atrophy **82**
collective memory 112, 316
collective perception: and behaviour
        change 87–89; cognitive heuristics
        and biases 80–81; of flood
        risks 79–87; influence of media
        narratives **82**, 82–84; organisational
        perceptions of risk 86–87; social
        and cultural factors 85–86; trust
        84–85
Collins, H.M. 111
Cologna, V. 81, 83

communication: with communities about
flood risk 164–168; and dispelling
myths 171–172; flood science for
172–176; technologies 316–317;
uncertainty 179–180
community activism: participation in 252–
254; role of flood groups 252–254
Community and Regional Resilience
Institute (CARRI) 17
Community-based Disaster Risk Reduction
(CBDRR) 123
community-based DRM 58, 60, 305–306;
characteristics of 246–248;
principles 247; principles of 246–
248; traditions 305–306, *306*
community-based learning 280
community capital 234–239, **236**, *237*
Community Capitals Framework 236, 259
community/communities 11–13, 249–259;
awareness and action 198–225;
and catchment-based solutions
202–214; complex 12; and complex
hazards 51–52; and compound
flooding 50–51; conscious 12–13;
as early flood warning systems
216–217, **218**; engaging with
buffers/managed retreat 205–208;
and flood insurance 219–220;
in flood management strategies
53–54, 53–61; flood-related
learning within 278–282; and flood
types 48–50; and flood warnings
214–216; inclusivity in 244–245;
issues in flood risk communication
with 180–182; and paradigms
for managing floods 199–200;
and polluted floodwaters 45–47;
resilience 52–53
community flood groups (CFGs) 249–254,
312, 317
community flood risk communication
186–189
community flood science **285**
community-focused FRM 5, 57–60; natural
flood management 60; place-
making 60–61; recovery-based
approach 61; spatial planning
60–61
community-implemented approaches to
measurement 329–331
community involvement: in disaster
management 232–233; international
context to 232–233

community learning for flood resilience
269–296
community-led DRM (CLDRM) 59–60;
characteristics of 248–249;
principles of 248–249
community-led flood risk management 225
community participation: auditing **245**,
245–246; in co-development of
learning strategies 295–296; in
disaster risk management 246–249;
equitable citizen participation
244–245; evaluating **245**, 245–246;
inclusivity in communities 244–
245; local flood risk management
(FRM) 240–246; models and
language of 240–242; processes
242–243; right tools, selecting
242–243
community resilience **17**, 17–19, **18**,
303–332; archiving as community
practice 152–156, **153**, **155**; beyond
'business as usual' 304–306;
changing risk 307–309; changing
society 307–309; communicating
flood science for 163–192;
community roles in DRM 311–
313; evidencing transformation
326–331; flood narratives 151–152;
and future floods 309–311; futures
for FRM 307–309; health and
well-being for 314–316; indices
with different approaches/scales/
users **328**; measuring change
326–331; opportunities for
systemic approaches 305–306;
participation 323–326; place-
based flood heritage approach
150–156; societal transformation
304–305; systemic flood impacts
314–316; technological innovation
316–318; transformative pathways
in governance 320–323, *321*; urban
designs 318–320
community risk communication 182–185
community science 290
complex communities 12
complex hazards: and community/
communities 51–52; floods
and droughts 51–52; floods
and pandemics 52; sequential
flood-heatwave events 52
compound flooding 50–51
conscious community 12–13

Construction Industry Research and Information Association (CIRIA) 187, 211
co-production of knowledge 119–120
Corner, A. 179, 180
COVID-19 pandemic 52, 168, 303, 308, 313
creative archiving practices 146–149
Creative Carbon Scotland 183
Cretney, R. 16
critical resilience 18–19
cross-cultural learning for resilience 313
Crowden, J. 139
cultural capital **236**
cultural flood archives 136–149; artistic and creative archiving practices 146–149, *147, 148, 150*; community memory 144–145; memorialisation 144–145; newspaper accounts of flood impacts 141–142; physical flood marking 138–139; visual media and flood artefacts 142–143
cultural meaning 78–79
cultural sector, role of 156
cultural theory of risk **73**
Cutter, S. L. 17, 48

dam burst floods 44
data: big 108, 257, 317–318; crowd sourced 134, 310; soft 257–258; thick 318
Davidson, J.L. 16
Davoudi, S. 16
'daylighting' rivers 204–205
de Groot, M. 203
de Groot, W.T. 203
Demeritt, D. 163, 165
democratising tendency 137
Derickson, K. 19
Derrida, J. 138
DeVerteuil, G. 18
De Vries, D.H. 21
dichotomies in flood risk management 308
DIKW knowledge pyramid 100, *100*
Dinnie, E. 13
disability 245, 312, 314
Disaster Education (report) 281
disaster management: community involvement in 232–233; international context 232–233
Disaster Resilience Framework (DFID) 14, *15*
disaster risk management (DRM) 5–6, *6*; community-based 246–248;

community-led 248–249; community participation in 246–249; transforming community roles in 311–313
disaster risk management (DRM) spiral 7, 18, 70, 84, 117, 123, 242, 259, 272, 275, 284, 292, 303, 313, 316, 329, 331
disaster risk reduction (DRR) 1, 104
dispositions in learning for flood resilience 285
diverse citizens 316–317
dominant flood narratives 287
dredging 44, 117
droughts 51–52
Dubois, D. 296
Dufty, N. 284, 290

Early Warning Systems (EWS) 217, **218**
ecological citizenship 239
ecological knowledge 105, 109
ecosystem services 203, *204*, 212
education: citizenship 276; flood (*see* flood education); sustainability 274–275
effective citizen 10
emotion 76–77
engagement with communities about flood risk 164–168
entrepreneurial knowledge 110
environmental humanities 131–132
environmental justice **20**, 21–22, 311–312; dimensions applied to FRM **48**; distributive **48**; procedural **48**; as recognition **48**; sense of justice/ legitimacy **48**
environmental monitoring 254–257, *255*, **256**
environmental risk 49
equitable citizen participation 244–245
ethics 140, 156; flood education 287
Eugene, N. 140
European Commission 209; Floods Directive 56
European Floods Database 134
evaluating participation tools 245, **245**
Evans, D. 149
Evans, R. 111
event-based science 289
experiential learning 271
exposure 19
Extreme Citizen Science (ExCiteS) 255

Falkenmark, M. 103
"Fatal Flood" 142
Filipe, A. 120
financial capital **236**
Fink, L.D. 276
Fink's *Taxonomy of Significant Learning*
    276
Fischer, A. 13
Flage, R. 175
Flanagan, S.M. 312
Flood and Water Management Act
    (UK) 5
flood archives: cultural 136–149; physical
    134–136; visual media 142–143
flood defence 8, 16, 22, 53–56, **54**, 78, 81,
    84, 200, 206, 219, 224
flood disadvantage 22
flood-drought continuum 173, **173**, 270
flood early warning 179
flood education: citizenship education
    276; defining 270–273; dominant
    flood narratives 287; ethics 287;
    facilitators of 272; focus of
    learning 271–272; formal settings
    for 272; groups for learning
    271; interdisciplinary learning
    278; language 287; learning for
    resilience 274–275; learning
    theories 272–273; levels of
    learning 276; need for 270–271;
    opportunities and barriers to 286–
    288; other framings for learning
    for resilience 276–278; pedagogies
    for *273*, 273–278; place-based
    288–290; process and timing 287;
    scale 288; significant learning
    276, *277*; sustainability education
    274–275; timings for learning 272;
    transdisciplinary learning (TL) 278;
    transformative learning 277–278
flood examples: 1953 storm surge,
    Netherlands/UK 4, 143–144,
    156; Hurricane Harvey 49, 145;
    Hurricane Ida 175; Hurricane
    Katrina 4, 16, 21, 49, 72, 75, 117,
    139, 143, 149; Hurricane Sandy
    296; Lynmouth floods 45, 46, **144**;
    Mississippi floods 4, 142, 146, 156,
    206; UK Summer floods 77
flood groups 252–254
flood-heatwave events 52
flood heritage 131–157, 288–290; within
    built environments 135–136;
    contexts 131–132; cultural flood

archives 136–149; framing 132–
    134; overview 131; physical flood
    archives 134–136; place-based
    150–156
flood histories 37, 134
flood-inspired theatre **294**
flood insurance 199, 214; access to 219;
    affordable 219; and communities
    219–220; in the USA 219
"Flood in the Highlands" 146, *147*
flood knowledge 99–124, **106–107**, 284;
    brokers and translators 113–116,
    116; controversies 117; local/lay
    knowledge of flooding 105–113;
    nature of 100–103; in practice
    122–123; scientific and technical
    103–104
flood learning: arts in learning for flood
    resilience 293–295; catchment-
    based flood models 293; game-
    based learning 291–293; learning
    approaches for 290–295; role-play
    as learning strategy 290
flood management strategies: catchment-
    based solutions and communities
    202–214; and community
    awareness/action 198–225;
    different paradigms for 199–200;
    infrastructural measures and
    communities 200–202; residual
    flood risk 214–224
flood maps 176, *177*, 178
flood marking: defined 138; physical
    138–139
flood materialisation *109*
The Flood Memorial (Grand Forks, North
    Dakota) 144
flood memorialisation 144–145
flood memory 12, 37, 53, 110, 112, 136,
    138, 142–143, 152–156, 175, 254,
    271, **285**
flood myths 165, 168, 171, 286,
    289
floodplain-based approaches 202–205;
    'daylighting' rivers 204–205;
    making space for water
    203–204
floodplains 4, 43, 48, 68, 103, 134–135,
    172–173, 202
flood poetry 142, 146
flood poor 37, 175
flood-related learning: child-centred
    learning 280–282; within
    communities 278–282;

community-based learning 280;
informal settings 279; social
learning for resilience 279
flood resilience: arts in learning for 293–
295; attitudes in learning for 285;
community learning for 269–296;
dispositions in learning for 285;
examples of skills acquisition in
learning for **286**; integration in
learning for 285–286; knowledge
in learning for 282–286; skills in
learning for 285, **286**
flood risk: collective perception of
79–87; communicating 173–174;
communication with communities
about 164–168; community
resilience to 303–332; defined 1;
engagement with communities
about 164–168; fluvial 43; mapping
and contestation 176–179; and
MEDCs 1; threshold concepts
173–174; visualisations of
176–179
flood risk communication **181–182**; with
communities 180–182; community
186–189; developing framework
for 186–189; issues in 180–182;
purposes of **169**; strategies **189**;
threshold concepts in **173**
flood risk management (FRM) 1–3,
5–6; changing citizens 308–309;
changing climate 307–308;
changing risk; changing society
307–309; changing society
308–309; children/young people
312–313; cognitive styles for
organising knowledge relevant
to **26**; concepts in local 10–22;
environmental justice dimensions
applied to **48**; future flood risk
307–308; futures for 307–309;
governance structures and
community roles in 233–234;
hydrocitizenship 308–309;
implications for 113–116, **116**,
119–120; implications for (new)
actors in 8; interdisciplinary
working in 22–25, **23**; knowledge
co-production in 121–122; role of
community in **53–54**, 53–61; social
shocks 307–308
flood science 37–47; myths 171
flood science communication 172–176;
arts links in 290; background

contexts 168–172; community
flood risk communication 186–189;
and dispelling myths 171–172;
flood risk and climate change
186; flood risk communication
with communities 180–182;
imperatives for effective **190–191**;
other approaches to 182–185;
troublesome knowledge/threshold
concepts 170–171; uncertainty in
practice 176–180
floods/flooding: application to researching
25–27; archiving 136–149; artefacts
142–143; causes of 37–44; in
climate change science 179–180;
droughts 51–52; fallacies based
on scientific understanding 172;
governance 7–8; high tide **39**;
impact experience 3–4; local/lay
knowledge of 105–113; losses **3**;
management 5–7; patterns 4; as
risk 69–72; and society's role 4–5;
as statistical events 174–176; types
38–42; warnings and communities
214–216; as wicked problem **9**,
9–10
flood stories 139–141
flood type: coastal flooding **40**, 43; dam
burst floods **41**, 44; fluvial flood
risks **38**, 43; groundwater flooding
39, 43, 47; ice-jam floods **42**;
jokulhlaup **42**; pluvial/flash floods
38; surface water flooding **38**, 43,
69, 75–76, 113, 211; tidal flooding
and storm surges **40**, 43; tsunami **41**
flood warning systems: communities
as 216–217, **218**; early 216–217,
218
fluvial flood risks **38**, 43
Funtowicz, S. 101
future stories 290

game-based learning 291–293
Ganje, L. 142
Geaves, L.H. 254
The Gloucestershire Floods 2007 (Thomas
and Wilson) 140
Golubchikov, O. 18
Gotham, K.F. 72, 75, 84
governance: community roles in
adaptive 322–323; structures,
and community roles in FRM
233–234; transformative pathways
in 320–323, *321*

Graham, I.D. 114
Greenhalgh, T. 115
Grineski, S. 50
Grothmann, T. 88
groundwater flooding **39**, 43, 47
Gunderson, L.H. 16
Gupta, A. 322

Hagemeier-Klose, M. 177
Hancock, B. 312
hard to reach groups 50, 311–312
Harries, T. 79
Haughton, G. 119
Hayek, F.A. 105
hazard research 8–9
health and wellbeing for community
        resilience 314–316
Hemming, H. 252
heuristics **82**; cognitive 80–81;
        defined 80
heutagogy 273, *273*
'High Water Everywhere' (Patton) 149
hobbyist knowledge 106, **107**
holistic thinking **26**, 203
Holmes, A. 152
Horst, C. 239
households 249–259
Hove, T. 70, 83
Hulme, M. 133
human capital 235, **236**
humanitarian aid agencies 317
human-made landscapes 135–136
Hurlbert, M. 322
Hurricane Harvey 49
Hurricane Katrina 16, 21, 49, 72, 75, 117,
        139, 143, 149
Hurricane Sandy 296
hybrid knowledge formations 119
hydro-citizenship 239, 308–309
hydrological cycle 27, 172–173
hydrosocial cycle 26–27

inclusion 58, 236, 242, 248
inclusivity in communities 244–245
indigenous knowledge 105, 118
inequalities 311–312
Information and Communications
        Technologies (ICT) 189, 316, 317
infrastructural measures and communities
        200–202
insurance *see* flood insurance
integration in learning for flood resilience
        285–286

interdisciplinary learning (IDL) 278
interdisciplinary research 22–25, **23**
Intergovernmental Panel on Climate
        Change (IPCC) 105
International Association for Public
        Participation 241
International Federation of Red Cross and
        Red Crescent Societies (IFRC) 217,
        218, 291, 296, 312
International Water Association (IWA)
        Principles for Water-Wise Cities
        319
intersectionality 86, 311
Islam, M.S. 236
ISM tool 90

Jacobson, S.K. 72, 186
Johnson, Lonnie 149
Jökulhlaups/megafloods **42**, 44

Kahneman, D. 80
Kaika, M. 16
Kais, S.M. 236
Karlqvist, A. 24
Kasperson, J.X. 81
Kasperson, R.E. 81
KEEPER (Knowledge Exchange
        Exploratory tool for Professionals
        in Emergency Response) 122
Kenney, L. 142
Keys, C. 215
Kinzig, A.P. 85
knowledge: explicit **102**; in learning for
        flood resilience 282–286; sharing
        257–259; tacit **102**; translation
        114–115
knowledge brokers/translators 113–116,
        116
knowledge co-generation 119–120
knowledge controversies 111, 117
knowledge exchange 25, 62, 103, 117, 122,
        164, 187, 198, 224, 259, 270, 310,
        318
knowledge integration 117–122; modes and
        scales of 118–119
knowledge justice 123
knowledge systems 100
Knowledge-to-Action (KTA) 114,
        115
knowledge transfer 122, **164**
Krasny, M.E. 296
Kuhn, T.S. 101
Kyne, D. 327

land: managing 208–214; and water
    208–214
Land, R. 171
Landseer, Sir Edwin 146
language: flood education 287
Lanza, T. 172
Laudan, J. 110
Lauder, T.D. 86
learning: approaches for flood learning
    290–295; from catchment-based
    flood models 293; transformative
    277–278
Learning for Resilience (LfR) 22, 274–280,
    283, 285–290, 295–296, 308,
    312–313, 318
Learning for Sustainability (LfS) 22
learning styles 24, 272–273, 288, 296
Lechowska, E. 72
Lehmann, R.J. 186
Leiss, W. 165
Leitch, A.M. 82
Less Economically Developed Countries
    (LEDCs) 2, 47, 198, 217, 243, 282,
    291
levels of learning 276
Liguori, A. 120
Linnér, B.-O. 305
lived experience: causes of flooding 37–44;
    flood intensifying factors 44–45;
    and flood science 37–47; local river
    flood histories 37
'Living laboratories' (LL) 310, 319
local flood risk management (FRM):
    agency in 231–260; background
    contexts 232–234; capacities
    234–239; civil agency 234–239;
    community capital 234–239;
    community participation 240–246;
    community participation in
    DRM 246–249; household and
    community 249–259; participation
    in resilience planning in practice
    249–259
local knowledge 111–112; place-based
    112–113
losses: direct 3; indirect 3
Lozano Nasi, V.L. 303
Luke, A. 178
Lyall, C. 24
Lynmouth flood (UK) (1952) 45–46

MacDonald, N. 131–132
Macias, T. 86

Mackinnon, D. 19
magic concept 120
Maiella, R. 76
making science public 167–168
*Making science public: challenges and
    opportunities* (Nerlich) 168
managed retreat 205–208
managing land and water 208–214
marginalised groups 61, 86, 156, 242, 296,
    311, 316
Marks, Kate 178
Maskrey, S.A. 246
Mason O'Connor, K. 280
Mazzocchi, F. 118
McEwen, L.J. 137, 152, 169, 171, 205, 280,
    286, 288, 289
meaningful participation 9, 69, 259,
    323
measuring change in resilience 326–329
media narratives, and collective perception
    82, 82–84
Meerow, S. 18
memorialisation 144–145
mental health 77, 220, 314
Meyer, J.H.F. 171
Meyer, M. 113
Miceli, R. 76
Millennium Ecosystem Assessment 203
misperception 75, 85, 167
Mississippi Delta 146
modelling risk communication 165–166
Mondino, E. 108
Monument Australia 144
More Economically Developed Countries
    (MEDCs) 1, 47, 61, 198, 214, 217,
    225, 233–234, 282, 303
Morse, R. 21
Mosher, Eve 185
Motoyoshi, T. 74
multidisciplinary learning **278**
multidisciplinary research **23**
Multi-Story Water project 141

Nakamura, S. 56
Nakashima, D.J. 105
Nanni, P. 77
narratives: counter 73, 108; dominant 71,
    73, 84, 287; sticky 108, 287
National Flood Forum (NFF) 252; *Flood
    Risk Communities' Charter* 308,
    323
National Flood Insurance Program (NFIP)
    219

National Graduate Institute for Policy
  Studies (GRIPS) 281
National Institute for Health and Clinical
  Excellence (NICE) 11
natural capital **236**
natural flood management (NFM) 45, 60,
  209, 211
naturebased solutions (NbS) 202, 209, 213
Negrete, A. 172
Nerlich, B. 168
Newell, J.P. 18
newspaper accounts of flood impacts
  141–142
Nicholas, G. 118
Nicholson, A. 111
Nobert, S. 163, 165
non-government organisations (NGOs) 8,
  13, 60, 120, 272, 317, 320
normalcy bias **82**
Norton, J. 123
Ntontis, E. 18
Nye, M. 239

observatories 317–318
O'Donnell, E. 212
Oki, T. 56
older people 37, 80, 244
O'Neill, E. 72, 75
opportunities and barriers to flood
  education 286–288
optimism bias 80
oral history, and flood stories 139–141
organisational perceptions of risk 86–87
Organisation for Economic Co-operation
  and Development (OEDC): 12
  Principles on Water Governance
  320–321; good governance 321

Paek, H.-J. 70, 83
Pagneux, E. 75, 76
pandemics 52
paradigm lock 99, 103, 114
participation: in community activism
  252–254; from household to
  community 249–259; participatory
  methods 25, 243, **244**; in resilience
  planning in practice 249–259;
  transformative pathways in
  governance 320–323, *321*
Participatory Action Research (PAR) 122
participatory design processes 10
Participatory Learning and Action (PLA):
  key principles of **244**
Patel, S.S. 17

Patton, Charlie 149
Paulsen, Erik 146
pedagogies: for flood education *273*,
  273–278; for place-based flood
  education 288–290
Pei, S. 52
Penning-Rowsell, E.C. 254
perception: climate change 75–76;
  collective 87–89; risk (*see* risk
  perception)
philosophy of science 101
physical exposure 19
physical flood archives 134–136; local
  flood histories 134; in physical
  landscape 134–135
physical flood marking 138–139
Pielke, R.A. 70, 172
place attachment 77–78
place-based: flood education 288–290;
  learning 288–289; local knowledge
  112–113
place-making 60–61
political capital **236**
polluted floodwaters: and communities
  45–47; risks from 45–47
Popper, K.R. 23
Post-Normal Science (PNS) 101
Prager, K. 88
preparedness 5–6, 17, 21, 51–52, 57, 68–
  70, 72, 76–77, 79, 84–85, 88, 164,
  182, 199, 223
Pressure and Release (PAR) model
  20, *21*
problem solving 10, 22, 101–102, 274–275,
  286–287, 289–290
Property Flood Resilience (PFR) 220, 320
property level protection (PLP) 16, 320;
  and communities 220–224
Protection Motivation Theory (PMT) 88
psychological distance 76
psychometric paradigm 72, **73**

Quinn, T. 78

Raaijmakers, R. 77
Rashid, H. 141
rationalist theory 72
Ravetz, J. 101
Raygorodetsky, G. 105
recovery-based approach 61
reflective communication cycle 187–188
reflective experiential learning 289
Research to Action 103
residual flood risk 214–224

residual risk 59
resilience 13–16; characteristics of
    a resilient community 238;
    community 52–53; cross-cultural
    learning for 313; defined 19;
    ecosystem **14**; engineering **14**;
    psychological **14**; social **14**; social
    learning for 279; socioecological
    14; sustainability education and
    learning for 274–275
resilience indices 327, **328**
resilience stories 224, 287, 295, 318
Resilientville (role-playing activity) 290
resistance 16, 89, 220–221, 310
resourcefulness 19
Retsö, D. 132
Reusswig, F. 88
Rezaie, A.M. 206
Rhine Delta Scheme **144**
risk communication 174–176; community
    182–185; conceptual models of
    165, **166**, *166*; methods 180, **191**;
    modelling 165–166; strategies
    81, 91
risk perception: cultural meaning
    78–79; emotion 76–77; floods
    69–74; importance of 69–70;
    individual factors 74–79; personal
    characteristics 74–79; place
    attachment 77–78; sensemaking
    78–79; sense of place 77–78; worry
    76–77
Risk Perception Paradox 70–72,
    71, 87
risks: cascading 50; compound 51–53;
    sequential 50
risk salience 75
risk society 198, 234
Rittel, H.W.J. 9
River Quaggie scheme, London, UK
    205
rivers: communicating, as complex
    systems 172–173; 'daylighting'
    204–205
Robertson, I. 133
Robinson, Eugene 49
Rogers, R.W. 88
role-play as learning strategy 290
Rose, C. 221

Šakić Trogrlić, R. 110
Sangster, H. 131–132
Sayers, P.B. 22, 58

scenario-based learning 290
Scherzer, S. 327
Schumann, R.L. 144
science: communicating, with the public
    167; different models for public
    engagement with **164**
science capital 163–164
Seebauer, S. 87
Selby, D. 24
self-efficacy 87
*Sendai Framework for Disaster Risk
    Reduction* 1–2, 214, 219, 232–233,
    280–281
sensemaking 78–79
sense of place 77–78
sequential flood-heatwave events 52
serious games 243, 290–292, **292**, 295–296
Shanahan, E.A. 182
The Sheffield Flood 1864 UK 44, 156
Sheftel, S. 139
Siders, A.R. 207–208
significant learning 276, *277*
Sills, E.O. 73, 79
Simpson, H. 119
Sitaraman, B. 12
skills in learning for flood resilience 285,
    286
Sloan, S. 140
Smith, K. 55
Smith, L. 133
social amplification of risk **73**
social capital 235, **235**, **236**
social disadvantage 308
social identity theory 80
social learning for resilience 279
social media 12, 72, 83–84, 137, 152, 171,
    186, 316
social norms 80, 85, 88, 91, 322
social shocks 307–308
social trust 84–85, 91
social vulnerability: inequalities and
    environmental justice 311–312;
    transforming 311–312
societal transformation 304–305
society, changing 308–309
socio-ecological resilience 18, 288, 303,
    332
socio-technical interfacing 320
soft data 257–258
Somerset Levels 111, 117, 136–137, 143
South, J. 11
South Carolina Flood Oral History
    Collection 140

Southern Road: Poems (Brown) 146
Southerton, D. 90
spatial justice 86, 311
spatial planning 60–61
specialist knowledge 75, 109, 117, 119
Spiekermann, R. 100, 104, 186
spontaneous volunteering 251–252
Stakeholder Competency Group (SCG)
        121–122
stakeholder engagement 321
stereotyping 50, 244
storying of science 182
structural measures 24, 56, 59, 200–202,
        220, 269
sunny day flooding 43
super tides 43
surface water flooding 43
sustainability **20**, 22, 56–58, 102, 234, 305
sustainability education and learning for
        resilience 274–275
Sustainability Science 101–102
sustainable drainage systems (SuDS) 211
Sustainable Flood Management (SFM) 56
Sustainable Flood Memories project 110
Sustainable Flood Memory (SFM) 152, **153**
sustainable flood risk management
        198, 269
Sutherland, P. 111
systemic flood disadvantage 22
systemic flood impacts 314–316
systems thinking 26–27, 53
Szreter, S. 235

'tactical urbanism' 319
Tait, J. 24
tame problem 10
technological innovation: developing
        citizen science and observatories
        317–318; diverse citizens 316–317;
        harnessing the power of stories 318;
        new communication technologies
        316–317; transformation through
        316–318; valuing big data 317–
        318; valuing thick data 318
temporal vulnerability **20**, 21
Theory of Planned Behaviour 88
Theory of Reasoned Action 88
thick data 108
thinking: systemic 53; systems 26–27
Thorne, C. 212
threshold concepts 168, 170–174, **173**, 284
Tierney, K. 4
Time of Flood (Crowden and Wright) 139
Tobin, G.A. 55

Total Flood Warning System (TFWS) 215;
        components of *216*
transdisciplinary learning (TL) 278
transdisciplinary research (TR) 25
transformation through technological
        innovation 316–318
transformative capacity 13
transformative learning 277–278
Transformative Learning Theory 277–278
transformative pathways 305, 320–322
Tree of Participation (ToP) *240*, 241
troublesome knowledge 170
trust: collective perception 84–85; and
        FRM 84–85; social 84
Tversky, A. 80
Twigger-Ross, C. 12, 17

UK Wetlands and Wildfowl Trust 211
uncertainty: communicating 179–180;
        communicating, in flood early
        warning 179; floods in climate
        change science 179–180
UNDRR: *Hyogo Framework for Action
        (2005–2015)* 232; *Sendai
        Framework for Disaster Risk
        Reduction (2015-2030)* 199, 232,
        269, 303
UNESCO 274, 291
United Kingdom (UK): DEFRA *Low
        cost flood resilience project*
        223; Environment Agency 179,
        249, 250; FloodRE scheme 219;
        National Flood Forum 285
United Nations (UN): 'Convention on the
        Rights of the Child' 280; Office
        for Coordination of Humanitarian
        Affairs 259; Sustainable
        Development Goals 305
United Nations Office for Disaster Risk
        Reduction 1
United States Agency for International
        Development (USAID) 19
urban design: transforming 318–320
US American Army Corps of Engineers
        142
US Federal Emergency Management
        Agency (FEMA) 219, 317
US Institute of Medicine 24

values and dispositions 282
Veer, E. 151
vernacular knowledge 119
Victoria, L. 247
Vignes, D.S. 149

Villarini, G. 52
visualisations of flood risk 176–179
visual narratives 146
volunteering 249–251; care of
        communities and place 249–251;
        as citizen science in environmental
        monitoring 254–257, *255*, **256**; civil
        agency in 249–251; developing
        roles of 313; spontaneous 251–252
Voss, M. 288
vulnerability: defined 19; social 20, **20**;
        temporal **20**, 21; types of **20**

Wachinger, G. 71, 72, 83, 84
Wagner, K. 177, 288
Walker, G. 49
Ward, N. 122
Wardman, J.K. 165, 167
water: making space for 203–204;
        managing land and 208–214
water cities: and citizens 318–320;
        designing 318–320
Water Sensitive Urban Drainage Systems
        (WSUD systems) in Greater
        Adelaide, South Australia 212
watery sense of place 110

Webber, M.M. 9
Weick, K.E. 78–79
Weitkamp, E. 80
well-being for community resilience
        314–316
Westcott, R. 88
Wetland: Life in the Somerset Levels
        (Sutherland and Nicholson) 111
White, Gilbert 72
Whitmarsh, L. 75
Whittle, R. 13
Wibeck, V. 305
wicked problem: and collaboration 10;
        defined 9; floods/flooding as **9**,
        9–10
Wickson, F. 25
Wieringa, S. 115
Woolcock, M. 235
worry 76–77
Wright, G. 139

Yerbury, H. 11

Zavar, E.M. 144
Zembrzycki, S. 139
Zhang, W. 52

Printed in the United States
by Baker & Taylor Publisher Services

Printed in the United States
by Baker & Taylor Publisher Services